ADVANCES IN
MEDICAL ONCOLOGY, RESEARCH
AND EDUCATION

Volume VI

BASIS FOR CANCER
THERAPY 2

ADVANCES IN MEDICAL ONCOLOGY, RESEARCH AND EDUCATION

Proceedings of the 12th International Cancer Congress,
Buenos Aires, 1978

General Editors: A. CANONICO, O. ESTEVEZ, R. CHACON and S. BARG, Buenos Aires

Volumes and Editors:

- I - CARCINOGENESIS. Editor: G. P. Margison
- II - CANCER CONTROL. Editors: A. Smith and C. Alvarez
- III - EPIDEMIOLOGY. Editor: Jillian M. Birch
- IV - BIOLOGICAL BASIS FOR CANCER DIAGNOSIS. Editor: Margaret Fox
- V - BASIS FOR CANCER THERAPY 1. Editor: B. W. Fox
- VI - BASIS FOR CANCER THERAPY 2. Editor: M. Moore
- VII - LEUKEMIA AND NON-HODGKIN LYMPHOMA. Editor: D. G. Crowther
- VIII - GYNECOLOGICAL CANCER. Editor: N. Thatcher
- IX - DIGESTIVE CANCER. Editor: N. Thatcher
- X - CLINICAL CANCER - PRINCIPAL SITES 1. Editor: S. Kumar
- XI - CLINICAL CANCER - PRINCIPAL SITES 2. Editor: P. M. Wilkinson
- XII - ABSTRACTS

(Each volume is available separately.)

Pergamon Journals of Related Interest

ADVANCES IN ENZYME REGULATION
COMPUTERIZED TOMOGRAPHY
EUROPEAN JOURNAL OF CANCER
INTERNATIONAL JOURNAL OF RADIATION ONCOLOGY, BIOLOGY, PHYSICS
LEUKEMIA RESEARCH

ADVANCES IN MEDICAL ONCOLOGY, RESEARCH AND EDUCATION

Proceedings of the 12th International Cancer Congress,
Buenos Aires, 1978

Volume VI

BASIS FOR CANCER THERAPY 2

Editor:

M. MOORE

Department of Immunology
Paterson Laboratories
Christie Hospital and Holt Radium Institute
Manchester

PERGAMON PRESS

OXFORD · NEW YORK · TORONTO · SYDNEY · PARIS · FRANKFURT

U.K.	Pergamon Press Ltd., Headington Hill Hall, Oxford OX3 0BW, England
U.S.A.	Pergamon Press Inc., Maxwell House, Fairview Park, Elmsford, New York 10523, U.S.A.
CANADA	Pergamon of Canada, Suite 104, 150 Consumers Road, Willowdale, Ontario M2J 1P9, Canada
AUSTRALIA	Pergamon Press (Aust.) Pty. Ltd., P.O. Box 544, Potts Point, N.S.W. 2011, Australia
FRANCE	Pergamon Press SARL, 24 rue des Ecoles, 75240 Paris, Cedex 05, France
FEDERAL REPUBLIC OF GERMANY	Pergamon Press GmbH, 6242 Kronberg-Taunus, Pferdstrasse 1, Federal Republic of Germany

Copyright © 1979 Pergamon Press Ltd.

All Rights Reserved. No part of this publication may be reproduced, stored in a retrieval system or transmitted in any form or by any means: electronic, electrostatic, magnetic tape, mechanical, photocopying, recording or otherwise, without permission in writing from the publishers.

First edition 1979

British Library Cataloguing in Publication Data

International Cancer Congress, 12th, Buenos Aires, 1978
Advances in medical oncology, research and education.
Vol.6: Basis for cancer therapy 2
1. Cancer - Congresses
I. Title II. Moore, M III. Basis for cancer therapy 2
616.9'94   RC261.A1   79-40064
ISBN 0-08-024389-4
ISBN 0-08-023777-0 Set of 12 vols.

In order to make this volume available as economically and as rapidly as possible the authors' typescripts have been reproduced in their original forms. This method unfortunately has its typographical limitations but it is hoped that they in no way distract the reader.

Printed and bound at William Clowes & Sons Limited Beccles and London

Contents

Foreword ix

Introduction xi

Immunity and Cancer

Immunology and cancer 3
   G. DELLA PORTA

Host immune responses in experimental tumor systems 13
   R. B. HERBERMAN

Studies of the immunogenic properties of a solubilized, partially purified tumor rejection antigen (TSTA) from a chemically-induced sarcoma 19
   L. W. LAW, E. APPELLA and G. C. DuBOIS

Study of macrophage receptors for allogenate antitumor cytophilic antibodies 29
   A. B. MAZZOLLI

Cancer immunotherapy: experimental findings involving lymphocytes and antibodies to the oncofetal antigen α-1-protein (AFP) 37
   A. J. L. MACARIO

Cancer Immunotherapy

Immunotherapy of cancer: developing concepts and clinical progress (Harold Dorn Memorial Lecture) 47
   R. W. BALDWIN

Symposium No.19 - Chairman's opening remarks 65
   D. PRESSMAN

Active specific immunotherapy: experimental studies 67
   R. W. BALDWIN, M. V. PIMM and R. A. ROBINS

Immunotherapy of lung cancer   77
   E. CARMACK HOLMES

Specific active immunotherapy and specific active immunoprophylaxis in lung cancer   85
   A. HOLLINSHEAD and T. STEWART

Specific antibodies as carriers of diagnostic or therapeutic doses of radioisotopes (or drugs)   95
   D. PRESSMAN

Overview of controlled prognostic evaluations of levamisole immunotherapy in clinical cancer   101
   W. K. AMERY

Problems of nonspecific immunotherapy in cancer   109
   R. L. IKONOPISOV

Chemoimmunotherapy: basic principles and clinical examples   121
   J. G. SINKOVICS, C. PLAGER and M. J. McMURTREY

Intrapleural Corybacterium parvum as adjuvant therapy in operable (stage I and II) bronchial non-small cell carcinoma: preliminary report   131
   THE LUDWIG LUNG CANCER STUDY GROUP

Symposium No.19 - Chairman's summary   137
   D. PRESSMAN

Lymphocyte subpopulations in tumor immunity   139
   A. J. L. MACARIO

Search for immunosuppressive effects of suppressor cells and alpha-fetoprotein in a mouse-hepatoma syngeneic system   143
   A. J. L. MACARIO and G. J. MIZEJEWSKI

Studies on natural killer cell-mediated cytotoxicity in normal individuals and cancer patients   149
   E. LOTZOVA, J. A. MAROUN, K. B. McCREDIE, B. DREWINKO and K. A. DICKE

Progress in Radiobiology in Cancer

Chairman's introduction to Symposium No. 8   157
   G. W. BARENDSEN and L. RÉVÉSZ

The influence of radiation induced recruitment of resting cells in tumours into the compartment of proliferating cells on the effectiveness of chemotherapeutic agents   159
   G. W. BARENDSEN

The relevance of radiobiological data on mammalian cells in culture for the prediction of tumor responses to fractionated radiation   165
   H. D. SUIT and L. E. GERWECK

Relevance of radiobiological data on mammalian cells in culture for the prediction of tumor responses to fractionated irradiation: effects of intercellular contact   173
   R. E. DURAND

Rationale for the selection of combined treatment schedules using
fractionated radiation and chemotherapy   183
  T. L. PHILLIPS

Possibilities and limitations in the use of hypoxic sensitizers and
hyperbaric oxygen for improvement of fractionated radiotherapy of
cancer   191
  J. F. FOWLER

The skin as a model for the analysis of radiation injury   201
  R. L. CABRINI

Summary of papers presented at Symposium No. 8   209
  L. RÉVÉSZ and O. C. A. SCOTT

Progress in Clinical Radiotherapy in Cancer

Clinical experience with negative pi mesons and other high-LET
radiations   215
  J. M. SALA and M. M. KLIGERMAN

Hyperbaric oxygen in radiotherapy   223
  R. SEALY

Hyperthermia and radiotherapy   235
  J. OVERGAARD

Combined hyperthermia and radiation *in vivo*   247
  J. OVERGAARD

Unconventional time-dose-fractionation in radiotherapy of cancer   253
  V. A. MARCIAL

Radiotherapy of subclinical disease: Are small amounts of radiation
effective for small amounts of cancer cells? Special reference to
the so-called prophylactic irradiation of lung and brain occult
metastases   263
  M. TUBIANA

Use of radiosensitizers in radiotherapy   271
  S. DISCHE

Chairman's closing remarks – Symposium No.9   275
  L. R. HOLSTI

Radiation therapy role in testicular germinomas   279
  M. A. BATATA, F. C. H. CHU, B. S. HILARIS, A. UNAL, W. F. WHITMORE, Jr.,
  H. GRABSTALD and R. GOLBEY

Results of 8,056 cases of carcinoma of cervix uteri treated by
irradiation: clinical use of Peking-type applicators   293
  CHINESE ACADEMY OF MEDICAL SCIENCES

Preoperative irradiation of T3-carcinoma in Bilharzial bladder   299
  H. K. AWWAD, H. ABD EL BAKI, M. N. EL BOLKAINY, M. V. BURGERS,
  S. EL BADAWY, M. A. MANSOUR, O. SOLIMAN, S. OMAR and M. KHAFAGY

The value of post-operative irradiation in stage II breast cancer: the pattern of appearance of metastases and the value of C.E.A. as a predictor of metastases 305
   S.A.S.I.B. BREAST STUDY GROUP

Comparison of three modalities of treatment for carcinoma of breast stage III. Results from a prospective randomized clinical trial 311
   A. CACERES, F. TEJADA, M. ZAHARIA, M. LINGAN, M. COTRINA, L. LEON, M. MORAN, O. CASTRO and A. SOLIDORO

Index 315

Foreword

This book contains papers from the main meetings of the Scientific Programme presented during the 12th International Cancer Congress, which took place in Buenos Aires, Argentina, from 5 to 11 October 1978, and was sponsored by the International Union against Cancer (UICC).

This organisation, with headquarters in Geneva, gathers together from more than a hundred countries 250 medical associations which fight against Cancer and organizes every four years an International Congress which gives maximum coverage to oncological activity throughout the world.

The 11th Congress was held in Florence in 1974, where the General Assembly unanimously decided that Argentina would be the site of the 12th Congress. Argentina was chosen not only because of the beauty of its landscapes and the cordiality of its inhabitants, but also because of the high scientific level of its researchers and practitioners in the field of oncology.

From this Assembly a distinguished International Committee was appointed which undertook the preparation and execution of the Scientific Programme of the Congress.

The Programme was designed to be profitable for those professionals who wished to have a general view of the problem of Cancer, as well as those who were specifically orientated to an oncological subspeciality. It was also conceived as trying to cover the different subjects related to this discipline, emphasizing those with an actual and future gravitation on cancerology.

The scientific activity began every morning with a Special Lecture (5 in all), summarizing some of the subjects of prevailing interest in Oncology, such as Environmental Cancer, Immunology, Sub-clinical Cancer, Modern Cancer Therapy Concepts and Viral Oncogenesis. Within the 26 Symposia, new acquisitions in the technological area were incorporated; such acquisitions had not been exposed in previous Congresses.

15 Multidisciplinary Panels were held studying the more frequent sites in Cancer, with an approach to the problem that included biological and clinical aspects, and concentrating on the following areas: aetiology, epidemiology, pathology, prevention, early detection, education, treatment and results. Proferred Papers were presented as Workshops instead of the classical reading, as in this way they could be discussed fully by the participants. 66 Workshops were held, this being the first time that free communications were presented in this way in a UICC Congress.

The Programme also included 22 "Meet the Experts", 7 Informal Meetings and more than a hundred films.

METHODOLOGY

The methodology used for the development of the Meeting and to make the scientific works profitable, had some original features that we would like to mention.

The methodology used in Lectures, Panels and Symposia was the usual one utilized in previous Congresses and functions satisfactorily. Lectures lasted one hour each. Panels were seven hours long divided into two sessions, one in the morning and one in the afternoon. They had a Chairman and two Vice-chairmen (one for each session). Symposia were three hours long. They had a Chairman, a Vice-chairman and a Secretary.

Of the 8164 registered members, many sent proferred papers of which over 2000 were presented. They were grouped in numbers of 20 or 25, according to the subject, and discussed in Workshops. The International Scientific Committee studied the abstracts of all the papers, and those which were finally approved were sent to the Chairman of the corresponding Workshop who, during the Workshop gave an introduction and commented on the more outstanding works. This was the first time such a method had been used in an UICC Cancer Congress.

"Meet the Experts" were two hours long, and facilitated the approach of young professionals to the most outstanding specialists. The congress was also the ideal place for an exchange of information between the specialists of different countries during the Informal Meetings. Also more than a hundred scientific films were shown.

The size of the task carried out in organising this Congress is reflected in some statistical data: More than 18,000 letters were sent to participants throughout the world; more than 2000 abstracts were published in the Proceedings of the Congress; more than 800 scientists were active participants of the various meetings.

There were 2246 papers presented at the Congress by 4620 authors from 80 countries.

The Programme lasted a total of 450 hours, and was divided into 170 scientific meetings where nearly all the subjects related to Oncology were discussed.

All the material gathered for the publication of these Proceedings has been taken from the original papers submitted by each author. The material has been arranged in 12 volumes, in various homogenous sections, which facilitates the reading of the most interesting individual chapters. Volume XII deals only with the abstracts of proffered papers submitted for Workshops and Special Meetings. The titles of each volume offer a clear view of the extended and multidisciplinary contents of this collection which we are sure will be frequently consulted in the scientific libraries.

We are grateful to the individual authors for their valuable collaboration as they have enabled the publication of these Proceedings, and we are sure Pergamon Press was a perfect choice as the Publisher due to its responsibility and efficiency.

Argentina
March 1979

Dr Abel Canónico
Dr Roberto Estevez
Dr Reinaldo Chacon
Dr Solomon Barg

General Editors

Introduction

The multifactorial nature of the immune response to malignant disease is stressed in several contributions dealing with the induction of tumour immunity and the factors which compromise the expression of effective resistance against tumour growth. Considerable progress is reported on the isolation and purification of a tumour rejection antigen of strong immunogenicity. The potential problems of non-specific- and specific active-immunotherapy are evaluated in experimental models and in certain clinical situations, notably carcinoma of the lung.

Progress in radiobiology related to cancer is herein restricted to some of the new developments in the treatment of different cancers, e.g. high-LET, chemo-radiotherapy and to the techniques developed for radiobiological studies involving responses of cultured cells, transplantable tumours and animal tissues. The emphasis throughout is upon fractionation regimes, and the extent to which their efficacy is improved by the use of hyperbaric oxygen and electron-affinic sensitizers. Papers on clinical radiotherapy take up the same themes and some additional ones, such as the interaction of radiation with hyperthermia.

March 1979

M. MOORE

Immunity and Cancer

Immunology and Cancer

Giuseppe Della Porta

*Division of Experimental Oncology A,
Istituto Nazionale per lo Studio e la Cura dei Tumori,
Via G. Venazian 1, 20133 Milan, Italy*

When I accepted Prof. Canonico's kind invitation to give this lecture, I did not realize immediately the difficulties of the task. Both Immunology and Cancer are extremely complex biologic entities: to put them together is a difficult operation, perhaps even in mathematical terms if we consider the number of parameters that must be taken into account and the number of unknown phenomena involved. I am certain that this complexity is the very reason of the slowness of the progress made. However, the last few years have seen a vast accumulation of data and reports. I know of a recent hand-book on tumor immunology in five volumes with a total of 1800 pages. To read aloud one out of three of those pages, would take us approximately 30 hrs. To present a comprehensive and detailed account of this material is impossible. So I had to take a decision, and I hope I won't disappoint you too much by my decision to present you my interpretation of the findings rather than detailed figures. This type of presentation is, of course, bound to be incomplete and sometimes superficial, and I apologize for this. As I apologize also to the many investigators for not mentioning their names. There are just too many individuals and laboratories involved, and to mention only a few of them would be meaningless. Fortunately, this is a field of science that, although relatively new, has stimulated many excellent reviews in which the details of the studies and their authors are given.

I shall start by considering the clinical results in immunotherapy and immuno-diagnosis; then I shall deal with the efforts to identify the tumor-specific or associated antigens and to discover their nature and it will be necessary to speak mostly of experimental tumors; finally I shall present some of the recent findings on the immune system and its modes of action. I wish to emphasize that the quantity and quality of the new information that we have been getting from basic immunology on the physiology of the immune system is of such value that it is there that our main hopes of a more correct, and valid, and useful application of immunology to the cancer problem lie.

IMMUNOTHERAPY

Many trials on immunotherapy have been, and are being, conducted throughout the world on patients with advanced cancer. Most of these studies use bacterial vaccines concomitantly with various forms of chemotherapy. Most of them have been set up not as formal trials but rather as crude but courageous attempts in face of an advanced disease. This approach makes an evaluation of the results most difficult. It is certainly impossible to reach the conclusion that any benefit has been achieved when the positive studies are on small groups of patients without adequate controls, but it is equally unfair to dismiss the case using similarly inadequate and negative studies. I appreciate the difficulties involved, but even a rapid perusal of the literature of the last 2 or 3 years indicates that 20 years after the discovery of BCG effect on experimental tumors, there is still no consensus of opinion on how, where and when, and at which dose BCG should be given, nor has its potency relative to other bacterial vaccines been established. Perhaps some of the early studies were too successful, as is still the dramatic local effect of BCG on cutaneous melanotic lesions, and therefore many studies have been set up in the expectation of clearly and easily discernible responses. The lesson has been learned now, that only little effect can be expected, particularly in advanced cancer, and that the trials must be designed accordingly. The final results of most of these new adequate trials have not been published yet, and I hope that in the course of this Congress we will hear formal reports of the latest data. So far, I think it is fair to say that the results have been inconsistent, although extension of survival over that achieved with chemotherapy alone has been observed.

A more convincing demonstration of a beneficial effect derives from the trials in which immunotherapy has been used as an adjuvant therapy after a solid tumor is removed by surgery or after a hematological cancer has been brought into remission by chemotherapy. Let us take as one of the best examples of this situation the trial with a single intrapleural injection of BCG in lung cancer patients stage I after surgical resection. At a median 2-year follow up the results showed a remarkable increase in the number of disease-free patients compared to an untreated control group, whereas there was no beneficial effect on stage II or stage III patients. These results emphasize that BCG may be beneficial when administered into the tumor region and when the amount of tumor left is minimal. Before a definite conclusion can be reached, I think that the final data should be available and that the trial should be duplicated with the same set-up with a larger group of patients. However, I don't know if this will be possible now that some benefit can be obtained also with combination chemotherapy.

It seems to me that the studies on malignant melanoma produce less convincing data, perhaps because they are more numerous and varied in their design and end points, with alternate positive and negative indications. The regression of cutaneous nodules injected with BCG or with Corynebacterium parvum is certainly quite out of the question, but this has been shown repeatedly to have no effect on untreated nodules and not to modify the evolution of the disease in any consistent way. The final results of the randomized prospective trials which are in progress to test adjuvant immunotherapy with or without chemotherapy after surgical resection are not known yet. The preliminary results of one study point to an extended disease-free interval, but with no ultimate improvement of the survival figures.

The information available with other types of solid tumors is too limited and needs extension and confirmation. The picture is not much more encouraging with leukemias and lymphomas. After the positive results reported by Mathé 15 years ago with BCG and allogeneic cells in acute lymphocytic leukemia, many other trials, with similar though not identical design, have not confirmed that original success. Studies with BCG on acute myelogenous leukemia gave suggestion of some prolongation of survival whereas a study on non-Hodgkin lymphoma showed a benefit on remission rate.

I think it is fair to say that in most instances where adjuvant immunotherapy is combined with chemotherapy, because chemotherapy is progressing more rapidly, immunotherapy is suffering from a sort of second rate classification, to the point that the exact repetition of a trial three or five years after the previous one was started, is unappealing and unethical. This could mean, perhaps, that we should not expect much more from BCG type of immunotherapy, unless we really understand its mechanism of action. A major effort should be made, at the level of the experimental animal, to develop the pharmacology of the immune system. Many other substances are already under test but it is my impression that it is too early to analyze the results. Very interesting information is coming from investigations on the mechanisms of action of interferons that, in addition to their antiviral action and to some effect on experimental and human tumors, have been found to have a regulating activity on the immune response which may explain the antiviral and antitumor effects. Moreover, a number of antigens, mitogens, bacteria, endotoxins, and double-stranded RNA forms have been found to induce the production of interferons by lymphocytes. It is evident, therefore, that we may have here a unifying hypothesis that may throw some light on the mechanisms of nonspecific immunotherapy.

Let me mention at this point that active specific immunotherapy is also continuing to attract a good deal of attention and efforts as the most direct approach, at least on theoretical bases, to the creation of a tumor immunity. However, we must first isolate and characterize tumor antigens and demonstrate their biologic relevance. The progress made in our knowledge of the nature of antigens and immune reactivity, and the advances of immunochemistry technology now permit of a more sophisticated approach than that of the vaccine made of a crude extract of tumor cells. Similarly, there is a growth in the quantity and quality of the studies on experimental animals aimed at increasing the immunogenicity of tumor antigens by altering either their structure or configuration or their presentation to the immune system.

The use of specific subcellular components such as transfer factor and immune RNA is very appealing and I believe will develop in the near future, particularly if the theoretical bases of this approach is clearly understood. Passive immunotherapy has not been abandoned but clinical data are too sparse to give any clear picture of the situation. It is worth noting, here, that a better knowledge of the nature of tumor antigens and of the methodology for in vitro sensitization of lymphoid cells may open up new developments of this approach.

If I had to summarize the present situation of immunotherapy in a couple of provocative sentences, I would say that in patients with advanced or disseminated cancer various types of immunotherapy have shown some effect, but that none has brought about a substantial benefit for the patient, at least at a level of confidence sufficient to suggest general use. I would add that, at least on theoretical bases, no benefit should be expected and that, perhaps, the advanced disease should be used for preliminary investigations on new types of immunotherapy. On the other hand, minimal cancer patients offer the best condition for immunotherapy. The results so far are not convincing, but almost everybody agrees that efforts should be continued, with properly controlled investigations, particularly in the direction of active specific immunotherapy.

Clinical evaluation of immunotherapy should include analysis of the immune response with the immediate scope of monitoring its efficacy and with the ultimate goal of correlating it with the antitumor response. This analysis has met with certain difficulties. Clearly, a modification of the general immune competence of a cancer patient cannot be interpreted with confidence as meaning that immunotherapy has indeed caused an immune reaction specifically against the tumor. I might add here that in general the so-called immune profile has been found to have little bearing in establishing the prognosis of the neoplastic disease. On the other hand, as we will see in a few minutes, analysis of the anti-tumor specific immune reactivity of the individual patient is still difficult and not reliable enough. A recent study, however, has shown that in a group of patients with acute leukemia in remission, who received active specific immunization in the form of pooled allogeneic leukemia cells or of soluble purified fractions from cell membrane extracts, 70% developed specific reactivity as measured by in vitro blastogenic response of the patient lymphocytes upon in vitro stimulation with the same extracts.

IMMUNODIAGNOSIS

Certainly I do not need to spend much time describing the potential applications of immunodiagnosis to cancer, which may comprise the screening of the population at large and of high risk groups, the aid in diagnosing or localizing a suspected tumor, the monitoring of the course of the disease with particular reference to the detection of occult metastases or recurrences, and the demonstration of the value of a therapy. This field of clinical investigation is of such importance that it has attracted the enthusiastic efforts of many investigators, and has caused many disappointments too. Nevertheless, it is progressing through two approaches, that is the detection of an immune response in the patient against tumor-associated antigens, and the detection by means of an immunologic procedure of tumor markers present in the serum or other fluids of the patients.

The immune response of the patient against the autochthonous tumor can be analyzed with assays for cell-mediated or humoral immunity. For the cell-mediated reactivity, the most logical and direct approach is to search for cytotoxicity by lymphoid cells on tumor cells. I really do not want to go into detail and describe the great efforts that have been put into this kind of study, which, unfortunately, have met with failure, because of lack of specificity and the presence of a frequent normal reactivity. Apart from technical difficulties in the assay, problems exist in both the target and effector cells. Cultured cells are difficult to characterize and may be either excessively sensitive or resistant to immune action. Effector cells are difficult to prepare and may contain various percentages of cells endowed with a natural killing capacity. This natural cell-mediated antitumor reactivity is per se a very interesting phenomenon that has also been observed in experimental animals, and whose significance is under study.

The same difficulties related to specificity, normal reactivity, and more often total absence of reactivity, have been encountered with the assays for detection, in the serum of the cancer patients, of antibodies against surface antigens of viable or intact tumor cells. These assays include indirect immunofluorescence, isotopic antiglobulins binding, immuno adherence, and complement dependent cytotoxicity. Among the most studied tumors are Burkitt's lymphoma, malignant melanoma and sarcoma. Particularly with Burkitt's lymphoma, the results are interesting and indicate that, at least in individual patients, the measurement of specific antibodies may contribute in monitoring the course of the disease. Moreover, the experimental application of serology to human tumors is making a valid contribution to the definition of surface antigens.

I could mention here that the difficulties of in vitro assays using viable cells and either lymphoid cells or serum can be due to the presence in the serum of tumor antigens from the cell membrane, of specific antibodies, and of immunocomplexes. It seems that their interference is not limited to the level of the in vitro test, but reaches the mechanisms of the generation of cytolytic T lymphocytes.

Another group of assays uses not cells but extracts of tumor cells from either surgical specimens or tissue cultures. These comprise, principally, the skin test for delayed hypersensitivity, the leukocyte migration inhibition test, and the leukocyte adherence inhibition test. Whereas the immunologic mechanism of delayed hypersensitivity is at least partially known, the mechanisms involved in the two in vitro tests are still not clearly understood. The inhibiting factor is released by stimulated lymphocytes upon contact with the antigen, and seems to be identical to the factor that causes the slowing down of electrophoretic mobility of macrophages, another assay that calls for wider appraisal.

A fourth in vitro assay of this group is the lymphocyte stimulation test that may have some advantages in the measurement of response but presents other problems particularly because of the interference of histocompatibility antigens when allogeneic combinations are tested.

Let me try to give you a comprehensive appraisal of these assays. They seem to be capable of measuring cancer-related immune reactivity in cancer patients. However, with a very large approximation, this reactivity is observed in little more than 50% of the patients tested with a good positive extract of a tumor of the same histologic type as that of the patient. Positive reactions with unrelated tumor extracts are seen in approximately 30% of the patients. Positive reactions with normal tissue extracts in the skin test in patients and in the in vitro tests with patient leukocytes, or with control leukocytes and tumor extracts in the in vitro tests, range from 5% to 20%. A large number of cancer patients do not respond at all, with no plausible clinical reason.

The assays, therefore, are not applicable as routine clinical tests. Moreover, they present problems in their reproducibility and in their reading. However, they are very useful as experimental tools for detailed studies of tumor antigens, of the procedures for their extraction from cells, and of their purification and characterization. In perspective, progress in investigations of tumor antigens will greatly help in improving the assays themselves, thereby facilitating their standardization.

We have also to consider the possibility of a purified tumor antigen being used for production in animal species of potent antisera which in their turn can be used for the detection of the same specificities in humans using the extremely sensitive radioimmunoassays. The same concept applies to other possible biochemical markers of neoplasia which even if not immunogenic in the patients, are highly so in other species. I need only mention that the carcino-embryonic antigen and alpha-feto-protein belong to this type of tumor marker. The CEA test is the only immuno-diagnostic assay now routinely used, and in spite of its limitations, gives benefit to the patient, chiefly by its contribution in monitoring the course of his disease. The problems of CEA specificity are too well known to be discussed here, but I would like to mention in passing that this is a good example of a substance which is not specific of the tumor cell since it is produced also by fetal tissues and, in much smaller quantities, also by adult normal cells, and which is not immunogenic, as far as we can detect, in the patient. Yet, a xenogeneic antiserum can be made and validly used. Other tumor biologic markers are similarly under investigation, some of them with encouraging preliminary results. Rather than give you a list of them, I want to stress that the proposed assays must be evaluated with very stringent criteria to establish their specificity and sensitivity, with studies that should be designed according to their potential applications. I am not saying this with the intention of discouraging small scale attempts in trying out new ideas, of which we do have indeed great need. But if an assay has to be suggested for a clinical use, the design of the study and the criteria for its evaluation should be made according to the suggested specific application. Clearly, an assay that has 20% false positives, or even far fewer, cannot be applied to the screening of the general population, but still could be most useful in the follow-up of a cancer patient and in the management of his disease. But in the latter case, the evaluation study must be set up properly and must take into account all the necessary clinical parameters, not more or less than in any good randomized clinical trial.

TUMOR ANTIGENS

We are coming now to the core of the problem, the tumor-specific or associated antigen, which is a proteic structure belonging to the cancer cell or produced by the same, in a different manner, qualitatively or quantitatively, from that of the corresponding normal adult cell. For the sake of clarity, let me repeat here again that we should consider two separate groups of tumor antigens with a different biologic reactivity. One group includes antigens which are capable of raising a detectable immune response, of whatever kind, in the autochthonous host, that is the patient carrying the tumor. I should specify that surface antigens seem to be the most relevant in the patient's immune reaction to his own tumor and may be the cause of an immune reaction involved in the control of the neoplastic growth. However, we must include in this group also the antigens defined by in vitro assays. The necessity of distinguishing between the two types of immunogenic tumor associated antigens according to the in vitro or in vivo procedure of their detection, is only operational in part since in humans we can rarely assume that a tumor associated transplantation antigen has been demonstrated. However, this is also due to the fact that in many instances experimental tumors have been found to be resistant to an immune reaction although they are capable of inducing an immune response.

In the second group are those structures which are not immunogenic in the cancer patient, in their native form, but are immunogenic in an allogeneic or xenogeneic host. I have already said that the latter group may include tumor markers of great importance because they can be exploited in immunodiagnosis.

I want to say that if our studies eventually prove that only the second group exists, and that the first, in humans, does not, we should stop talking of tumor immunology but we could continue to speak of immunologic methodologies applied to tumors.

If we analyze in the most schematic manner (as in Fig. 1) the antigens which may be present on tumor cells and may be relevant to our discussion, we should start with the conventional major histocompatibility alloantigens, which are certainly the best known antigens and the best example of transplantation antigens. Near to them we may locate the antigens coded by the I-region of the major histocompatibility complex that, as we will see later, are of great importance in regulating immune responses. Then, we have the series of normal fetal and differentiation antigens which may be present with their normal structure and specificities on tumor cells although in different amounts than on normal adult cells and for this reason they and the group of the tumor associated antigens overlap.

I have signed a question mark near the neoplastic transformation specific antigen because we do not even know whether it exists. Naturally, its importance goes far beyond its immunogenicity, involving an essential part of the mechanism of transformation. Instead, a series of antigens that are present on and within cells of tumors induced by oncogenic viruses are quite well known. They must be divided in two categories, that consisting of structural proteins of the viruses and that directly or indirectly linked to viral gene expression, as we will see in a moment. I should say here that they are characteristically cross-reacting in the sense that tumors induced by the same virus bear the same antigens.

I will proceed now to the antigens of chemically induced tumors, which were the tumors, induced by methylcholanthrene in mice, on which tumor immunology was founded, just little more than 20 years ago. They are characteristically individual transplantation antigens. An animal immunized against one tumor will reject that tumor only. This is the general rule which in spite of some exceptions, has on one hand greatly favored the studies aimed at establishing the dynamics of the anti-tumor immune reactions and at discriminating between transplantation tumor antigens and other tumor antigens, and, on the other hand, however, has indicated that is extremely unlikely that many individually specific antigens, found on the same type of tumors induced in the same animal by the same carcinogen, can represent a proteic structure essential for neoplastic transformation.

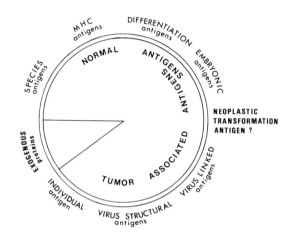

Fig. 1. A schematic representation of antigens on the surface of a tumor cell.

I want to point out here that a variety of microbial antigens can be present on somehow infected tumor cells, and also that components of fetal calf serum used in culturing tumor cells may be bound intimately to the cell surface. Both can interfere considerably in our studies by causing uncontrolled cross-reactivities.

Going back to the experimental tumors induced by oncogenic viruses, I will use only a small part of the available data, that which may help me in delineating what can happen on the surface of the transformed cell.

The C-type oncornaviruses are now well studied in several species where certain differences have been found. However, the scheme of their gene products as observed in the murine and in the avian systems could serve as a model in our present discussion. As you can see in Table 1, the first three viral genes code for several proteins which are fairly well known and which constitute the structural proteins of the virus and the enzyme essential for the transcription of the viral RNA into the cellular DNA. We should put the question of the relationship between these viral gene products and the cell surface antigens which have been recognized in virus-induced tumors and thought to be the virus-induced expression of cellular genes. At present I believe that we should come to the conclusion that all of them are closely linked to viral products and that some of the previous observations indicating a specific expression of cellular genes should be ascribed to polymorphism of C-type viruses and to the simultaneous presence in a given tumor of non cross-reacting viruses. However, studies in the avian system with the Rous sarcoma virus have demonstrated an additional gene, the src gene, that is associated with neoplastic transformation and its maintenance through the production of a protein with kinase activity, does not code for structural proteins of the virus, is not required for viral replication, and seems responsible for the appearance of a virus associated surface antigen, perhaps directly or by means of the activation of cellular genes.

TABLE 1   Type C Oncornavirus Antigens

Viral genes		Virionic proteins	Cell surface virus related antigens
gag	(murine)	p15, p12, p30, P10	GCSA MCSA
pol		Reverse transcriptase	
env	(murine)	gp70, p15E	G_{IX} $G(RADA_1)$ FMR
src	(avian)		VCSA (transformation protein?)

The importance of this kind of study with oncogenic RNA viruses is manifold and goes beyond the understanding of the surface makeup of virus transformed cells. In the meanwhile this model is used for a systematic search for similar viruses in human tumors.

I will bypass entirely the subject of oncogenic DNA viruses, but I wish to remind you that a neoplastic disease of widespread diffusion, caused by a DNA virus, the Marek disease of poultry, has been brought under control by simple vaccination. I mean to say that, if either RNA or DNA viruses are found to be the causal agents of some human tumors, the studies conducted in experimental situations to specify many immunologic details of virus-cell interactions will take on a practical meaning.

Interest in the immunologic aspects of chemically induced tumors has been renewed in the last few years by the evidence that they can express besides the individual antigen, also alien histocompatibility antigens, that is antigens which are normally expressed by cells of murine strains with a different haplotype. The available experimental data may give support to different hypotheses that can be stated as follows: 1) the alien antigens are the result of a mutation at the structural genes of the major or minor histocompatibility antigens;   2) they are the result of a derepression of genes coding for alloantigens, that are normally silent in a given haplotype and can be activated by the chemical carcinogen or by neoplastic transformation by itself;   3) they are the result of a close interaction, at the cell surface, of the normal histocompatibility antigens with undefined structures related to the neoplastic process.

Whichever hypothesis will turn out to be exact, the experimental evidence which was restricted to murine tumors, fibrosarcomas, lung adenomas and lymphomas all studied with quite different approaches, has recently found a further development in the human hairy cell leukemia. In vitro sensitisation of the patient lymphocytes with normal cells from a large pool of unrelated individuals generated cytotoxic T lymphocytes that lysed autologous leukemia cells but not normal lymphocytes. I have little doubt that this area of tumor immunology, though still very obscure, may soon give rewarding contributions.

THE IMMUNE SYSTEM

On the other hand, the importance of the major histocompatibility complex in immunology is demonstrated by another series of experimental results.

In recent years there has been a great improvement in knowledge of the genetics of the major histocompatibility complex, and it has been recognized that it is important not only because it codes for the histocompatibility antigens but also for its fundamental role in the regulation of immune response. There are many similarities between murine and human MHC regions and their study is progressing in parallel. The I region in the mouse and the region around the HLA-D locus in man have the greatest importance in determining the capacity of immune reaction of an individual. A series of excellent genetic investigations taking advantage of the polymorphism of the genes involved, has established that in the mouse the products of the genes represented in the I region are involved in most of the known interactions between the subpopulations of T cells, between T and B cells, and between macrophages and T cells, and control their functions. The Ia products are particularly important in immune response to cell surface antigens, and seem to play a role in the phenomenon known as associative recognition, by which the specific reactivity to different membrane antigens can be greatly enhanced.

Also our knowledge of the cellular components of the immune system, and their interactions, is developing rapidly. I will single out only a few issues that seem to me more relevant to our discussion.

Macrophages are intensly investigated and their central role in immunity has been well established. Their major function is the uptake of the antigen and its presentation to lymphocytes either indirectly by release of antigen, perhaps complexed to Ia products, or directly by cell-to-cell physical interaction. However, macrophages are involved in other aspects of specific and nonspecific immune functions through the secretion of various products and may acquire the capacity to kill.

The understanding of the pathway of maturation and differentiation of T cells has been favored by a better knowledge of specific markers at cell surface. At the same time, the functions of the various subpopulations and their specific interactions are becoming better known. For our problem, there are two aspects which to my mind are of prime importance. First, the discovery of the subclass of suppressor T cells, which are capable of stopping specific immune reactions. Their induction seems to be regulated by a T-T cell interaction system which is responsible for a regulatory feedback mechanism of the immune response: antigen-specific, activated helper T cells can induce not only B cells to produce antibody, but also prekiller T cells to differentiate to killer cells and resting T cells to exert a potent suppressive activity on the same helper T cells. The importance of this T-T feedback mechanism is that it could constitute an obstacle to immunotherapy very difficult to remove.

It has also been suggested that the suppressive function may become overwhelming in the presence of an excess of free antigens and in the absence of an appropriate macrophagic function. Again here, specific genes of the I region play a regulatory role.

The second point is in the area of cytotoxic T cells where the major recent development is the recognition that in certain situations in the mouse they can lyse only target cells that carry not only the specific antigen but also the same H-2K or H-2D determinants as the original stimulating cells. This phenomenon has been called H-2 restriction and has been found to take place in cytotoxicity against virus-infected cells, hapten-modified cells and minor histocompatibility antigens. Similar restrictions involving I region products are operational in T-B lymphocyte cooperation and in lymphocyte-macrophage interactions. We do not know yet whether tumor antigens fall within the restricted systems. If they do, one implication would be that also in vitro assays proposed for immunodiagnosis could be restricted, and this could offer another explanation for the failure of cytotoxicity tests that are often performed in allogeneic combinations. Moreover we must keep in mind that genetic restrictions could also hinder passive immunotherapy manipulations.

The last issue I would like to consider is that of the natural antitumor immune reactivities. In the peripheral blood of normal individuals there are Fc receptor-positive lymphoid cells which can kill tumor cells in in vitro assays. Similar natural killer cells have been found in mice and rats where the natural activity decreases with age and can be augmented by inoculation of tumor cells or of non-specific immunogens all of which share the property of being interferon inducers. Moreover, there are indications that natural killer cells may play a role also in vivo, by conditioning resistance to tumor growth. The nature of the specificities involved and the significance of the phenomenon are not yet clear. It has recently been suggested that it may be a rather selective defense mechanism where interferon plays an important role in protecting the normal cells from the natural killers and leaving virus-infected cells and tumor cells unprotected. At any rate, I think that natural reactivity can be taken as a further support to the reality of tumor immunology and, particularly, to the concept of immunosurveillance.

Besides a cellular natural response, also a humoral one against tumor cells has been observed in man and mice. Its role in the control of tumor growth is still undefined. On the other hand, also specifically raised antibodies that in vitro show a complement-dependent cytotoxicity and are able to mediate a cellular cytotoxicity, in vivo have been thought to block antitumor immunity and even to stimulate tumor growth. However, a possible benefit has been observed in leukemias, and, additionally, specific antibodies have been postulated as carriers of antineoplastic drugs. If a positive role of antibodies is eventually demonstrated, then the new technology of hybridomas with massive production of antibodies may prove useful also in the field of cancer.

I must come to an end. I hope I have given you a balanced picture of the situation, pointing out the difficulties in the clinical field and the new discoveries of experimental studies. May I stress in closing that only close interaction between applied and basic research will avoid mistaken and dangerous application to the patient of concepts not even proven with certainty in the experimental animal, and will allow successful achievements to be made.

Host Immune Responses in Experimental Tumor Systems

R. B. Herberman

*Laboratory of Immunodiagnosis, National Cancer Institute,
Bethesda, Maryland, U.S.A.*

ABSTRACT

Host immune responses to tumors are multifaceted, and dependent upon a diversity of interrelated components, including the immunogenicity of the tumor cell, induction and effector mechanisms, and upon factors which may compromise their effective function. Several parameters (e.g. amount, form, duration of exposure) may determine the immunogenicity of tumor antigens, and in some systems there is the additional requirement for appropriate histocompatability complex (MHC) antigens.

Host factors include general immunocompetence and genetically-determined ability to recognise particular antigens; and the capacity for co-operation between the various cellular components of the immune response, and antibodies. Most important are those intra-tumor interactions, where in the MSV system cytotoxic T cells and macrophages have been implicated in successful resistance. Other possibly relevant mechanisms are those mediated by natural killer (NK) cells. These may be of two types: lymphoid (weakly theta-antigen positive in the mouse) or monocyte-macrophages (mouse and man). The activity of lymphoid NK cells in rodents may be boosted by bacterial immunostimulants and poly I:C, a phenomenon mediated by interferon. Their *in vivo* relevance to tumor resistance can be evaluated by assessment of the rate of clearance of isotopically-labelled tumor cells, studied in irradiated mice in which T cell responses are absent. Further dissection of these reactions and of the factors which interfere with their expression, should facilitate the design of more effective immunotherapy, and methods for assessing prognosis or monitoring therapy.

There is considerable evidence for a major role of immune responses in resistance against a variety of experimental tumors, but it has been very difficult to relate the results of a given in vitro assay to in vivo resistance. This has led to some skepticism regarding the in vivo relevance of these assays. However, a likely explanation for the lack of apparent correlation is that the host immune responses to tumors are quite complex. To adequately understand the role of immune responses, we probably need to be aware of a variety of factors which may be involved in the induction and expression of anti-tumor immunity. Such factors include: a) antigenicity and immunogenicity of the tumor cell; b) the mechanisms involved in induction of tumor immunity; c) the effector mechanisms for expression of immune responses against tumors; and d) the mechanisms which can interfere with effective anti-tumor immunity.

In regard to immunogenicity of tumor antigens, several major factors have been identified (Herberman, 1977): a) the ability to induce transplantation resistance. It is important to note that not all cell surface tumor associated antigens, even those which are able to induce detectable immune responses, can function as tumor associated transplantation antigens. b) the amount, form and route of antigen presented to the host. In many tumor systems, the amount of antigen presented, its form and the duration of exposure have been critical factors in determining whether resistance, no effect, or even enhancement of tumor growth are induced. c) the presence of appropriate major histocompatibility complex (MHC) antigens. As will be discussed below, there is some restriction of T cell immune responses to antigens. Immunization with allogeneic tumor cells or tests of hosts against allogeneic tumor materials may be unsuccessful either because of lack of shared tumor antigens or because of a lack of shared MHC antigens.

In addition to considerations related to the antigens themselves, a variety of host factors may be involved in successful induction of tumor immunity: a) responsiveness of the host. This includes general immunologic competence and also the genetically-determined ability to recognize the particular tumor antigens; b) the frequent need for cell cooperation in anti-tumor responses. Instances for cooperation between macrophages and T cells, T cells and B cells and T cells with other subpopulations of T cells have all been described, c) relevance of a particular in vitro assay to in vivo resistance. The effector mechanisms that may be involved in responses to tumors and which may be measured by in vitro assays are quite heterogeneous and these can include cytotoxic, proliferative and lymphokine responses by T cells; production of anti-tumor antibodies by B cells which can directly affect tumor cells or can interact with K cells or macrophages to produce antibody-dependent cell-mediated cytotoxicity; and cytotoxicity by macrophages and natural killer (NK) cells.

In regard to the importance of MHC compatibility for inducing effective immune responses against tumors, it is commonly assumed that the MHC restriction, which has been shown for responses by cytotoxic T cells to viral antigens (Doherty et al, 1976) or chemical haptens, also applies to responses to tumor antigens. However, although some preference for histocompatibile targets has generally been seen in cytotoxicity assays, the restriction has been found not to be complete. In a murine sarcoma virus (MSV) tumor system, which has been extensively studied in my laboratory, T cells have been shown to respond to tumor associated antigens on appropriate allogeneic as well as syngeneic tumor cells (Holden et al, 1977). In cytotoxicity assays performed for 18 hrs or more instead of 4 hrs, good killing of allogeneic targets was seen. In addition, it has been possible to induce secondary responses in vivo and in vitro with allogeneic stimulator cells, to adsorb cytotoxic effector cells on monolayers of allogeneic cells, to competitively inhibit cytotoxicity with cold targets in a 4 hr ^{51}Cr release assay, and to produce migration inhibitory factor in response to allogeneic tumor cells.

Although MHC-incompatible tumor antigens can be recognized, the MHC has been shown to play a central role in the immune response to soluble tumor antigens. Such soluble antigens need to be presented to immune lymphocytes by macrophages (Landolfo et al, 1977) and these macrophages need to be MHC-compatible with the immune lymphocytes. In the MSV tumor system, the genetic control for this MHC restriction of T cell-macrophage interaction has been shown to be in the I-A region of the H-2 complex (Landolfo et al, 1978).

Most immune responses to tumors are measured in peripheral lymphoid organs, with spleen, blood or peritoneal exudate cells usually being used. One possible explanation for a lack of clear relationship between detectable immune responses by these peripheral lymphoid cells and tumor growth is that such tests may not adequately reflect the immune responses actually taking place in the site of tumor growth. Strongly reactive anti-tumor effector cells would not be expected to be useful if they were unable to enter into the tumor. Therefore attention has recently been focused on immune responses occurring in situ. H. Holden and others in my laboratory have extensively studied the nature of the host cells in MSV tumors in mice (Holden et al, 1976, 1978). After mechanical and enzymatic disaggregation of tumors, it has been possible to separate various cell types by 1g velocity sedimentation. Regressing MSV tumors contained a large number of T cells, macrophages of varying sizes, and a small number of B cells. In such regressing tumors, the T cells had strong, specific cytolytic activity and macrophages have strong, non-specific cytostatic and cytolytic activity against tumor cells. At least two and usually three peaks of macrophages of various sizes were cytotoxic. In contrast, progressor MSV tumors, induced by a variant strain of MSV, had only one peak of macrophage-mediated cytolytic activity, associated with the smallest sized macrophages. Development of cytolytic activity, but not cytostatic activity, appeared dependent on T cells since in progressively growing MSV tumors in nude, athymic mice, no cytolytic macrophage activity has been detected.

Recently there has been increasing recognition of natural cell-mediated cytotoxicity as a potentially important anti-tumor effector mechanism in addition to that of specifically immune T cells and of activated macrophages. Although natural cellular cytotoxicity was first recognized only a few years ago, there has already been extensive research in many laboratories on the nature of the effector cells, the possible mechanisms of cytotoxicity, the factors regulating the levels of reactivity, and the relevance of natural immunity to in vivo resistance against tumor growth and immune surveillance. A principal component of natural cell-mediated cytotoxicity in rodents and man has been found to be a particular subpopulation of lymphocytes which have been termed NK cells (Herberman and Holden, 1978). However, it is important to consider the possible role of other natural effector mechanisms. Macrophages or monocytes of normal individuals have been found to have potent cytostatic and cytolytic activity, as measured in long-term in vitro assays (Mantovani et al, 1979a; Tagliabue et al, 1979). Monocytes isolated by adherence from PBL of normal human donors caused appreciable lysis of human and mouse tumor cell lines. Transformed cell lines were quite susceptible to lysis, whereas little or no cytotoxicity was seen against untransformed cells (Mantovani et al, 1979b). Peripheral blood monocytes or peritoneal macrophages of normal mice gave a similar pattern of cytotoxicity.

The cytolytic effects of macrophages or monocytes only became evident after incubation with target cells for 48-72 hr. In contrast, NK cells display very rapid cytolytic effects and represent the main natural effector cells in short-term assays. Many investigators have attempted to characterize NK cells and this has led to considerable controversy. However, in recent studies, we have found that most of the discrepancies are technical in nature and that NK cells in rodents and man share many common features (Herberman and Holden, 1978). On the basis of initial cell separation studies, NK cells appeared to be null cells, i.e., lacking characteristic markers of either T cells or B cells. In mice, high levels of NK activity were found in nude, athymic mice and this seemed to support the non-T

cell nature of NK cells. However, it has been recently shown that treatment with
high concentrations of anti-Thy 1 serum plus complement, or repeated treatments,
eliminated most mouse NK activity (Herberman et al, 1978). Use of congenic antiserum and of mice congenic for Thy 1 gave the appropriate results. Therefore it
now appears that mouse NK cells have a low density of Thy 1 antigen.

An area of substantial interest has been the factors influencing the levels of NK
activity. Some clues regarding regulation of activity were obtained when it was
found that inoculation of NK-susceptible target cells, viruses, or bacterial
immunoadjuvants such as BCG or Corynebacterium parvum caused a rapid increase in
cytotoxicity by cells with the characteristics of NK cells. When it was found
that rat NK activity could also be boosted by poly I:C, a potent interferon in-
ducer, the possible mediation of the boosting by interferon was seriously considered. We have performed an extensive series of experiments in mice (Djeu et al,
1979) and have found that inoculation of interferon, as well as poly I:C and other
interferon inducers, can appreciably augment NK activity. The effects were particularly noticeable in mice older than 12 weeks of age, whose spontaneous levels
of NK activity had already declined to negligible levels. The kinetics of boosting with the various agents was consistent with the reported kinetics of interferon
production. Gidlund et al (1978) have obtained similar results and showed that
the boosting could be blocked by administration of anti-interferon. After overcoming some of the problems associated with in vitro lability of mouse NK cells,
we have been able to incubate mouse spleen cells with poly I:C or interferon and
produce rapid augmentation of NK activity (Djeu et al, 1979). Similarly, these
effects were also inhibited by anti-interferon. The degree of boosting in vitro
by interferon has varied substantially among strains, with high NK strains being
more sensitive to boosting. This suggests that the differences in NK activity
among strains may be due, at least in part, to differential susceptibility to
interferon produced by various in vivo stimuli. As in the in vivo experiments,
spleen cells from older mice, with low or absent NK activity, underwent at least
as great an increase in cytotoxicity as did cells from young mice.

There already have been a number of suggestions or indications of mediation of in
vivo resistance against tumors by NK cells. The findings of high NK activity in
nude and neonatally thymectomized mice provide a very plausible explanation for
the observations of resistance to tumors in such animals (Herberman and Holden,
1978). Most of the data available on the in vivo role of NK cells involve correlations between in vivo resistance of mice to challenge with long transplanted
tumors and in vitro NK activity. Recently, it has been possible to perform in
vivo studies of anti-tumor resistance in lethally irradiated mice and thereby
eliminate the possible contribution by immune T cells. Furthermore, to obtain a
quantitative measure of resistance within a relatively short period of time, survival of tumor cells within the host was determined by the use of radioisotopes.
Mice have been inoculated intravenously with tumor cells prelabeled with ^{125}I-
iododeoxyuridine and then the amount of isotope in various organs, mainly the
lungs, spleen and liver, was measured after 4 hr (Riccardi et al, 1979). In young
mice of strains with high NK activity, there was a major decrease in radioactivity
recovered at 4 hr in the spleen, liver and lungs, when compared to the amount
present at 15 minutes in the same group of mice or to the amount recovered at 4 hr
from older mice or from low NK strains. In these experiments, pretreatment of the
mice with lethal irradiation had no effect on in vitro or in vivo reactivity,
pyran, poly I:C and C. parvum increased both types of reactivity, and silica and
carrageenan inhibited both activities. Thus, this in vivo assay correlates very
well with NK cell reactivity and should provide a very useful and rapid tool for
assessment of the in vivo role of NK cells against a variety of tumors and under a
wide range of experimental conditions.

A further important factor to consider in regard to tumor immunity is that despite
the frequent involvement of one or more aspects of the immune system, usually

resistance is ultimately ineffective and tumors grow progressively and kill the host. Therefore one needs to consider possible mechanisms for ineffective tumor immunity: a) tolerance. Currently this is not thought to be a frequent explanation for lack of resistance to tumors. b) inadequate levels of immune responses. There may simply be a question of the balance between the rapidity of development of immune responses, the extent of this reactivity, and the rapidity of tumor growth. c) interference or suppression. This is being focused on increasingly as a likely explanation for many instances of ineffective anti-tumor immunity. Interference with immune responses could be mediated by specific or nonspecific serum blocking factors, suppressor T cells or macrophages, or by immunosuppressive materials in tumor cells.

We have observed, in a variety of tumor systems, that macrophages from tumor-bearing individuals can interfere with immune responses against the tumors (Herberman et al, 1979). A predominant activity of these suppressor macrophages has been inhibition of lymphoproliferative responses to tumor antigens as well as to other antigens and to mitogens. However, recent evidence has indicated that proliferation-independent immune responses can also be inhibited. This has been observed with in vitro generation of secondary cytotoxic responses (Klempel and Henney, 1978) and with production of migration inhibitory factor (Varesio et al, 1979).

As illustrated by the above examples, the host immune response is multifaceted. The recent progress in dissecting out several of the major factors involved in induction and expression of tumor immunity is encouraging. It is anticipated that further advances in our understanding of this important area will be forthcoming. As we develop more insight into the details of the immune response and the factors interfering with effective resistance against tumor growth, we should be able to rationally manipulate the systems and devise better approaches to immunotherapy of cancer. In addition immunologic assays which are found to correlate well with extent of disease or with immune manipulations should be useful for assessing prognosis and for monitoring of therapy.

REFERENCES

Djeu, J. Y., J. A. Heinbaugh, H. T. Holden, and R. B. Herberman (1979). Augmentation of mouse natural killer cell activity by interferon and interferon inducers. J. Immunol., in press.
Doherty, P. C., R. V. Blanden, and R. M. Zinkernagel (1976). Specificity of virus-immune effector T cells for H-2K or H-2D compatible interactions: Implications for H-antigen diversity. Transpl. Rev., 29, 89-124.
Gidlund, M., A. Orn, H. Wigzell, A. Senik, and I. Gresser (1978). Enhanced NK cell activity in mice injected with interferon and interferon inducers. Nature, 273, 759-761.
Herberman, R. B. (1977). Immunogenicity of tumor antigens. Biochem. Biophys. Acta Rev. Cancer, 473, 93-119.
Herberman, R. B. and H. T. Holden (1978). Natural cell-mediated immunity. In G. Klein and S. Weinhouse (Eds.), Advances in Cancer Research, Vol. 27, Academic Press, New York. pp. 305-377.
Herberman, R. B., M. E. Nunn, and H. T. Holden (1978). Low density of Thy 1 antigen on mouse effector cells mediating natural cytotoxicity against tumor cells. J. Immunol., 121, 304-309.
Herberman, R. B., H. T. Holden, J. Y. Djeu, T. R. Jerrells, L. Varesio, A. Tagliabue, S. L. White, J. R. Oehler, and J. H. Dean (1979). Macrophages as regulators of immune responses against tumors. In Proceedings of the 8th International Congress of the Reticuloendothelial Society, in press.
Holden, H. T., J. S. Haskill, H. Kirchner, and R. B. Herberman (1976). Two functionally distinct anti-tumor effector cells isolated from primary murine sarcoma virus-induced tumors. J. Immunol., 117, 440-446.

Holden, H. T., S. Landolfo, and R. B. Herberman (1977). T cell dependent reactivity against tumor associated antigens on allogeneic target cells. Transpl. Proc., 9, 1149-1152.

Holden, H. T., L. Varesio, T. Tanyama, and P. Puccetti (1979). Functional heterogeneity and T cell-dependent activation of macrophages from murine sarcoma virus (MSV)-induced tumors. In Advances in Experimental Medicine and Biology, in press.

Klimpel, G. R. and C. S. Henney (1978). A comparison of the effects of T and macrophage-like suppressor cells on memory cell differentiation in vitro. J. Immunol., 121, 749-754.

Landolfo, S., R. B. Herberman, and H. T. Holden (1977). Stimulation of mouse migration inhibitory factor (MIF) production from mouse MSV-immune lymphocytes by soluble tumor associated antigen: Requirement for histocompatible macrophages. J. Immunol., 118, 1244-1248.

Landolfo, A., R. B. Herberman, and H. T. Holden (1978). Macrophage-lymphocyte interaction in migration inhibition factor (MIF) production against soluble or cellular tumor-associated antigens. I. Characteristics and genetic control of two different mechanisms of stimulating MIF production. J. Immunol., 121, 695-701.

Mantovani, A., T. R. Jerrells, J. R. Dean, and R. B. Herberman (1979a). Cytolytic and cytostatic activity on tumor cells of circulating human monocytes. Submitted for publication.

Mantovani, A., A. Tagliabue, J. H. Dean, T. R. Jerrells, and R. B. Herberman (1979b). Cytolytic activity of circulating human monocytes on transformed and untransformed human fibroblasts. Submitted for publication.

Riccardi, C., P. Puccetti, A. Santoni, and R.B. Herberman (1979). Rapid in vivo assay of mouse NK cell activity. Submitted for publication.

Tagliabue, A., A. Mantovani, M. Kilgallen, R. B. Herberman, and J. L. McCoy (1978). Natural cytotoxicity mediated by normal mouse monocytes and macrophages and its responsiveness to lymphokines. Submitted for publication.

Varesio, L., R. B. Herberman, and H. T. Holden (1979). Suppression of a proliferation-independent immune response by macrophages infiltrating a murine tumor. Submitted for publication.

Studies of the Immunogenic Properties of a Solubilized, Partially Purified Tumor Rejection Antigen (TSTA) from a Chemically-Induced Sarcoma

L. W. Law, E. Appella and G. C. DuBois

Laboratory of Cell Biology, National Cancer Institute, Bethesda, Maryland 20014, U.S.A.

ABSTRACT

The tumor rejection antigen, TSTA, has been solubilized with NP40 from plasma membranes of the Meth A sarcoma. The antigen assayed by an in vivo rejection assay in syngeneic BALB/c mice was released in good yield and partially purified by a series of chromatographic procedures. The isolated TSTA immunized specifically against Meth A in exceedingly low amounts (0.5 to 5 μg providing greater than 90% tumor inhibition). A linear dose response was obtained with the soluble antigen and with the chromatographed fractions. H-2, β-2 microglobulin and the viral structural proteins gp70 and p30 were separated from TSTA. The antigen was not degraded and appreciable loss as determined by specific activity did not occur during purification or storage. The solubilized, partially purified TSTA retained all the immunogenic and biologic characteristics of the intact sarcoma cell. No deviations of the immune responses of the host were observed following treatment with solubilized antigen.

INTRODUCTION

Individually distinct tumor rejection antigens (TSTAs) are associated with most chemically induced neoplasms regardless of species. The antigens which reside on the membranes of tumor cells have the capacity to elicit tumor rejection in the syngeneic or autochthonous host. Efforts to solubilize and purify these TSTAs, with retention of tumor rejecting capacity, have not been too successful (Natori, Law & Appella, 1977; Law & Appella, 1975). We have reported that TSTA from the membranes of a methylcholanthrene-induced sarcoma, Meth A, could be isolated, solubilized with the detergent NP40 and partially purified by conventional chromatographic procedures including gel filtration, lectin affinity chromatography and column electrophoresis (Natori, Law & Appella, 1977).

Meth A was maintained in ascitic form and was developed also into a tissue culture-passaged line. Most of the work in this laboratory involving isolation of the antigen from membranes and tumor challenge was performed with the in vivo passaged line; the TD50 was regularly 10^2 cells. Meth A has a strong TSTA such that complete protection in syngeneic BALB/c mice against a challenge of 2×10^4 cells (200 x TD50) is achieved through a single immunization with 10^6 irradiated Meth A cells. Cross-reactivity in vivo has not been observed with any other neoplasms. The TSTA elicits specific cellular immune responses by T cells as observed in adoptive transfer studies, but we have been unable to develop syngeneic antisera to the Meth A TSTA. Recently, DeLeo and coworkers (1977) have reported the induction of specific syngeneic antisera against Meth A as detected in vitro in a humoral cytotoxic assay; it was hoped that this antiserum could be used to establish a reliable competition in vitro assay to better quantitate TSTA in the several purification steps; however the antigen producing specific cytotoxic antisera against Meth A appears not to be TSTA (Appella, to be published).

Some of the salient findings of our earlier work may be summarized as follows:

1. Following solubilization from isolated membranes (CM) with NP40, this material (CS) was concentrated and applied to an Ultragel AcA 22 column. Major tumor rejection activity was found in Fractions I and III; 45% of the activity of the NP40-solubilized antigen was recovered in Fraction III.

2. The majority of H-2 and β-2 microglobulin activities, also assayed in the AcA 22 fractions were separated from tumor rejection activity.

3. Fraction III material was subjected to Lens culinaris lectin chromatography; 97% of the TSTA activity passed through the column but 97% of the H-2 activity was recovered in the bound material. The unbound LcH(u) material was found to induce 50% tumor inhibition at a dose of approximately 12 µg of protein.

4. Both the CS and LcH(u) materials were effective in immunizing in a specific manner against Meth A.

5. Recovery of TSTA from the CM following NP40 solubilization was variable but a yield as high was 30-40% was obtained. There was an increase in specific activity of TSTA however in succeeding fractionations and this is shown in Table 1; recovery in this preparation was only 6% however.

TABLE 1   Recovery of TSTA from Membranes of Sarcoma Meth-A

Type of preparation	Protein(mg)	TSTA(units[a])	Specific activity (units/mg)	H-2.4(units[b])	β_2-Microglobulin (units[b])
CM	1,710(100)[c]	1,899,810(100)	1,111	434,782(100)	2,631,550(100)
NP40-CS	710(42)	118,570(6.2)	167	311,550(72)	1,914,190(73)
Fraction III	134(7.8)	53,600(2.8)	400	11,056(2.5)	73,024(2.7)
LcH-u	130(7.6)	52,000(2.7)	400	308(0.07)	37,290(1.4)

[a] One unit corresponds to the amount of protein that induces 50% inhibition of tumor growth compared to that of control mice.

[b] One unit corresponds to the amount of protein that inhibits 50% of the binding.

[c] Numbers in parentheses, percentages.

6. A linear dose-response TSTA activity was found in the several fractions assayed as well as in CM and CS.

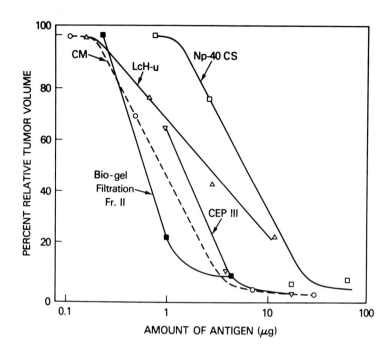

Fig. 1. In vivo assay for tumor rejection antigen (TSTA) using crude membrane (CM), NP40 CS, the solubilized antigen and several chromatographic fractions: LcH-u, the lectin unbound fraction, fraction III from column electrophoresis and fraction II from rechromatography. Immunization done twice at 10 day intervals and challenge with 2×10^4 Meth A cells 10 days after last immunization. At least 8 mice for each point on curve (Natori, Law & Appella, 1977).

7. LcH(u), the unbound fraction, with high TSTA activity when subjected to column electrophoresis showed only 3 major fractions, one of which retained most of the TSTA activity; this was 10-fold higher than that of the original NP40 CS. This fraction coincided with the position of a standard size marker protein of 70,000 daltons.

We report here in continuation of this work some biologic and immunogenic properties of a new preparation of Meth A TSTA released from the membranes of dissociated sarcoma cells by NP40 followed by further fractionation on an AcA 22 column, LcH (lens culinaris) affinity column, WGA (wheat germ agglutinin) column and AcA 54 chromatography. Specific details of the procedures and problems in solubilization and fractionation have been published by Appella and coworkers (1978).

The striking immunity provided by the unbound material (LcH-u) obtained following Lens culinaris lectin affinity chromatography in syngeneic BALB/c mice challenged with 2×10^4 Meth A sarcoma cells at 200 X the TD50 is shown in Table 2. Immuniza-

TABLE 2 Immunogenic Potential of LcH Unbound Solubilized TSTA from Meth A

Group	Immunized with these doses of antigen x 2:						Controls (mean tumor volume)
	0.5μg	1.0	5.0	10.0	25	50	
1	-	-	-	0.006[a]	0.04	0.011	(520 mm^3)
2	0.016	0.08	0.005	0.0	0.0	-	(225 mm^3)
3[b]	-	-	-	0.02	-	0.0	(360 mm^3)
4[c]	-	0.21	0.12	-	-	0.15	(434 mm^3)
5	-	-	0.16	-	-	-	(288 mm^3)

[a] Immunogenic index = $\frac{\text{mean tumor volume of immunized group}}{\text{mean volume of controls}}$, thus, 1.0 = no inhibition of tumor growth and 0.0 = complete inhibition. At least 8 BALB/c mice in each group; mice immunized twice, 10 days apart and challenged at 10 days after last immunization with 2 x 10^4 Meth A ascites cells.

[b] Sloan-Kettering subline of Meth A, negative for MuLV and MuLV-related antigens, (See DeLeo and coworkers (1977)).

[c] Only one preparation used throughout this study but LcH-u material for groups 4 and 5 had been stored for 4 months at -20°C.

tion in group 1 was accomplished with the lectin-unbound material, freshly prepared, while groups 4 and 5 were immunized with the same preparation stored at -20°C for 4 months. There is some indication of loss of immunogenic potential of the stored material, although a complete dose-response study was not done. Greater than 95% tumor inhibition was achieved with as little as 1.0 μg (0.5 μg x 2) of the lectin-unbound fraction (group 2). On a μg/protein basis this activity is far greater than that achieved with the LcH fraction reported earlier (Natori, Law & Appella, 1977), where approximately 12 μg was required for 50% tumor inhibition. The Sloan-Kettering subline of Meth A, found to be negative for MuLV and MuLV-related antigens (DeLeo and coworkers, 1977), was also strikingly inhibited (group 3). LcH-u, as well as the NP40-CS preparations, showed specificity in the induction of immunity; this is discussed later. LcH-u still contained 5% H-2 activity and presented many bands, but principally those of low MW, on SDS gel.

Chromatography of the LcH-u material on WGA (wheat germ agglutinin) column completely removed H-2 as assayed by a sensitive radioimmunoassay (RIA) employing broadly-reacting sheep anti-H2 antiserum but tumor rejection activity was found to be present in both the bound and unbound fractions. These results with WGA chroma-

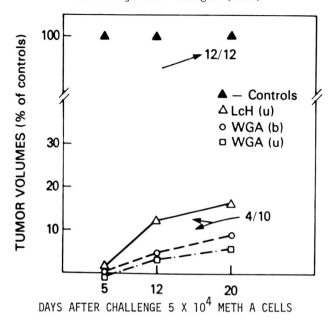

Fig. 2. Immunogenic activity of the fractions bound (b) and unbound (u) following wheat germ agglutinin (WGA) chromatography. BALB/c mice were immunized x 2 with 10 µg protein; 12/12 with progressively growing tumors in control group but only 4/10 positive in each of the immunized groups.

tography are at variance with the findings of Sikora and coworkers (1978) who did find binding of TSTA from chemically induced sarcomas of C57BL origin to WGA columns. The detergent was removed from the unbound fraction following chromatography on the WGA column and the material was then chromatographed on an ACA 54 column. The material in the 60,000 dalton range, representing the major peak was assayed for tumor rrejection activity at 0.5, 1.0 and 5.0 µg doses. This fraction did not contain the viral structural protein gp70 nor p30. Two immunizations, 10 days apart, followed by challenge with 2×10^4 Meth A cells were done. The results are shown in Table 3. Activity was retained in this major peak.

TABLE 3 Immunogenicity of Materials Following WGA and AcA-54 Chromatography

Groups		Immunization with[a]:	Immunogenic index
1	a	0.5µg x 2 WGA (unbound)	0.27
	b	1.0µg x 2 WGA (unbound)	0.21
	c	5.0µg x 2 WGA (unbound)	0.07
2		10.0µg x 2 WGA (unbound)	0.06
3	a	0.5µg x 2 AcA-54[b] (II)	0.33
	b	1.0µg x 2 AcA-54 (II)	0.44
	c	5.0µg x 2 AcA-54 (II)	0.77

[a] See Table 2 for details of immunization and challenge. 8 mice used for each experimental and control group.

[b] Details of AcA54 preparation in text.

Specificity of Response to Solubilized TSTA from Meth A

Previously, it was shown that the crude membrane (CM) and NP40 solubilized (CS) materials from Meth A induced immunity only against Meth A and not against other syngeneic neoplasms such as mKSA, an SV40 induced sarcoma and LSTRA, an M-MuLV-induced lymphoma. Table 4 records the results of specificity studies against mKSA and 2 additional methylcholanthrene-induced sarcomas CI-3 and CI-4 of BALB/c strain mice; all of these neoplasms developed in this laboratory have their own strong individually specific TSTA. Immunizations here were done with the material that

TABLE 4 Specificity of the Immunity Induced by Solubilized Meth A Tumor Rejection Antigen (TSTA)

Tumor Challenge	Immunization	$\frac{\text{No. Tumorous}}{\text{No. Challenged}}$ (immunogenic[a] index)
Meth A (2×10^4 cells)	$10\mu g \times 2$[b] $20\mu g \times 2$ $100\mu g \times 2$ none	1/8, 1/10 (0.016) 0/8 0/8 (0.0) 16/16 -
mKSA[c] (2×10^4 cells)	$100\mu g \times 2$ none	14/14 (1.2) 12/12 -
CI-3 (1×10^5 cells)	$20\mu g \times 2$ none	10/10 (1.0) 10/10 -
CI-4 (1×10^5 cells)	$20\mu g \times 2$ none	8/8 (1.0) 8/8 -

[a] Immunogenic index = $\frac{\text{mean tumor volume treated group}}{\text{mean tumor volume of controls}}$

[b] Two immunizations subcutaneously 10 days apart with s.c. challenge 10 days after last immunization.

[c] mKSA = SV40-induced sarcoma; CI-3 and CI-4 are methylcholanthrene-induced sarcomas, all in BALB/c mice and having their own strong TSTA. Solubilized membranes from mKSA sarcoma provided complete protection against mKSA when immunized with $1.0\mu g$ protein.

was not bound by the lectin column. The chromatographed unbound fraction (LcH-u) induced strong immunity but only against the Meth A sarcoma, behaving in this respect as did intact Meth A cells, and the CM and CS preparations.

Immune Reactivities of the Host to Solubilized TSTA

Several investigators have reported immune deviations of the host following inoculations of solubilized tumor rejection, or tumor associated antigens. (See particularly Price and Robins, 1978). The development of a "null" state following large doses of antigen, inhibition of growth attained only over a very restricted antigen dose range (Pellis and Kahan, 1975), a rapid decay of the immune state, facilitation (rather than inhibition) of tumor growth, Paranjpe and Boone (1975); Rao and Bonavida (1977), specific inhibition of cell-mediated immune responses, Baldwin and Price (1977); Baldwin, Embleton & Moore (1973), etc. have been described. These immune "deviations" have been attributed to qualitative changes in the antigen resulting from processing or to inappropriate responses on the part of the host as a result of the differing form in which the antigen is presented, resulting in the production of tumor specific antibody rather than the appropriate cell-mediated response. Most of these crude soluble or solubilized preparations studied however did not produce tumor rejection immunity, were feebly immunogenic under restricted conditions or indeed were not assayed for immunogenic potential.

It was of interest therefore to look for inappropriate or modified immune responses in BALB/c mice with our NP40-solubilized antigen and the partially purified lectin-unbound material (LcH-u) described in this study and both shown to be strikingly immunogenic.

We did not observe "overloading" with antigen nor the development of a "null" state.

TABLE 5 Inhibition of Growth of Meth A Sarcoma with Varying Doses of NP40-Solubilized Meth A TSTA

Dose (in µg x 2)	No. tumorous mice / No. challenged[a] (% control tumor volume)
1.25[b]	6/8 (25%)
5	1/8 (0.7%)
20	1/8 (0.9%)
50	0/8
100	0/8
500	0/8
1000	0/8
None	24/24 (420 mm^3)

[a] Challenge at 10 days after the second immunization with 2×10^4 Meth A ascites cells.

[b] The LcH chromatographed material (LcH-u) behaved similarly when assayed at 5, 50 and 500 µg dose levels; that is, complete inhibition of growth was achieved.

Total doses of antigen ranging from 2.5 µg to 2000 µg protein per mouse all produced immunity to a challenge with 2×10^4 Meth A cells; nor was there an optimal or a restricted tumor dose at which the antigen was most, or only, effective.

TABLE 6   Influence of Challenge Load in BALB/c Mice Immunized with NP40 Solubilized Antigen (50µg x 2)

No. of cells in challenge	No. of tumorous mice Total challenged	Latent period (in days)
10^3	0/5 (2/5)	15
10^4	0/5 (4/5)	11
10^5	1/5 (5/5)	9
5×10^5	3/5 (5/5)	9
10^6	5/5[a] (5/5)	8

[a] Latent period = 15 days compared with 8 days in controls.

BALB/c mice immunized with 50 µg x 2 of the NP40-solubilized material (CS) were shown to be solidly immune when challenged 60 days after the last immunization indicating a stable immune state induced by soluble antigen. Spleen cells from mice,

TABLE 7   Stability of Immunity Produced by Solubilized Meth A TSTA

BALB/c mice immunized with 50 µg x 2 of NP40 (CS). Challenged with 2×10^4 Meth A sarcoma cells at day:		
10	30	60
	No. positive/No. inoculated	
0/8 (10/10)[a]	0/4 (4/4)	0/4 (3/4)

[a] Non-immunized controls shown in parentheses.

similarly immunized, were found to maintain the capacity to neutralize Meth A sarcoma cells as measured in the Winn assay, 150 days after immunization, without an intervening challenge of tumor cells (data not shown).

We did not observe any modification of the tumor rejection response through pretreatment of BALB/c mice with NP40 solubilized material prior to immunization as reported by Baldwin, Embleton and Moore (1973) who used soluble or membrane preparations of a rat hepatoma. (The preparations of Baldwin and coworkers did not however have tumor rejection activity.

Thus, in the limited studies we have now carried out it is clear that the immune responses of the BALB/c host to solubilized and to the partially purified TSTA of Meth A do not differ qualitatively from the responses to TSTA presented on tumor cells. We have been unable to employ other in vitro assays such as inhibition of fluorescence or inhibition of cell-mediated cytotoxicity, used by others in studies of soluble tumor antigens; clear-cut specificity of response in these assays was never attained in our hands employing Meth A cells.

REFERENCES

Appella, E., G. DuBois, T. Natori, M. J. Rogers, and L. W. Law (1978). Histocompatibility antigens and tumor specific transplantation antigens of a methylcholanthrene induced sarcoma. In R. W. Ruddon (Ed.), Biological Markers of Neoplasia, Elsevier Press, Amsterdam.

Baldwin, R. W., M. J. Embleton, and M. Moore (1973). Immunogenicity of rat hepatoma membrane fractions. Brit. J. Cancer, 28, 389-399.

Baldwin, R. W. and R. A. Robins (1977). Induction of tumor-immune responses and their interaction with the developing tumor. In Comtemporary Topics in Molecular Immunology, Vol. VI, pp. 177-207.

DeLeo, A. B., H. Shiku, T. Takashi, M. John, and L. J. Old (1977). Cell surface antigens of chemically induced sarcomas of the mouse. J. Exp. Med., 146, 720-734.

Law, L. W., and E. Appella (1975). Studies of soluble transplantation and tumor antigens. In F. F. Becker (Ed.), Cancer, Vol. 4, Plenum Press, New York. pp. 135-154.

Natori, T., L. W. Law, and E. Appella (1977). Biological and biochemical properties of nonidet P40-solubilized and partially purified tumor specific antigens of the transplantation type from plasma membranes of a methylcholanthrene induced sarcoma. Cancer Research, 37, 3406-3413.

Paranjpe, M. S., and C. W. Boone (1975). Specific depression of the antitumor cellular immune response with autologous tumor homogenate. Cancer Research, 35, 1205-1209.

Pellis, N. R., and B. D. Kahan (1975). Specific tumor immunity induced with soluble materials: Restricted range of antigen dose and of challenge tumor load for immunoprotection. J. Immunol., 115, 1717-1722.

Price, M. R., and R. A. Robins (1978). Circulating factors modifying cell-mediated immunity in experimental neoplasia. In J. E. Castro (Ed.), Immunological Aspects of Cancer, University Park Press, Baltimore. pp. 155-182.

Rao, V. S. and B. Bonavida (1977). Detection of soluble tumor-associated antigens in serum of tumor bearing rats and their immunologic role in vivo. Cancer Research, 37, 3385-3389.

Sikora, K (Personal communication).

Study of Macrophage Receptors for Allogenate Antitumor Cytophilic Antibodies

A. B. Mazzolli

Segunda Catedra de Hustologia, Facultad de Medicina,
Universidad de Buenos Aires, Buenos Aires, Argentina
Seccion Leucemia Experimental, Instituto de Investigaciones Hematologicas,
Academia Nacional de Medicina, Buenos Aires, Argentina

ABSTRACT

The object of this paper was to investigate the presence of cytophilic activity in the serum of BALB mice inoculated with an allogeneic AKR lymphoma. This model makes tumor-bearing (progressor) mouse and tumor rejecting (regressor) animals simultaneously available. When mouse lymphoma cells were incubated with guinea pig peritoneal cells, a background value of "natural cytophilic activity" was observed within a range of 20 to 50 tumor-bearing macrophage per thousand; this value was not altered by pre-incubation of the macrophages with normal mouse serum. Incubation of the guinea pig macrophages with regressor serum significantly increased the number of the macrophages bearing lymphoma cells (106 $\pm$ 30). On the contrary, incubation with progressor serum gave results similar to that of the control. Thymectomy in this model does not alter either *in vivo* incidence or humoral cytophilic activity. These results showed guinea pig macrophage Fc receptors are not species specific. On the other hand it is postulated that AKR lymphoma cells are capable of restoring T-immunocompetence to a thymectomized allogeneic BALB host.

Key words: Cytophilic activity, allogeneic AKR lymphoma, thymectomy, guinea pig macrophages.

INTRODUCTION

In the last few years, several authors (Evans and Alexander, 1972; Evans and Grant, 1972; Alexander, 1976; Russell and Mc Intosh, 1977) have attributed to the macrophage, an important role in tumour rejection, in spite of which the mechanism involved remains obscure. On the other hand, the presence of anti-tumour cytophilic activity has been reported by Hoy and Nelson (1969) and Mitchell and Mokyr (1972), using murine tumours and either allogeneic or syngeneic peritoneal macrophages. Mitchell and Mokyr (1972) also demonstrated cytophilic activity with human leukemia cells using mouse peritoneal macrophages.

The first indication of the existence of macrophage-associated immunoglobulins was given by Girard and Murray (1954). Boyden (1964) using mouse or rabbit spleen macrophages, concluded that there was an antibody which had a strong avidity not only for specific antigen but also for macrophages, describing this antibody as "cytophilic".

The object of the present work was to investigate the participation of cytophilic activity in tumour immunity both in intact and thymectomized mice, applying an indirect technique developed previously in our laboratory for an autoimmune model (Mazzolli and Barrera, 1974). This was made possible by the use of an experimental model in which an AKR lymphoma is conditioned to grow in BALB mice, leading to a reproducible allogeneic tumour incidence which makes tumour-bearing (progressor) an tumour-rejecting (regressor) animals simultaneously available (Saal and colleagues, 1972; Pasqualini, 1976).

MATERIALS AND METHODS

Animals. Male and female, 2-4 month-old BALB and AKR mice raised in our colony were used in an AKR tumour-BALB host combination. Adult outbred albino guinea pigs of both sexes, weighing between 500 and 700 g were used as peritoneal cell donors.

Donor tumour. It consisted of an AKR lymphoma of spontaneous origin, maintained as a subcutaneous syngeneic cell line, denominated L15.

Allogeneic tumour. Glass cylinders made out of neutral glass tubing, 1 cm in diameter and 1.5 cm in length, were implanted under the skin of BALB mice. Two days later, a solid fragment of L15, containing about 1.5×10^5 cells, was loaded into a 15-gauge trocar and discharged into the cylinder with 0.2 ml of physiological saline. In this model, about half of the animals develop an allogeneic tumor, which kills them in an average of 39 days. This makes it possible to obtain serum simultaneously from progressor and regressor animals, 20-25 days after tumor challenge (Saal and colleagues, 1972; Pasqualini, 1976).

Control tumor. Sarcoma 180 growing in Swiss mice was used as a source of control tumor cells.

Cytophilic test. The technique used, described elsewhere (Mazzolli and Barrera, 1974; Barrera, Mazzolli and Mancini, 1976), consisted of the following steps (fig. 1): 1) The introduction in a glass chamber of 0.2 ml of a guinea pig peritoneal cell suspension obtained with Hanks solution, without the use of irritants; 2) 1 h incubation of the glass-adherent cells with either normal, progressor or regressor serum, pre-heated at 56°C for 30 min; 3) the addition of a tumor cell suspension, 2×10^6 cells/ml, which remained in the chamber for 1 h; 4) the addition of a neutral red solution, 300 µg/ml, which permitted the identification of macrophages by the presence of the red granules in their cytoplasm; 5) the placing of a coverslip on the chambers which were then inverted for microscopic observation of the glass-adherent cells. All procedures were carried out at room temperature with repeated washings with Hanks solution between each step. The presence of cytophilic activity was detected by the adherence of one or more tumor cells to the surface of the macrophage. The results were expressed as the number of macrophages bearing tumor cells per thousand.

Thymectomy. In BALB mice the thymus was removed either during the first 24 h of life (neonatal) or at 1 month of age (adult thymectomy). In this strain runting led to the death of approximately 25% of the neonatally thymectomized animals within the first month of age while all animals survived adult thymectomy. The animals were used when 2 months old. Only animals which could be autopsied and in which no thymic remnants were encountered, were included in the results.

In vivo experiments. The incidence of lymphoma in BALB mice bearing a glass cylinder was compared in 2 groups of animals: those which had undergone a neonatal thymectomy and those which had been submitted to an adult thymectomy; sex and age-matched intact animals were included in each case.

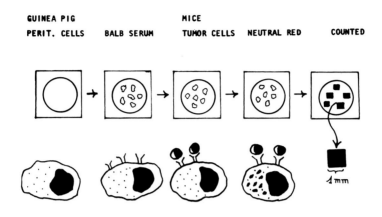

Fig. 1   Anti-tumour cytophilic technique

RESULTS

When mouse tumour cells were incubated with guinea pig peritoneal cells, some of them adhered to the surface of the macrophages; this background value representing the "natural cytophilic activity" of the guinea pig, varied from one animal to the other, within a range of 20 to 50 tumour-bearing macrophages per thousand; this value was not altered by pre-incubation of the macrophages with normal mouse serum, which was the control used in each experiment (table 1).

TABLE 1 Natural Cytophilic Activity

EXPERIMENT	No. tumour-bearing macrophages per thousand							Average
	Without serum	With preheated normal BALB serum						
A	22#	22	20	21				21
B	50	45	45	67	55	50	50	52
C	32	18	22	22	20	20	32	20
D	50	52	50	50	52			51

each value was obtained with a different serum with the same guinea pig peritoneal cells in each experiment.

As can be seen in Table 2 incubation of the macrophages with serum of mice which had rejected a tumour (regressor serum) significantly increased the number of macrophages bearing lymphoma cells (fig. 2). On the contrary, incubation with serum from tumour-bearing mice (progressor serum) gave results similar to that of the controls. It is interesting to note that similar results were obtained when either lymphoma or L15 donor tumour cells were used as target.

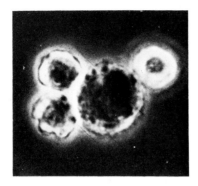

Fig. 2 Normal peritoneal guinea pig macrophage incubated with regressor mouse serum, with three tumour cells adhered on the surface (x 1800).

TABLE 2 Anti-Tumour Cytophilic Activity

	Mouse sera		
Experiment	Regressor	Progressor	Normal
1	108 ± 30^a (6)^b	52 ± 10 (6)	52 ± 8 (6)
2	81 ± 20 (5)	26 ± 8 (5)	20 ± 6 (7)

\# No. tumour-bearing macrophages per thousand

a Average ± SE

b Number of serum mice

That the positive results obtained with regressor serum were tumour-specific was demonstrated by replacing lymphoma cells with Sarcoma 180, in which case no increase in cytophilic activity was observed (Table 3).

TABLE 3 Tumour specificity of cytophilic activity of regressor serum

	No. tumour-bearing macrophages per thousand	
Tumour target cells	Lymphoma regressor serum	Normal serum
AKR Lymphoma	152 ± 38^a (6)^b	29 ± 2 (5)
Sarcoma 180	43 ± 6 (7)	56 ± 4 (6)

a Average ± SE

b Number of serum mice

In vivo experiments with tumour incidence, yielded very similar results in thymectomized animals as compared to the simultaneous controls, both after neonatal and adult thymectomy (Table 4). All progressor animals died in an average of 37 days. Our allogeneic model at times yields a lower tumour incidence, which accounts for the difference in tumour incidence between the control intact groups of animals in the two experiments and is the reason why simultaneously controls are always included. The few animals with thymic remnant were discarded and only autopsied animals were considered.

TABLE 4 Lack of Effect of Thymectomy on the Incidence of an Allogeneic AKR Tumor (Lymphoma P) Growing in BALB Mice Bearing a Glass Cylinder.

Groups	Lymphoma lethal tumors/No. mice	%
Neonatal thymectomy	14/39	36
Simultaneous intact controls	14/36	39
Thymectomy at 1 month of age	17/28	61
Simultaneous intact controls	16/26	61

As for cytophilic activity the serum obtained from thymectomized animals did not alter the background values. As can be seen in Table 5, incubation of the macrophages with serum of mice which had rejected the allogeneic tumour (regressor serum), significantly increased the number of macrophages bearing tumour cells, both in the neonatally thymectomized and in the intact controls in two different experiments in each case. On the contrary, incubation with serum of tumour-bearing animals (progressor serum), either thymectomized or not, gave results similar to that obtained with normal serum.

TABLE 5 Anti-Tumour Cytophilic Activity in Neo-Thymectomized Mice

Groups	Serum mice Normal	Progressor	Regressor
Intact Mice	1° 52# ± 3^a (6)^b 2° 22 ± 1 (6)	65 ± 6 (5) 26 ± 3 (5)	106 ± 22^c (6) 71 ± 12^c (5)
Thymec tomized	1° 53 ± 3 (7) 2° 96 ± 8 (6)	36 ± 5 (7) 103 ± 15 (9)	115 ± 15^c (8) 242 ± 30^c (4)

\# % macrophages bearing adherent lymphoma cells

a Average ± SE

b Number of mice

c Significantly different from the normal serum group, P< 0.01, as calculated by Student's T test.

DISCUSSION

The results obtained indicate the presence of anti-tumor cytophilic activity in regressor serum and its absence in progressor serum; the specificity of the reaction was demonstrated by the negative results obtained with cells from a different tumor. These experiments were made possible by 1) the use of an experimental model in which an AKR tumor-BALB host combination yields a reproducible tumor incidence, making a progressor and regressor animals available simultaneously (Saal and colleagues, 1972; Pasqualini, 1976); 2) the demonstration that the receptor for cytophilic antibodies present on the macrophage is not species-specific, permitting the adherence of mouse cytophilic antibodies on guinea pig macrophages detectable by specific tumor cell adherence.

It is noteworthy that no cytophilic activity was detectable in tumor-bearing animals. This is comparable to results obtained in this experimental model with lymphocyte phytohemagglutinin stimulation where no blast formation could be observed in progressor animals (Conesa, Rumi and Pasqualini, 1976) and with the leukocyte migration inhibition test in the presence of tumor antigen, which was not altered in progressors (Braun, Saal and Pasqualini, 1974); both reactions became positive after tumor rejection. Together with the increased cytophilic activity, it can be postulated that these three immune parameters are related to cell-mediated immunity leading to tumor rejection.

The results obtained indicate that thymectomy, performed either at birth or at 1 month of age, does not alter tumor incidence in an AKR lymphoma-BALB host combination. Furthermore, the increase in specific anti-tumor cytophilic antibodies associated with tumor rejection was also observed in neonatally thymectomized mice.

There is one interesting explanation for this phenomenon and that is, the possibility that the allogeneic AKR lymphoma transplant behaved as a thymus graft: the AKR lymphoma proved to be of T-cell origin since it could be destroyed by anti-Thy serum in the presence of complement and since both the donor syngeneic and the allogeneic lymphomas could form E-rosettes (Conesa, Pasqualini and Astaldi, 1978). According to other authors (Stutman, Yunis, and Good, 1970; Stutman, 1975) mouse newborn tissues contain cells that have already received some thymic influence and are sensitive to the humoral activity of the thymus; it may be that an AKR neoplastic T-cell is capable of providing the necessary factor, stimulating the host lymphocytes to become T-immunocompetent. Further experiments will be carried out in order to confirm this hypothesis.

REFERENCES

Alexander, P. (1976). The functions of the macrophage in malignant disease. Ann. Rev. Med., 27, 207.
Barrera, C. N., Mazzolli, A. B., and Mancini, R. E. (1976). Cytophilic activity in experimental immunologic orchitis in guinea pigs. Fert. Steril., 27, 21.
Boyden, S. V. (1964). Cytophilic antibody in guinea pigs with delayed type hypersensitivity. Immunol., 7, 474.
Braun, N., Saal, F. and Pasqualini, C. D. (1974). Aspectos inmunológicos del crecimiento tumoral murino alogeneico. Sangre, 19, 156.
Conesa, L. C. G., Rumi, L., and Pasqualini, C. D. (1976). Phytohemagglutinin unresponsiveness of spleen cells of mice bearing an allogeneic lymphoma. Bol. Ist. Sieroterap. Milanese, 55, 308.
Conesa, L. C. G., Pasqualini, C. D., Astaldi, G. (1978). E-Rosette forming lymphocytes in thymus lymph nodes and in tumor from mice bearing syngeneic or allogeneic lymphoma. Boll. Ist. Sieroter. Milan., in press.
Evans, R., and Alexander, P. (1972). Role of macrophages in tumor immunity. I. Cooperation between macrophages and lymphoid cells in syngeneic tumor immunity. Immunol., 23, 615.
Evans, R., Grant, C. K. (1972). Role of macrophages in tumor immunity. III. Cooperation between macrophages and lymphoid factors in an in vitro allograft situation. Immunol., 23, 677.

Girard, K. F., Murray, E. G. D. (1954). The presence of antibody in macrophage extracts. Can. J. Biochem. Physiol., 32, 14.

Hoy, W. E. and Nelson, D. S. (1969). Studies on cytophilic antibodies. V. Allo-antibodies cytophilic for mouse macrophages. Austr. J. Exp. Biol. Med. Sci., 47, 525.

Mazzolli, A. B. and Barrera, C. N. (1974). A method for detecting cytophilic activity in a homologous system. J. Immunol. Methods, 4, 41.

Mitchell, M. S. and Mokyr, M. B. (1972). Specific inhibition of receptors for cytophilic antibody on macrophages by isoantibody. Cancer Res., 32, 832.

Pasqualini, C. D. (1976). Why does a tumor grow? Alergol. et Immunopathol., 4, 449.

Russell, S. W., McIntosh, A. T. (1977). Macrophages isolated from regressing Moloney sarcomas are more cytotoxic than those recovered from progressing sarcomas. Nature, 268, 69.

Saal, F., Colmerauer, M. E. M., Braylan, R. C. and Pasqualini, C. D. (1972). Tumor growth in allogeneic mice bearing a plastic cylinder. J. Natl. Cancer Inst., 49, 451.

Stutman, O., Yunis, E. J. and Good, R. A. (1970). Studies on thymus function. II. Cooperative effect of newborn and embryonic hemopoietic liver cells with thymus function. J. Exp. Med., 132, 601.

Stutman, O. (1975). Humoral thymic factors influencing postthymic cells. Ann. N.Y. Acad. Sci, 249, 89.

Cancer Immunotherapy: Experimental Findings Involving Lymphocytes and Antibodies to the Oncofetal Antigen α-1-Protein (AFP)

Alberto J. L. Macario

(Work done in collaboration with G. J. Mizejewski)
Division of Laboratories and Research, New York State Department of Health,
Albany, New York 12201, U.S.A.

ABSTRACT

Cancer immunotherapy encompasses maneuvers which promote humoral and cell-mediated immunity by specific and nonspecific means. Specific immune responses require a tumor-specific antigen, or an antigen that, although not restricted to the tumor, is expressed predominantly in the malignant cell as opposed to all other contemporary cells in the host. These antigens can be membrane-bound or released into the circulation, as for example monoclonal immunoglobulins and AFP of myelomas and hepatomas, respectively. This report reviews the immunotherapeutic effects of antibodies and lymphocytes raised against AFP, and focus upon the hepatoma BW7756 in C57L/J mice. Two experimental approaches are considered: 1. Passive immunotherapy with heterologous anti-AFP antiserum and syngeneic lymphocytes; and 2. Active immunotherapy with allo- and xenogeneic AFP. This is associated with nonspecific means aiming at the breaching of tolerance to autologous AFP and, thereby to the hepatoma with which it shares antigenic determinants.

The immunotherapeutic effects are measured in vivo (tumor incidence, growth rate and histology of tumor and host's tissues) while the interaction between antibodies (and/or lymphocytes) with AFP and target tumor-cells are also studied in vitro. To elucidate the immunotherapeutic mechanisms the following points are examined: a) binding of antibodies and lymphocytes to the tumor-cell membrane; b) demonstration of surface and intracellular AFP determinants; c) effect of antibodies upon tumor-cell metabolism; d) cytopathogenic effect of antibodies and lymphocytes in hepatoma cultures; e) antibody-mediated tumor-cell lysis and relationship to the complement system; and f) immunosuppressive potential of AFP and role of estrogen-binding in this suppression.

KEY WORDS

Cancer immunotherapy; mouse hepatoma BW7756; lymphocyte immunotherapy; alpha-fetoprotein; anti-alpha-fetoprotein antibodies; tolerance to alpha-fetoprotein, breaching of.

INTRODUCTION

Cancer is a major health problem in our era. To deal with it the following means are important: prevention, early diagnosis of incipient tumors and treatment of overt disease. Immunology can contribute to all three of them by providing vaccines, diagnostic reagents and anti-tumor antibodies and cells. Thus immunotherapy includes several maneuvers which promote humoral and cell-mediated immunity. This can be achieved by specific and nonspecific means. The former are those which utilize tumor-specific antigens to elicit anti-tumor antibodies and/or lymphocytes. Nonspecific means of immunotherapy are those which favor initiation of immunity and/or amplify an ongoing immune response, such as adjuvants and agents that remove suppressor cells.

Elicitation of a specific immune response requires a tumor-specific antigen, or an antigen which although not restricted to the tumor, is expressed predominantly in the tumor cell as opposed to other contemporary cells of the tumor-host. These antigens can be sessile, on the cell surface, or can be soluble molecules released by the cancer cells into the extracellular space. Examples of the latter are idiotypic determinants of monoclonal immunoglobulins secreted by myelomas and AFP produced by hepatomas (Eisen, Sakato and Hall, 1975; Ruoslahti, Pihko and Seppälä, 1974).

This report focuses upon immunity to AFP. The following points can be regarded as advantages to the utilization of AFP as a tumor antigen in the elicitation of immune responses against hepatomas: __Firstly__, several normal and pathological conditions, including tumors, which occur in humans and are characterized by elevated synthesis and secretion of AFP, have well defined animal counterparts. Thus questions which cannot be asked experimentally in humans can be investigated in a matched animal model. __Secondly__, soluble tumor antigens, and AFP in particular, can be obtained in workable amounts from several fluids: serum of tumor-bearing individuals, amniotic fluid and spent medium of cultures of hepatoma cells. __Thirdly__, since AFP is in solution, it can be purified and characterized more readily than cell-membrane structures. It can, therefore, be prepared without contaminants for immunization _in vivo_ and for immunological assay _in vitro_. __Fourthly__, the probability of contamination with oncogenic virus particles is smaller for AFP than for other tumor antigens which have to be extracted from malignant cells. This point becomes clearer if one thinks that AFP can be obtained from the amniotic fluid of tumor-free individuals. Thus AFP can be considered for immunization, as a vaccine, without the fear of administering oncogenic particles along with the antigen. __Lastly__, from another angle and since AFP purportedly displays immunosuppressive ability (Alpert and co-workers, 1978; Auer and Kress, 1977; Murgita and Tomasi, 1975; Peck, Murgita and Wigzell, 1978), it is interesting to investigate whether neutralization of tumor-derived AFP by specific antibody releases the immune system of the tumor-host from a status of suppression, thereby promoting an effective anti-cancer response.

A prerequisite for the antibodies and/or lymphocytes directed to a soluble, extracellular tumor product, to be cytotoxic-cytolytic, is that the antigenic determinants of the soluble molecule are expressed, in whole or in part, on the tumor cell surface. Thus a direct interaction is possible between antibodies (and/or lymphocytes) and the cancer cell. This requirement is fulfilled by the hepatoma BW7756, which in its host of origin, the C57L/J mouse, produces large amounts of AFP. As we shall discuss below, AFP and the hepatoma cell surface share epitopes which allow anti-AFP antibodies to interact with the tumor cells, and bring about cytotoxicity and lysis.

In this report the immunotherapeutic effects of antibodies and lymphocytes on the hepatoma BW7756 are described. Reference to other similar experimental systems is made only to a limited extent, as required for the understanding and interpretation of the data obtained with the hepatoma BW7756, _in vitro_ and _in vivo_ (in C57L/J mice). Two main experimental approaches are discussed:

1. Passive immunotherapy with heterologous (rabbit) anti-mouse AFP antiserum and

syngeneic lymphocytes; and 2. Active immunotherapy with xenogeneic AFP, together with nonspecific means which aim at breaching the tolerance to mouse AFP, and consequently to the hepatoma. The immunotherapeutic effects of antibodies and lymphocytes are measured in vivo, while the interaction between antibodies and/or lymphocytes with the target tumor-cell and AFP are also studied in vitro. The in vivo measurements include tumor incidence, growth rate, macroscopic pathology and histopathology of the tumor and organs of the host. Parallel in vitro assays examine a number of parameters necessary for the elucidation of the mechanism whereby anti-AFP immunotherapy affects the hepatoma.

IN VIVO STUDIES

I. Passive Immunotherapy

Perhaps the simplest approach to anti-BW7756-hepatoma immunotherapy is the utilization of anti-AFP antibodies. Serotherapy attempts have been published (Mizejewski and Allen, 1974; Mizejewski and Dillon, 1978). The results suggest that heterologous anti-AFP antiserum has some effect on tumor growth and AFP production. In an earlier preliminary work (Mizejewski and Allen, 1974), rabbit anti-AFP antiserum was given in various ways: multiple small doses over four weeks beginning the day of tumor challenge; a single large dose 2 days after tumor challenge; and, incubation of the tumor inoculum with antiserum for 30 minutes prior to implantation in mice. Control mice were given normal rabbit serum or were left untreated. At the end of the experiment, 28 days after tumor challenge, a reduction in tumor size was observed in all three antiserum-treated groups. The smallest tumors were found in the mice which had been implanted with cells preincubated with the antiserum in vitro. Moreover, the serum AFP levels did not increase in the antiserum-treated mice as much as in the control groups. It should be mentioned that the control mice receiving normal rabbit serum also had smaller tumors than untreated controls, but the degree of growth inhibition due to normal serum was only about one-half of that due to antiserum, and AFP levels were not significantly diminished. In addition, examination at autopsy revealed that antiserum-treated mice had a tumor with massive hemorrhage, necrosis, and poor capsulation, whereas control groups did not show these destructive lesions. These results were confirmed in a subsequent publication (Mizejewski and Dillon, 1978), which also indicated that a large dose of antiserum, repeated several times, is more effective to diminish tumor growth than a similar sequence of injections of small but increasing doses over several weeks. The same effect was produced by large constant doses of normal rabbit serum but not by the other administration schedule involving several gradually increasing doses. Thus the data are difficult to interpret and the mechanism whereby anti-AFP antiserum brings about its antitumor effect hard to figure it out. It has been postulated (Mizejewski and Dillon, 1978) that formation of immune complexes could occur between rabbit proteins and mouse antibodies against them, produced during the therapeutic period, with deposition near and/or inside the tumor and subsequent development of allergic tissue reactions. The latter would cause cell lysis, vascular damage and tissue destruction.

We have undertaken the same kind of approach but instead of using antiserum we utilized spleen cells from mice which had or had had tumors. We found (Macario and Mizejewski, 1978) that spleen cells from mice bearing tumors implanted 14 and 21 days before excision of the spleens and injected into syngeneic recipients two days prior to tumor challenge diminish tumor incidence (Table 1). Spleen cells obtained after 21 days post-tumor implantation do not display this protective capacity. We also investigated whether a still more dramatic defensive result could be obtained by allowing the spleen cells to get in contact with the tumor cells at the time of implantation. Cells from tumor-bearing mice proved to be suppressive rather than defensive, in the sense that they enhance tumor

TABLE 1 Spleen Cells from Mice Bearing a Hepatoma Transplanted
14 and 21 Days Before Excision of the Spleen
Protect Syngeneic Recipients[a]

Mice transfused with spleen cells[b]	Tumor incidence in spleen-cell recipients Days after tumor challenge[c]		
	14	18	28
yes	4/32 (13)[d]	7/32 (22)	15/32 (47)
no	35/63 (56)	40/63 (63)	46/63 (73)

a: Data extracted from Macario and Mizejewski, 1978. b: Obtained from mice bearing a hepatoma transplanted either 14 or 21 days before excision of the spleen. c: 10^5 hepatoma cells inoculated subcutaneously 48 hours after intraperitoneal transfusion of 50×10^6 spleen cells. d: Mice with palpable tumor/mice studied (percent of mice with palpable tumor). Days 14 and 18 $p<0.001$; day 28 $p<0.01$.

incidence and growth. However, spleen cells from tumor-free mice, which had been inoculated at least twice with hepatoma cells, did show a pronounced defensive potential (Table 2). While spleen cells from normal mice injected prior to tumor implantation did not show protective or enhancing ability (Macario and Mizejewski, 1978), they did protect when administered along with the hepatoma cells, but to a lesser degree than spleen cells from tumor-free mice (Table 2). These results were confirmed at various spleen to hepatoma cell ratios (Macario and co-workers, 1978). The mechanisms of these enhancing and protective effects are under investigation. They most likely involve suppressor as well as effector cells.

TABLE 2 In situ Immunotherapy of a Mouse Hepatoma with Spleen
Cells from Syngeneic Donors Preimmunized to the Tumor[a]

Spleen cell donors	Tumor incidence in recipients of a mixture of spleen and hepatoma cells[b] Days after the cell-mixture inoculation			
	14	18	28	35
preimmunized[c]	0/3 (0)[d]	0/3 (0)	0/3 (0)	0/3 (0) 0[e]
normal	1/3 (33)	2/3 (66)	2/3 (66)	2/3 (66) 2.04
-[f]	1/3 (33)	2/3 (66)	3/3 (100)	3/3 (100) 3.33

a: From Macario and Mizejewski, 1978. b: Five and 0.4×10^6 spleen and hepatoma cells, respectively. c: Tumor-free mice which had been inoculated twice with hepatoma cells prior to spleen excision. d: Mice with palpable tumor/mice studied (percent of mice with palpable tumor). e: Mean tumor weight (grams) at autopsy on day 35. f: No spleen cells were given to the mice in this group.

II. Active Immunotherapy

Since AFP is a normal component of the body, which reaches high levels during fetal life, tolerance develops as it happens for any other self-component. Therefore, a tumor-host does not react against tumor-derived AFP. Attempts have been made to break tolerance to AFP, the rationale being that in adulthood the only tissues that might eventually be damaged by anti-AFP antibodies are the tumor ones. Normal adults produce very little, if any AFP. Rat AFP emulsified in Freund's complete adjuvant was used for immunization of C57L/J mice (Ruoslahti and co-workers, 1976). These mice, however, had the same tumor incidence after challenge with hepatoma cells as did control groups preimmunized with rat albumin. This lack of anti-tumor protection occurred despite the fact that AFP-immunized mice produced anti-AFP antibodies reacting with rat as well as mouse AFP. Several explanations were proposed accounting for the absence of anti-tumor protection in AFP-immunized mice: a) AFP does not stay on the tumor cell surface for a time long enough to allow antibody binding and cell lysis; b) anti-AFP antibodies exhibit low affinity for the autologous (tumor-derived) AFP; lower than its combining affinity for the xenogeneic (rat) AFP used for immunization. The association constant is too low to trigger the chain of reactions leading to membrane damage, and c) the control antigen, rat serum albumin, is also immunogenic and anti-albumin antibodies have the same anti-tumor effect as anti-AFP antibodies. This consideration is specially relevant because the hepatoma cells secrete albumin and this molecule has several similarities with AFP (Ruoslahti and co-workers, 1976).

Explanation (a) is considered below (in vitro studies), explanation (b) awaits testing, and explanation (c) was ruled out by recent work (Engvall and co-workers, 1977) which demonstrated that nonimmunized mice behave identically to mice immunized with rat AFP or albumin.

We have followed the same approach but instead of utilizing only one heterologous AFP we used a mixture of three (Mizejewski and Macario, 1978). The species origin of the AFP used are phylogenetically at various distances from the mouse, namely, human, rat and guinea pig. We hypothesized that it would be easier to deceive the self-recognition mechanisms by administering such a mixture of widely phylogenetically different AFPs. Breaching of tolerance to autologous AFP was expected to ensue the administration of these foreign AFPs and consequently anti-tumor effects would be observed. Table 3 shows the data and demonstrates that only a slight retardation in tumor growth was caused by preimmunization with the mixture of xenogeneic AFPs, as well as with a control mixture of sera. Our results agree with the ones discussed above (Engvall and co-workers, 1977; Ruoslahti and co-workers, 1976), and show that active immunization leading to production of autoantibodies specific for AFP does not affect significantly hepatoma growth.

IN VITRO STUDIES

As reviewed above, there is some evidence that anti-AFP antibodies have anti-BW7756 hepatoma effects in vivo. This observation, although inconclusive, has led to the investigation of the direct interaction of antibodies to AFP with hepatoma cells in vitro. First of all it was attempted to establish a correlation between anti-tumor effect in vivo and cytotoxicity in vitro (Mizejewski and Allen, 1974). It was found that rabbit anti-mouse AFP antiserum displays a low cytotoxic titer on hepatoma target cells. Although addition of a source of complement enhances cytotoxicity, significant cytolysis occurs also in the absence of complement. These results were confirmed and extended in a later publication (Mizejewski, Young and Allen, 1975). Anti-AFP antiserum excerted cytopathogenic effects in cultures of hepatoma cells which contrasted with

TABLE 3 Frequency of Hepatoma Growth in Mice Preimmunized With a Mixture of AFP from Three Different Species of Mammals[a]

Mice pretreated with	Tumor incidence in pretreated mice Days after tumor challenge[b]			
	14	18	28	35
AFP mixture[c]	4/15 (27)[d]	6/15 (40)	6/15 (40)	6/15 (40)
Serum mixture[e]	3/15 (20)	6/15 (40)	6/15 (40)	6/15 (40)
Phosphate buffered saline	7/15 (47)	9/15 (60)	9/15 (60)	9/15 (60)
Nothing	7/15 (47)	8/15 (53)	9/15 (60)	10/15 (67)

a: From Mizejewski and Macario, 1978. b: 10^5 hepatoma cells inoculated subcutaneously. c: Human, rat and guinea pig AFP. d: Mice with palpable tumor/mice studied (percent of mice with palpable tumor). e: Same species as for AFP.

cultures exposed to normal rabbit serum. The latter cultures did not show the severe sloughing observed in the antiserum-exposed cultures. Some degree of cell suffering was produced also by the normal serum, but the cells returned to normal after serum removal and refeeding of cultures with regular medium. On the contrary, the lesions produced by the antiserum were irreversible. It was also demonstrated that this cytopathogenic effect mediated by anti-AFP antiserum was specific for the hepatoma cells in as much as it did not occur in normal liver and muscle cells equally treated with the antiserum. Furthermore, the cytopathogenic and cytolytic effects could be abolished by preincubation of the antiserum with AFP. This most likely means that AFP determinants on the hepatoma cell surface were involved in the antiserum mediated cytopathogenesis and cytolysis. At this point, it is relevant to ask the question of whether AFP is an integral part of the cell membrane for a length of time enough to permit antibody binding and all the subsequent reactions determining membrane damage, or it forms only a putative coat on the cell surface. Moreover, it is also important to understand the consequences of the antibody-cell membrane interaction upon the cellular physiology, even under conditions nonconducive to cytolysis, for example, after a short exposure to antiserum. Using synchronized target cells it was shown that anti-AFP antiserum is most cytotoxic at the time in the cell cycle (late S and early G_2) which coincides with AFP synthesis and secretion (Allen and Ledford, 1977). Blockade of AFP synthesis rendered the hepatoma cells insensitive to the cytotoxic potential of the anti-AFP antiserum. Moreover, anti-mouse albumin antiserum was not cytolytic although the target hepatoma cells were actively synthesizing albumin. Both antisera, anti-AFP and anti-albumin, however, inhibited the synthesis of the protein for which they are specific, but they did not affect synthesis of the other protein. These studies indicate that AFP remains in the cell membrane as an integral part of it, sufficiently exposed for a period of time long enough to allow interaction with antibody and eventually lysis, even in the absence of added complement. In support of this, recent investigations (Mizejewski, 1978) have shown that anti-AFP antibodies bind to the membrane and to the cytoplasm of hepatoma cells. This was demonstrated by indirect immunofluorescence of nonfixed and formalin-fixed cells, respectively.

CONCLUSIONS AND PERSPECTIVES

From the data reviewed thus far it appears that although a considerable amount of information has already been gathered with regard to immunotherapy of the mouse hepatoma BW7756, the most exciting experiments have still to be done. Elicitation of cellular immunity by immunization with AFP has not yet been reported and certainly deserves attention. Furthermore, AFP seems to either suppress or enhance immunological functions (Alpert and co-workers, 1978; Auer and Kress, 1977; Murgita and Tomasi, 1975; Peck, Murgita and Wigzell, 1978) and therefore it is relevant to explore the possibility that this oncofetal protein, as well as others, suppresses anti-tumor immune reactions and thereby favors tumor growth. We have begun an investigation on the role of AFP in the regulation of the immune reaction against the BW7756 hepatoma and found that it amplifies the immunosuppressive ability of spleen cells from tumor bearing animals (Macario and Mizejewski, 1978). This amplification is however not very pronounced and it requires more experimentation.

A promising road to an analysis of the cellular and molecular struggle between immuno-surveillance mechanisms and tumor tricks to avoid and/or overcome them, is supplied by the model system discussed in this report (hepatoma BW7756 and C57L/J mice). The recent observation that AFP binds to other molecules, such as esteroid hormones (Keller, Galvanico and Tomasi, 1976; Mizejewski and co-workers, 1978; Uriel and co-workers, 1975), and the suggestion that standard sources of AFP contain other suppressive factors (Tomasi, 1978) add a new dimension to the system. Thus new perspectives are opened, not only regarding the possible role of AFP in immune regulation and its dependency upon binding to other molecules, but also on other biological functions that AFP might have in the regulation of gene expression via hormone binding. In this regard, the observation that AFP preparations exhibit post-synthetic heterogeneity (Lester, Miller and Yachnin, 1977) and contain fatty acids (Parmelee, Evenson and Deutsch, 1978) indicate that there is still much to be done to identify the active form(s) of AFP and to elucidate the molecular mechanisms of its interaction with other molecules, and with cells, that brings about immune regulation.

REFERENCES

Allen, R. P., and B. E. Ledford (1977). The influence of antisera specific for α-fetoprotein and mouse serum albumin on the viability and protein synthesis of cultured mouse hepatoma cells. Cancer Res., 37, 696-701.
Alpert, E., J. L. Dienstag, S. Sepersky, B. Littman, and R. Rocklin (1978). Immunosuppressive characteristics of human AFP: Effects on tests of cell mediated immunity and induction of human suppressor cells. Immunol. Commun., 7, 163-186.
Auer, I. O., and H. G. Kress (1977). Suppression of the primary cell-mediated immune response by human α_1-fetoprotein in vitro. Cell Immunol., 30, 173-179.
Eisen, H. N., N. Sakato, and S. J. Hall (1975). Myeloma proteins as tumor-specific antigens. Transplant. Proc., 7, 209-214.
Engvall, E., H. Pihko, H. Jalanko, and E. Ruoslahti (1977). Effect of specific immunotherapy with preimmunization against alpha-fetoprotein on a mouse transplantable hepatoma: Brief communication. J. Nat. Cancer Inst., 59, 277-280.
Keller, R. H., N. J. Calvanico, and T. B. Tomasi, Jr. (1976). Immunosuppressive properties of AFP: Role of estrogens. In W. H. Fishman and S. E. Sell (Ed.), Onco-Developmental Gene Expression, Academic Press, New York. pp. 287-295.

Lester, E. P., J. B. Miller, and S. Yachnin (1977). A postsynthetic modification of human α-fetoprotein controls its immunosuppressive potency. Proc. Nat. Acad. Sci. USA, 74, 3988-3992.

Macario, A. J. L., and G. J. Mizejewski (1978). Search for immunosuppressive effects of suppressor cells and alpha-fetoprotein in a mouse-hepatoma syngeneic system. This Proceedings.

Mizejewski, G. J. (1978). Studies of tumor-associated antigen cytotoxicity in mouse hepatoma BW7756. Clin. Immunol. Immunopathol. (in press).

Mizejewski, G. J., and R. P. Allen (1974). Immunotherapeutic suppression in transplantable solid tumors. Nature, 250, 50-52.

Mizejewski, G. J., and W. R. Dillon (1978). Immunobiologic studies in hepatoma-bearing mice passively immunized to alpha-fetoprotein. Arch. Immunol. Exp. Therap. (in press).

Mizejewski, G. J., and A. J. L. Macario (1978). (in preparation).

Mizejewski, G. J., J. M. Plummer, K. A. Blanchet, M. Vonnegut, and H. I. Jacobson (1978). Alpha-fetoprotein: Immunoreactivity of the major estrogen-binding component in mouse amniotic fluid. Immunology (in press).

Mizejewski, G. J., S. R. Young, and R. P. Allen (1975). Alpha-fetoprotein: Effect of heterologous antiserum on hepatoma cells in vitro. J. Nat. Cancer Inst., 54, 1361-1367.

Murgita, R., and T. B. Tomasi, Jr. (1975). Suppression of the immune response by α-fetoprotein. II. The effect of mouse α-fetoprotein on mixed lymphocyte reactivity and mitogen-induced lymphocyte transformation. J. Exp. Med., 141, 269-286.

Parmelee, D. C., M. A. Evenson, and H. F. Deutsch (1978). The presence of fatty acids in human α-fetoprotein. J. Biol. Chem., 253, 2114-2119.

Peck, A. B., R. A. Murgita, and H. Wigzell (1978). Cellular and genetic restrictions in the immunoregulatory activity of alpha-fetoprotein. I. Selective inhibition of anti-Ia-associated proliferative reactions. J. Exp. Med., 147, 667-683.

Ruoslahti, E., E. Engvall, H. Jalanko, and H. Pihko (1976). Immunization of mice against autologous alpha-fetoprotein (AFP) - reduction of serum AFP during tumorigenesis but lack of effect on incidence of transplanted hepatomas. In W. H. Fishman and S. E. Sell (Ed.), Onco-Developmental Gene Expression, Academic Press, New York. pp. 349-353.

Ruoslahti, E., H. Pihko, and M. Seppälä (1974). Alpha-fetoprotein: Immunochemical purification and chemical properties. Expression in normal state and in malignant and non-malignant liver disease. Transplant. Rev., 20, 41-60.

Tomasi, T. B., Jr. (1978). Suppressive factors in amniotic fluid and newborn serum: Is α-fetoprotein involved? Cell. Immunol., 37, 459-466.

Uriel, J., C. Aussel, D. Bouillon, F. Loisillier, and B. de Nechaud (1975). Liver differentiation and the estrogen-binding properties of alpha-fetoprotein. Ann. N. Y. Acad. Sci., 259, 119-130.

Cancer Immunotherapy

Immunotherapy of Cancer: Developing Concepts and Clinical Progress
(Harold Dorn Memorial Lecture)

R. W. Baldwin

Cancer Research Campaign Laboratories, University of Nottingham, University Park, Nottingham NG7 2RD, U.K.

INTRODUCTION

Development of the concept of immunological recognition of malignant cells must be viewed as one of the major advances in tumour biology over the past two decades. I well remember some 25 years ago embarking upon a programme of research into the immunology of chemically induced animal tumours at a time when the discipline of tumour immunology was barely respectable. Today there is hardly a single aspect of cancer research where immunological factors are not held to be relevant. But, it is in the development of immunological approaches to cancer diagnosis and treatment where one looks for progress. Unfortunately, this is one area of investigation where translation of experimental animal studies to the clinic is so difficult and one has to recognise that in many cases the optimism of the investigators has clouded reality.

IMMUNE MECHANISMS IN TUMOUR REJECTION

Cell mediated immunity is generally recognised as being important in tumour rejection (Herberman, 1974), but within this broad classification, the contribution of the various subpopulations of lymphocytes has yet to be resolved (Fig. 1). By analogy with the effector mechanisms involved in allograft rejection, it has been argued that immunological rejection of tumour cells may also be mediated by cytotoxic T lymphocytes (T killer cells). Experiments with animal tumours have established that well defined T lymphocyte populations from tumour immune donors are cytotoxic *in vitro* for tumour cells (Cerottini and Brunner, 1974). Lymphoid cell populations are also able to adoptively transfer immunity from immunized animals to normal recipients (Klein and Sjögren, 1960). Moreover, T lymphocytes sensitized *in vitro* by co-cultivation with tumour cells can suppress *in vivo* growth of tumour cells (Treves, 1978). T cells recognise antigens with the same discriminatory precision as B cells involved in antibody production so that the induction of T killer cells requires that the tumour cell expresses a tumour associated 'rejection' antigen. But, as discussed later, this is not always the case since, for example, naturally occurring animal tumours are only infrequently able to induce immunity to naturally-occurring (spontaneous) tumours (Baldwin, Embleton and Pimm, 1978). These findings with spontaneous tumours have brought full circle the argument concerning the existence of human tumour antigens especially in relation to their capacity to induce immunological rejection of malignant cells. So after more than a decade of research it still cannot be stated unequivocally that human tumours are able to induce T killer cells following recognition of specific tumour antigens.

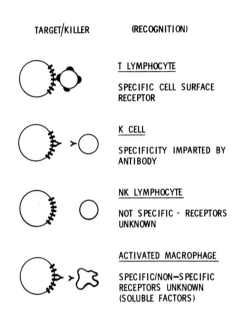

Fig. 1. Cell mediated immune cytolysis of tumour cells (Baldwin and Byers, 1978).

K cells involved in antibody mediated reactions (Fig. 1) do not fall into the two major subclasses of lymphocyte, namely T and B lymphocytes, since they lack the typical markers of these cell types (Perlmann, P., 1976). The main surface characteristic of K cells is the presence of receptors specific for the Fc portion of the IgG molecule and the cytolytic activity of K cells requires antibody molecules binding to either the target tumour cell or the effector cell. Therefore even though K cells are formed and persist, e.g. in peripheral blood and spleen, independent of a specific recognition effect, antibody directed towards tumour associated antigens is required. Therefore the limitations and restrictions described for T killer cells similarly apply to K cell responses.

A third subpopulation of lymphocytes currently engaging the attention of tumour immunologists is the natural killer or NK cell (Herberman and Holden, 1978). Natural killer cells constitute a distinct subpopulation of cells within the immune system being found in the lymphoid organs of several species including humans. The basic observation which drew attention to natural cell mediated immunity arose out of quite extensive investigations comparing the *in vitro* cytotoxicity of peripheral blood lymphocytes for cultured human tumour cells. These showed that lymphocytes from normal individuals, who would not be expected to have been exposed to tumour associated antigens, were frequently highly cytotoxic for tumour cells. There are also several correlations in animal systems between *in vivo* resistance to tumours and the levels of NK cell activity which suggest that these cells play a role in immunosurveillance. For example, the tumour rejecting capacity of mice following T cell depletion by thymectomy and whole body irradiation and reconstitution with bone marrow cells depended upon the NK cell activity of the bone marrow donors

(Haller and co-workers, 1977). NK cells are non-adherent, non-phagocytic lymphocytes which, like K cells, do not express cell surface markers for either B or mature T cells and they do not appear to require specific recognition signals for their induction or interaction with tumour cells (Herberman and Holden, 1978). The relevance of this subpopulation of lymphocytes in tumour immunotherapy is emphasized by the finding that NK cell activity is greatly enhanced by treatment with bacterial immunostimulants such as bacillus Calmette Guérin (BCG) and *C. parvum* (reviewed by Herberman and Holden, 1978; Baldwin and Byers, 1978).

Finally, macrophages activated by various means may be cytotoxic for tumour cells (reviewed by Baldwin and Pimm, 1978; Baldwin and Byers, 1978). This activation can be brought about by a variety of non-specific agents such as BCG, hence their potential in immunotherapy. Activation may also be brought about in a specific fashion through interaction with products released by sensitized lymphocytes, but it is generally accepted that the cytotoxic effects are mediated by non-specific factors.

ESCAPE FROM IMMUNOLOGICAL CONTROL

From the above considerations of effector mechanisms potentially operating against a developing tumour, one might expect that 'escape' from immunological restraint in the host would be a rare event. This is not the case, unfortunately, and several factors have been proposed to account for the failure of immunological control.

Firstly, one has to consider whether tumour cells do elicit immune rejection responses as frequently as was originally proposed. In recent years it has become widely accepted that many types of experimental animal tumours induced with extrinsic agents such as chemical carcinogens and oncogenic viruses exhibit neoantigens capable of eliciting tumour immune rejection responses (Baldwin, 1973; Herberman, 1977). In comparison, tumours arising naturally, i.e. without deliberate exposure to extrinsic agents are less frequently immunogenic when assessed by this criterion (Baldwin, Embleton and Pimm, 1978). For example, in a comprehensive analysis of spontaneous rat tumours, these being predominantly mammary carcinomas, sarcomas and kidney tumours, it was found that only 7 of 30 were able to elicit immunity against tumours transplanted into syngeneic rats (Table 1) (Baldwin, Embleton and Pimm, 1978; Middle and Embleton, 1978). In comparison sarcomas induced with 3-methylcholanthrene and hepatomas induced with aminoazo dyes in the same rat strain are most frequently able to initiate tumour rejection responses against the homologous tumour (Baldwin, 1973). There are still far too few studies with tumours of different aetiologies and appropriate histological types, e.g. mammary, lung and colorectal carcinomas, to make valid judgements upon the relative merits of experimentally-induced versus spontaneous tumours as models for human cancer. Nevertheless it must be recognised that human cancers are not necessarily able to elicit a rejection response, especially where this involves T killer cells generated following recognition of tumour associated antigens and in this case a specific immunological approach to therapy will be pointless.

Interference Factors

Where tumours do elicit immunological reactions capable of producing tumour rejection, their effectiveness in the tumour-bearing host may be influenced by a number of factors actually generated by the developing tumour. This was initially formulated in terms of blocking antibody interfering with cell mediated immunity (Baldwin and Robins, 1977; Hellström, Hellström and Nepom, 1977). The effects are

Table 1 Rejection Responses to Transplanted Subcutaneous Tumours in Preimmunized Syngeneic WAB/Not Rats

Tumour	Site of origin	Tumour type	Min. cell inoculum growing in controls	Max. cell inoculum rejected by immunized[1] rats	Immunogenicity Index[2]
Sp1	Skin	Squamous ca.	10^3	10^5	100
Sp2	Subcutaneous	Sarcoma	10^5	$<10^5$	0
Sp7	Subcutaneous	Sarcoma	5×10^4	5×10^6	100
Sp20	Subcutaneous	Sarcoma	2×10^5	$<2 \times 10^5$	0
Sp24	Subcutaneous	Sarcoma	10^3	10^3	1
Sp25	Subcutaneous	Sarcoma	5×10^4	$<5 \times 10^4$	0
Sp41	Subcutaneous	Sarcoma	5×10^4	2×10^5	4
Sp3	Breast	Carcinoma	10^3	10^3	1
Sp4	Breast	Carcinoma	10^3	10^5	100
Sp6	Breast	Carcinoma	10^3	$<10^3$	0
Sp9	Breast	Carcinoma	10^3	$<10^3$	0
Sp11	Breast	Carcinoma	10^2	$<10^2$	0
Sp14	Breast	Carcinoma	10^4	$<10^4$	0
Sp15	Breast	Carcinoma	10^3	10^3	1
Sp21	Breast	Carcinoma	5×10^4	$<5 \times 10^4$	0
Sp22	Breast	Carcinoma	10^3	$<10^3$	0
Sp45	Kidney	Nephroblastoma	10^6	$<10^6$	0
Sp63	Kidney	Nephroblastoma	5×10^4	$<5 \times 10^4$	0
Sp78B	Kidney	Nephroblastoma	10^4	$<10^4$	0
Sp71	Skin	Chemodectoma	5×10^5	$<5 \times 10^5$	0

[1] Rats were immunised by implantation of irradiated (15,000 R) tumour or by excision of growing tumour.
[2] Ratio of cell growth inocula rejected by immune rats to minimum inoculum growing in controls.

much more complex, however, so that as illustrated in Fig. 2, a series of
responses, primarily cell mediated, can lead to immunological rejection of the
tumour. On the other hand, the actual growth of the tumour and its interaction
within the immunological network of the host can and does generate a powerful set
of suppressor effects. This includes release of tumour antigen and tumour specific
immune complexes which can specifically inhibit T killer cells. Circulating immune
complexes may also interfere with the cytotoxic effects mediated by K cells through
binding tumour specific antibody as well as inhibiting antibody synthesis. Interference with macrophage mediated cytotoxic effects may also be produced since
tumours release factors which inhibit chemotaxis of macrophages into the tumour
bed (James, 1977).

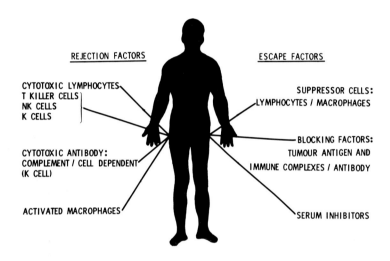

Fig. 2. Immune processes modifying tumour development.

Host responses to tumour associated antigens may also lead to the expansion of T
lymphocyte clones functioning as specific suppressor cells, so limiting the induction of T killer cells and antibody (Broder, Muul and Waldmann, 1978). Similarly
there may be the induction of non-specific 'suppressor' macrophages which inhibit
several arms of the immune response (Baldwin and Byers, 1978).

IMMUNOTHERAPY

From the above considerations, it is easy to see that it is still not possible to
design rational approaches to immunotherapy. Consequently, most clinical trials
have adopted an empirical approach involving treatment of patients with immunological adjuvants. These include various bacterial vaccines but bacillus Calmette
Guerin (BCG), the attenuated strain of *Mycobacterium bovis* and killed *Corynebacterium parvum* have been widely used (Milas and Scott, 1978; Baldwin and Pimm,
1978). Various subcellular preparations of BCG including the methanol extraction
residue (MER) and the synthetic muramyl dipeptide are also being evaluated

(Baldwin and Pimm, 1978). In addition synthetic immunostimulants such as levamisole have been tested.

From the numerous experimental animal and clinical trials, it is possible to identify three modes of presentation of these immunological adjuvants.

1. General immunostimulation where agents are administered systemically.

2. Active specific immunotherapy, this being defined as treatment with vaccines containing the appropriate tumour antigen so as to enhance tumour specific immune responses.

3. Regional immunotherapy. Here agents are administered so as to localize within tumour deposits and so initiate specific as well as non-specific immunity.

Clinical Studies

One of the original trials of the clinical potential of immunostimulation was presented by Mathé and his colleagues (1969) on the treatment of acute lymphoblastic leukaemia. Here patients received Pasteur BCG by skin scarification with or without radiation-attenuated allogeneic leukaemic cells, following chemotherapy induced remission. This study demonstrated that all of the untreated patients relapsed within 130 days whereas half of the 20 immunotherapy patients remained in remission for greater than 295 days and 7 have been in remission for 7 to 12 years (Table 2). Several attempts have been made to confirm the value of treatment with BCG alone in ALL during remission, but as summarized in Table 2, none have proved entirely successful. Unfortunately, these trials have not been entirely comparable with the original trial of Mathé and co-workers (1969). For example, the Strains of BCG used have varied from Pasteur to Glaxo and Tice preparations, whilst the doses and mode of administration have not been standardized. So the value of immunostimulation with BCG alone still remains unresolved and the original trial did not establish whether addition of irradiated allogeneic cells improved the response to BCG treatment. In view of the success of intensive combination chemotherapy with prophylactic treatment of the CNS, it has been concluded that chemotherapy is the treatment of choice in ALL (Powles, 1978). But, this does not take into account the differential toxicities of the two modalities of treatment and the supportive role of immunotherapy for remission maintenance deserves further consideration.

TABLE 2 Immunotherapy for Acute Lymphoblastic Leukaemia (Powles, 1978)

Study	Induction chemo. (months)	Immunotherapy	Median remission (months)		
			Immuno	No treatment	MTX
Mathé	24	BCG Pasteur + allog. ALL cells	7	2	—
MRC, UK	5½	BCG Glaxo	6	4	14
Heynes	3½	BCG Tice	4	4	8
Heynes	11½	BCG Tice	5	6	14
Poplack	—	BCG Pasteur + allog. ALL cells	21	—	35

The role of immunotherapy in the treatment of acute myeloblastic leukaemia (AML) has been evaluated more critically in the trials at St. Bartholomew's Hospital, London comparing the response to chemotherapy alone, with chemotherapy together with immunotherapy for the maintenance of chemotherapy-induced remission (Powles, 1978). The immunotherapy here was BCG Glaxo by intradermal puncture technique together with radiation attenuated allogeneic AML cells at separate sites. Analysis of overall survival in the two groups indicated a positive response, but when this was broken down into two summative components, remission duration and survival after relapse, immunotherapy essentially influenced the latter component (Fig. 3). There are now several trials completed or in progress attempting to confirm the St. Bartholomew's Hospital studies either using exactly the same protocol or with variations (Powles, 1978). The general conclusion to be drawn from these studies is that immunotherapy essentially prolongs survival after relapse. It also appears that treatment with BCG and irradiated allogeneic AML cells is superior to BCG alone. This conclusion still requires confirmation, however, and evidence that this reflects a specific immune response is lacking.

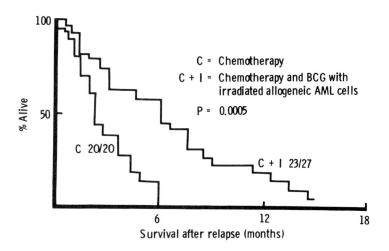

Fig. 3. Immunotherapy in acute myelogenous leukaemia: Survival after relapse (from Powles, 1978).

Solid Tumours

Turning next to solid tumours, there is no difficulty in assembling a whole series of immunotherapy trials, especially in malignant melanoma, where treatment with BCG or *C. parvum*, with or without tumour cells as a source of putative tumour antigen have been reported to give beneficial responses (reviewed by Goodnight and Morton, 1978). To review these many trials is an almost impossible task, made even more difficult by variations in trial design. Not the least of these problems has been the use of historical controls and in most cases, the value of immunotherapy as a component in the treatment of solid tumours awaits confirmation from properly conducted randomized trials. Here one must further recognise that immunotherapy at best will only be able to deal with small amounts of tumour. This is well established with experimental animal tumours such as carcinogen-induced guinea pig hepatomas (Zbar and co-workers, 1972) and rat sarcomas (Baldwin and Pimm, 1971) and from this it can be predicted that treatment of advanced disease will show no benefit. This has to be balanced against the knowledge that treatment with agents such as BCG and *C. parvum* is not without an associated toxicity.

In the past, treatment of malignant melanoma has often been cited as the model for other solid tumours in view of supportive evidence of cell mediated and humoral immune responses to melanoma associated neoantigens. Here the role of immunotherapy in the post-surgical treatment of malignant melanoma can be illustrated by the trials of Morton and his colleagues (1978) where patients following surgery for primary melanoma who had histologically proven evidence of regional node metastases received repeated intradermal injection of BCG or BCG plus allogeneic cells from a cultured melanoma cell line. The data shown in Fig. 4 compares survivals in patients having no post-operative treatment with those receiving BCG or BCG and cultured melanoma cells. Immunotherapy patients are showing better survival, but there is only a small benefit. Also, because of the small numbers of patients, the trial does not resolve whether the addition of allogeneic melanoma cells increases the response to that achieved with BCG alone. Moreover not all immunotherapy trials in Stage II malignant melanoma have shown a beneficial response. For example, in a small controlled study at Nottingham (McIllmurray and co-workers, 1977), on patients rendered clinically tumour-free by surgery, those treated with BCG and irradiated autologous cells had a more rapid recurrence rate and early deaths when compared to controls receiving surgery only.

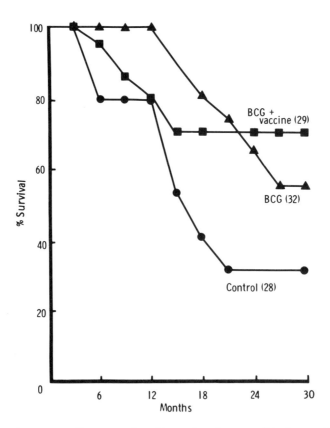

Fig. 4. Immunotherapy of malignant melanoma (Morton and colleagues, 1978). Patients following surgery for primary melanoma who had histologically proven evidence of regional node metastases (Stage II) received repeated intradermal injections of BCG or BCG + allogeneic cells from a cultured cell line.

There are also a number of trials where various forms of immunostimulation are being evaluated in the post-operative treatment of lung cancer (Table 3).

TABLE 3   Immunotherapy in the Post-Operative Treatment of Lung Cancer

Trial	Immunotherapy
Herberman, 1978	BCG Pasteur: $3-5 \times 10^8$ organisms ± allogeneic cells
Roscoe, 1977	BCG Glaxo: $5-25 \times 10^7$ units
Yamamura, 1978	BCG cell wall ± autologous cells
EORTC, 1978	BCG Pasteur
Stewart, 1978	Freund's adjuvant + lung tumour extracts
Takita and co-workers, 1978	BCG Glaxo + autologous cells
Amery, 1978	Levamisole

None of these have conclusively established the value of immunotherapy, but it is of interest to compare two recent trials where BCG alone or together with irradiated allogeneic cells are being given. In the trial reported by Roscoe and his colleagues (1977) no difference in survival between control patients with those treated post-operatively with BCG either by intradermal injection or multi-puncture technique. In comparison, another similar trial (Herberman, personal communication) suggests that repeated immunostimulation with Pasteur BCG increases survival in post-operative Stage I lung cancer patients whether or not allogeneic tumour cells were included in the treatment. These two trials illustrate the dilemma of current clinical trials since the type and dose of BCG were so different. In the trial by Roscoe and his colleagues (1977), 6 to 12 injections of BCG (Glaxo) were given, whilst in the other trial, BCG (Pasteur) was administered at a higher dose of organisms and repeatedly at two-week intervals. One may speculate that repeated treatment with BCG is important, but this in itself poses logistic problems. As discussed later a single intrapleural injection of BCG post-operatively may be at least as effective and the advantages and disadvantages of these two approaches need to be considered.

It may also be that there are qualitative differences in the responses to different BCG preparations although in animal tumour trials it has proved extremely difficult to establish this point (Willmott, Pimm and Baldwin, 1978). Nevertheless one is attracted by the possibility of using cell free extracts of BCG or even more defined agents. In this context trials reported by Yamamura (1978) indicate that repeated intradermal injection of BCG cell wall skeleton preparations produced a prolongation of survival in Stage III lung cancer patients.

Chemoimmunotherapy

As already commented upon, immunotherapy basically will only be effective against small amounts of tumour and in recognition of this, Hersh and Gutterman at the M. D. Anderson Hospital and Tumor Institute, Houston, have pioneered the concept of incorporating non-specific immunostimulation with chemotherapy in the treatment of disseminated tumours (Hersh, Gutterman and Mavligit, 1977). These approaches have been developed in large part from studies on malignant melanoma where it has been reported that the response to DTIC chemotherapy is improved by repeated intradermal treatment with Pasteur BCG (Gutterman and colleagues, 1978). But these trials have been evaluated without concurrent controls and not surprisingly this has led to some questioning of the conclusions when similar responses have not been obtained in other trials (Goodnight and Morton, 1978). One now expects many of these trials to be re-evaluated and perhaps this is the time to question the fundamental basis of these approaches since the original proposition that tumour specific immune responses were being enhanced has not been established with any degree of confidence.

Mode of Action

In reviewing the current status of clinical trials in the leukaemias as well as with various types of solid tumour, I am not convinced that there is compelling evidence to support the view that clinical benefit accrues from stimulation of tumour specific immune responses. Indeed apart from the trials in acute myeloblastic leukaemia, it does appear that responses induced by BCG alone are just as effective as when attenuated tumour cells are also given. In this case one needs to ask what is the immunological basis for treatment with various immunostimulating agents. Here one might have expected guidance from the many studies on the immune responses elicited against experimentally-induced animal tumours. However the responses are much more complex than was originally envisaged and there is still no clear understanding of the host responses that are critical for tumour rejection (Baldwin and Byers, 1978).

One of the principal responses to the treatment with bacterial immunostimulants is the induction of activated macrophages, these cells being cytotoxic *in vitro* for cultured tumour cells (Baldwin and Byers, 1978). Indeed this is such a uniform response, that many investigators have concluded that activation of macrophages constitutes a principal component of the non-specific host defence system against malignant cells (Baldwin and Pimm, 1978; Milas and Scott, 1978). But surprisingly many of the *in vivo* studies with transplanted animal tumours have failed to establish conclusively that macrophages activated with agents such as BCG and *C. parvum* can suppress tumour growth when injected together with tumour cells (Table 4). It is controversial, therefore, whether activated macrophages directly control tumour growth or play a secondary role such as the promotion of lymphocyte infiltration through the release of soluble mediators. Here it is pertinent to note that activated macrophages secrete interferons since these substances enhance natural killer cell activity (Herberman and Holden, 1978). Indeed it has been postulated that interferon production may represent a common pathway by which BCG and *C. parvum* generate natural killer cell activity, the suggestion being that NK cells are one of the principal effector cells in 'non-specific' tumour resistance (reviewed by Baldwin and Byers, 1978).

TABLE 4 Cytotoxicity of Activated Macrophages for Tumour Cells

Tumour	Peritoneal cells stimulated by:	Cytotoxicity for tumour cells *In vitro*	*In vivo*
Mouse			
Sarcoma	BCG	+	−
	Pyran copolymer	+	+
	C. parvum	+	+[1]
Leukaemia	BCG	+	−
	Pyran copolymer	+	−
Hamster			
Lymphoma	BCG	NT	−
Rat			
Sarcoma	BCG	+	±

[1] Requires host cell cooperation.

When one considers the complexities of the immune responses modified by treatment with bacterial immunostimulants, it becomes even more difficult to provide a rational interpretation of combination chemotherapy and immunotherapy. Many chemotherapeutic agents are cytotoxic for lymphocytes so that immunostimulating agents may simply be repairing immunological damage. On the other hand immunotherapy could be acting synergistically with immunotherapy through the generation of specific or non-specific resistance. Here it should be recognised that as long ago as 1970 Currie and Bagshawe demonstrated that treatment with *C. parvum* after cyclophosphamide enhanced the response to chemotherapy. This approach has been extended by other investigators, notably Fisher and his colleagues (1978) in studying treatment of murine mammary carcinomas, to show that combination of chemotherapeutic agents with *C. parvum* is more effective than with single agents, but the rationale of these manipulations is as yet undefined.

From the many experimental tumour studies in animals it must be concluded that there is little support for the use of bacterial immunostimulants in the treatment of primary tumours. But immunotherapy can be used in conjunction with conventional therapies for the treatment of early recurrent or metastatic disease. This can be illustrated by the studies summarized in Fig. 5 on the treatment of transplanted tumours derived from spontaneously arising rat mammary adenocarcinomas (Greager and Baldwin, 1978). These tumours when implanted into the mammary pad develop locally and metastasize to the regional lymph nodes and then more widely, particularly to the lungs. Treatment of rats at a stage when the mammary tumours were only just palpable (stage I) by repeated intravenous injection of BCG or more effectively *C. parvum*, followed by surgical resection of the primary tumour significantly decreased the incidence of regional lymph node metastases. But if treatment was delayed until regional lymph node metastases were well established, then systemic immunostimulation was ineffective.

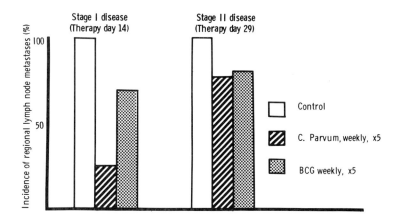

Fig. 5. Immunotherapy of post-surgical metastases of rat mammary carcinoma Sp4 (Greager and Baldwin, 1978). Rats received intravenous injections of BCG or *C. parvum* starting either at days 14 or 29. The primary tumours developing in mammary pad tissue were then surgically resected on day 36.

REGIONAL IMMUNOTHERAPY

Whilst bacterial adjuvants may be used as general immunostimulants in the treatment of malignant disease, infiltration of these agents directly into tumour deposits will often produce a more marked therapeutic effect. This form of treatment, which can be referred to as *regional immunotherapy* was originally demonstrated in the treatment of cutaneous metastatic malignant melanoma where lesions injected with BCG underwent regression (Morton and colleagues, 1970). Comparably, animal studies initiated by Rapp and his colleagues (Zbar, Bernstein and Rapp, 1971) established that intralesional injection of viable BCG organisms into intradermal grafts of the guinea pig line 10 hepatoma induced regression of the local tumour. Following on from these studies it has been established that administration of BCG or *C. parvum* under conditions allowing their localization within tumour deposits suppresses the growth of several types of experimentally-induced and spontaneous animal tumours (Milas and Scott, 1978; Baldwin and Pimm, 1978). Animal studies have also indicated that regional immunotherapy can be used for treating metastatic deposits and this indicates its greatest clinical potential. Injection of BCG into intradermal grafts of the guinea pig line 10 hepatoma caused the local tumour to regress but more importantly, treatment also prevented the development of regional lymph node metastases (Zbar, Bernstein and Rapp, 1971). In a comparable manner, growth of intramammary gland implants of a spontaneous rat mammary adenocarcinoma could be controlled by intralesional injection of *C. parvum* (Fig. 6). Repeated treatment with *C. parvum* starting when tumours were already established but before overt metastases were present prevented the progressive growth of the local tumour. Associated with this, there was a marked reduction in the incidence of rats developing regional lymph node metastases (Greager and Baldwin, 1978). In animal studies, regional immunotherapy has been most widely studied for its effect on pulmonary tumour growth and both BCG and *C. parvum* are effective in this respect (Milas and Scott, 1978; Baldwin and Pimm, 1978).

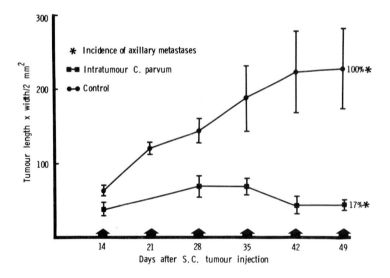

Fig. 6. Influence of multiple intratumour injections of *C. parvum* on growth and spread of mammary adenocarcinoma Sp4 (Greager and Baldwin, 1978).

This approach is illustrated in Fig. 7 showing the response to treatment of transplanted rat squamous cell carcinoma. When subcutaneous growths of this tumour are surgically resected and no further treatment given, all of the rats develop pulmonary metastases. Surgery followed by intravenous injection of BCG so that the organisms become localized in pulmonary tissue significantly decreases the incidence of pulmonary metastases with a concomitant increase in survival times (Baldwin and Pimm, 1973). This approach has subsequently been substantiated in studies showing that intravenous BCG following complete excision of primary canine osteosarcoma delayed the development of pulmonary metastases (Owen and Bostock, 1974) and the method has subsequently been applied to the treatment of canine mammary cancer (Bostock and Gorman, 1978).

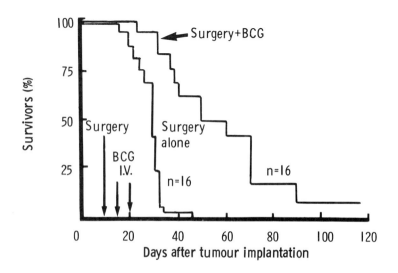

Fig. 7. Post-surgical treatment of pulmonary metastases of rat epithelioma Sp1 by intravenous injection of BCG.

Mechanisms of Action

The mechanisms involved in regional immunotherapy are still poorly understood, since it is only within recent years that serious attention has been given to the nature of host cells within a tumour mass. It is now firmly established, however, that at least with experimental systems the actively growing as well as the regressing tumour may contain a range of host cells including macrophages and monocytes, T lymphocytes, B lymphocytes, null cells which include natural killer and K cells and finally granulocytes. Therefore, when agents such as BCG and *C. parvum* are administered in a manner allowing their localization within a tumour, they may elicit a multiplicity of responses including stimulation of macrophages and lymphocytes (Fig. 8).

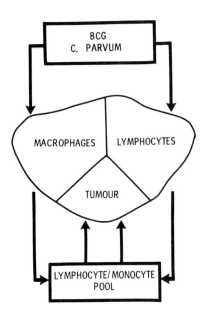

Fig. 8. Regional immunotherapy: host responses (Baldwin and Byers, 1978).

Clinical Studies

As already described, one of the initial approaches to regional immunotherapy was by Morton and his colleagues (1970) in the treatment of metastatic lesions from malignant melanoma by intralesional injection of BCG. Undoubtedly, however, the greatest potential of regional immunotherapy is revealed by recent trials in lung cancer treatment. In the trial established by McKneally and his associates (1976, 1977), patients received post-operatively a single intrapleural injection of BCG (Tice), this being administered through the chest tube just before its removal in patients undergoing lobectomy and by thoracentesis in patients undergoing pneumonectomy. Fourteen days after BCG treatment, patients received antituberculous chemotherapy. The trials as presently designed only indicate a clinical benefit in Stage I patients, that is tumours where there is no obvious metastatic spread or small tumours involving the hilar lymph nodes. Here as illustrated in Fig. 9 there is a significant difference in the survivals of patients receiving BCG post-operatively where 2/30 have died when compared with surgery controls (14/36 deaths). A similar trial was initiated in May 1976 at Nottingham, the only variant being the use of BCG (Glaxo) rather than the Tice Strain (Iles and colleagues, 1978). Again comparing Stage I non-anaplastic lung cancer, 2/22 BCG-treated patients have died within a 24 month period of the trial compared with 7/23 patients receiving surgery alone.

Several collaborative trials have now been established (Ludwig Lung Cancer Co-operative Group, 1978), to evaluate the benefit or otherwise of intrapleural bacterial vaccines in the post-operative treatment of lung cancer. Even if these prove to have some value it is likely that this will be limited to patients with minimal residual disease. So immunotherapy still has some way to go to offer benefit to the majority of lung cancer patients.

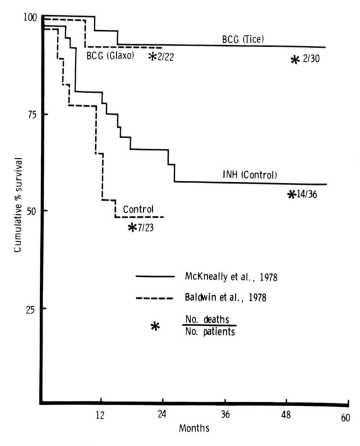

Fig. 9. Intrapleural BCG in lung cancer, Stage I : non-anaplastic. Data summarized from McKneally and colleagues (1977) and Iles and colleagues (1978) compares survival of patients receiving BCG postoperatively with surgery only controls.

CONCLUSIONS

To conclude then, what is the current status of immunotherapy for the treatment of human malignant diseases? Undoubtedly where tumours express rejection-type antigens, procedures designed to enhance specific responses such as those mediated by T killer lymphocytes will have the greatest potential. Underlying all of these approaches to immunotherapy, however, is the inescapable fact that the immunological system is only capable of dealing with limited amounts of tumour. In consequence immunotherapy basically will be potentially useful for the treatment of residual or recurrent disease or micrometastases. A more critical reservation is the question of whether human tumours do or do not express tumour rejection antigens. Here one must recognise the disappointing finding that unlike tumours induced with chemical carcinogens or oncogenic viruses, most naturally occurring tumours in animals do not express tumour rejection antigens. Added to this we still lack generally acceptable methods for detecting immune responses to human tumours. So, apart from a few special cases such as the EBV-related antigens in Burkitt's lymphoma, it is still not possible to decide with any degree of confidence which types of human tumour, if any, express tumour associated antigens with the capability of evoking immune responses in the patient.

If it turns out that human tumours, like the naturally occurring animal tumours, do not consistently express neoantigens functioning as rejection antigens, then one will be limited to the use of non-specific forms of therapy. Although it has not been possible so far to identify the cell types responsible for this type of anti-tumour response, attention has been focused upon the macrophage and the natural killer cell since both populations are stimulated by the bacterial immunostimulants currently favoured in immunotherapy. In this case, it is unlikely that *general* immunostimulation will be very effective when compared with the responses elicited following treatment with bacterial adjuvants under conditions leading to their localization in tumour deposits. But this does not mean that direct intralesional injection of agents has any significant potential except, perhaps at the research level. Rather efforts should be directed towards elucidating procedures where administration of agents allows their localization in tumour deposits. There is already support for the latter approach in the post-operative treatment of lung cancer by intrapleural injection of BCG. Before instilling bacterial suspensions into almost every body cavity, however, some regard should be given to the principles of this treatment and recognition given to its potential toxicity.

ACKNOWLEDGEMENTS

The author's work is supported by grants from the Cancer Research Campaign.

REFERENCES

Baldwin, R. W. (1973). Immunological aspects of chemical carcinogenesis. *Adv. Cancer Res.*, *18*, 1-75.

Baldwin, R. W., and V. S. Byers (1978). Immunoregulation by bacterial organisms and their role in the immunotherapy of cancer. *Seminars in Immunopathology*, 1. (In press).

Baldwin, R. W., and M. V. Pimm (1971). Influence of BCG infection on growth of 3-methylcholanthrene-induced rat sarcomas. *Europ. J. Clin. Biol. Res.*, *16*, 875-881.

Baldwin, R. W., and M. V. Pimm (1973). BCG immunotherapy of local subcutaneous growths and post-surgical pulmonary metastases of a transplanted rat epithelioma of spontaneous origin. *Int. J. Cancer*, *12*, 420-427.

Baldwin, R. W., and M. V. Pimm (1978). BCG in tumor immunotherapy. *Adv. Cancer Res.*, *28*, 91-147.

Baldwin, R. W., and R. A. Robins (1977). Induction of tumor-immune responses and their interactions within the developing tumor. *Contemp. Topics Mol. Immunol.*, *6*, 177-207.

Baldwin, R. W., M. J. Embleton, and M. V. Pimm (1978). Host responses to spontaneous rat tumours. In *Antiviral Mechanisms in the Control of Neoplasia*. Plenum Press, New York. (In press).

Bostock, D. E., and N. T. Gorman (1978). Intravenous BCG therapy of mammary carcinoma in bitches after surgical excision of the primary tumour. *Europ. J. Cancer*, *14*, 879-883.

Broder, S., L. Muul, and T. A. Waldmann (1978). Suppressor cells in neoplastic disease. *J. Natl. Cancer Inst.*, *61*, 5-11.

Cerottini, J-C., and K. T. Brunner (1974). Cell mediated cytotoxicity, allograft rejection and tumor immunity. *Adv. Immunol.*, *18*, 67-132.

Currie, G. A., and K. D. Bagshawe (1970). Active immunotherapy with Corynebacterium parvum and chemotherapy in murine fibrosarcomas. *Br. Med. J.*, *1*, 541-544.

Fisher, B., J. Linta, J. Hanlon, and E. Saffer (1978). Further observations on the inhibition of tumor growth by *Corynebacterium parvum* with cyclophosphamide. V. Comparison of the effects of tilorone hydrochloride, levamisole, methanol-soluble fraction of *Mycobacterium butyricum*, BCG and a nonviable aqueous ether extract of *Brucella abortus* preparation in the treatment of mice with tumors. *J. Natl. Cancer Inst.*, *60*, 391-399.

Goodnight, J. E., and D. L. Morton (1978). Immunotherapy for malignant disease. *Ann. Rev. Med.*, *29*, 231-283.

Greager, J. A., and R. W. Baldwin (1978). Influence of immunotherapeutic agents on the progression of spontaneously arising rat mammary adenocarcinomas of varying immunogenicities. *Cancer Res.*, *38*, 69-73.

Gutterman, J. U., E. M. Hersh, G. M. Mavligit, M. A. Burgess, S. P. Richman, M. Schwarz, V. Rodriguez, and M. Valdivieso (1978). Chemoimmunotherapy of disseminated malignant melanoma with BCG. Follow up report. In W. D. Terry and D. Windhorst (Eds.). *Immunotherapy of Cancer: Present Status of Trials in Man*. Raven Press, New York. pp. 103-111.

Haller, O., M. Hansson, R. Kiessling, and H. Wigzell (1977). Role of non-conventional natural killer cells in resistance against syngeneic tumour cells *in vivo*. *Nature*, *270*, 609-611.

Hellström, K. E., I. Hellström, and J. T. Nepom (1977). Specific blocking factors - are they important? *Biochim. Biophys. Acta (CR)*, *473*, 121-148.

Herberman, R. B. (1974). Cell-mediated immunity to tumor cells. *Adv. Cancer Res.*, *19*, 207-263.

Herberman, R. B. (1977). Immunogenicity of tumor antigens. *Biochim. Biophys. Acta (CR)*, *473*, 93-119.

Herberman, R. B., and H. T. Holden (1978). Natural cell-mediated immunity. *Adv. Cancer Res.*, *27*, 305-377.

Hersh, E. M., J. U. Gutterman, and G. M. Mavligit (1977). BCG as adjuvant immunotherapy for neoplasia. *Ann. Rev. Med.*, *28*, 489-515.

Iles, P. B., D. F. Shore, M. J. S. Langman, and R. W. Baldwin (1978). Intrapleural BCG in operable lung cancer. In *Recent Results in Cancer Research*, Springer-Verlag. (In press).

James, K. (1977). The influence of tumour cell products on macrophage function *in vitro* and *in vivo*: a review. In K. James, B. McBride and A. Stuart (Eds.), *The Macrophage and Cancer*. Edinburgh. pp. 225-246.

Klein, E., and H. O. Sjögren (1960). Humoral and cellular factors in homograft and isograft immunity against sarcoma cells. *Cancer Res.*, *20*, 452-461.

Ludwig Lang Cancer Study Group (1978). Search for the possible role of 'immunotherapy' in operable bronchial non-small cell carcinoma (Stage I and II): a Phase I study with *Corynebacterium parvum* intrapleurally. *Cancer Immunol. Immunother.*, *4*, 69-75.

Mathé, G., J. L. Amiel, L. Schwarzenberg, M. Schneider, A. Cattan, J. R. Schlumberger, M. Hayat, and F. De Vassal (1969). Active immunotherapy for acute lymphoblastic leukaemia. *Lancet*, *I*, 697-699.

McIllmurray, M. B., M. J. Embleton, W. G. Reeves, M. J. S. Langman, and M. Deane (1977). Controlled trial of active immunotherapy in management of stage IIB malignant melanoma. *Br. Med. J.*, *1*, 540-542.

McKneally, M. F., C. Maver, and H. W. Kausel (1976). Regional immunotherapy of lung cancer with intrapleural BCG. *Lancet*, *I*, 377-379.

McKneally, M. F., C. Maver, and H. W. Kausel (1977). Intrapleural BCG immunostimulation in lung cancer. *Lancet*, *I*, 1003.

Middle, J. G., and M. J. Embleton (1978). Immunogenicity of spontaneously arising tumours in inbred rats. *Br. J. Cancer*, *38*, 181-182.

Milas, L., and M. T. Scott (1978). Antitumor activity of *Corynebacterium parvum*. *Adv. Cancer Res.*, *26*, 257-306.

Morton, D. L., F. R. Eilber, R. A. Malmgren, and W. C. Wood (1970). Immunological factors which influence response to immunotherapy in malignant melanoma. *Surgery*, *68*, 158-164.

Morton, D. L., E. C. Holmes, F. R. Eilber, F. R. Sparks, and K. P. Ramming (1978). Adjuvant immunotherapy of malignant melanoma: preliminary results of a randomized trial in patients with lymph node metastases. In W. D. Terry and D. Windhorst (Eds.). *Immunotherapy of Cancer: Present Status of Trials in Man*. Raven Press, New York. pp. 57-72.

Owen, L. N., and D. E. Bostock (1974). Effects of intravenous BCG in normal dogs and in dogs with spontaneous osteosarcoma. *Europ. J. Cancer*, *10*, 775-780.

Perlmann, P., H. Perlmann, B. Wahlin, and S. Hammarström (1977). Quantitation, fractionation and surface marker analysis of IgG- and IgM-dependent cytolytic lymphocytes (K cells) in human blood. In P. A. Miescher (Ed.), *Immunopathology* VIIth International Symposium. Schwabe, Basel. pp. 321-334.

Powles, R. L. (1978). The application of immunotherapy to the treatment of cancer. In *International Encyclopedia of Pharmacology and Therapeutics*. Pergamon Press, New York. (In press).

Roscoe, P., S. Pearce, S. Ludgate, and N. W. Horne (1977). A controlled trial of BCG immunotherapy in bronchogenic carcinoma treated by surgical resection. *Cancer Immunol. Immunother.*, *3*, 115-118.

Treves, A. J. (1978). In vitro induction of cell-mediated immunity against tumor cells by antigen-fed macrophages. *Immunol. Rev.*, *40*, 205-226.

Willmott, N., M. V. Pimm, and R. W. Baldwin (1978). Comparison of BCG and *C. parvum* preparations for adjuvant contact therapy of a rat sarcoma. *Develop. Biol. Stand.*, *38*, 39-43.

Yamamura, Y. (1978). Immunotherapy of lung cancer with oil-attached cell wall skeleton of BCG. In W. D. Terry and D. Windhorst (Eds.). *Immunotherapy of Cancer: Present Status of Trials in Man*. Raven Press, New York. pp. 173-179.

Zbar, B., I. D. Bernstein, and H. J. Rapp (1971). Suppression of tumor growth at the site of infection with living Bacillus Calmette-Guérin. *J. Natl. Cancer Inst.*, *46*, 831-839.

Zbar, B., I. D. Bernstein, G. L. Bartlett, M. G. Hanna, and H. J. Rapp (1972). Immunotherapy of cancer: regression of intradermal tumors and prevention of growth of lymph node metastases after intralesional injection of living *Mycobacterium bovis*. *J. Natl. Cancer Inst.*, *49*, 119-130.

Symposium No. 19 — Chairman's Opening Remarks

David Pressman

Roswell Park Memorial Institute, Buffalo, New York 14263, U.S.A.

It is over eighty years since the first report of the use of Immunology in the treatment of cancer. In 1895, J. Herricourt and Charles Richet published a report that antiserum prepared against a cancer is effective in the treatment of cancer (Comptes Rendus Acad. Sci. 107: 748 (1895). On the basis that bacterial diseases namely diphtheria had recently been cured by use of antisera against diphtheria, they reasoned that a similar antiserum prepared against cancer might be effective against cancer. Indeed, Richet claims to have been the first to use anti-diphtheria antibody in the treatment of human diphtheria. However, in 1901, Emil von Behring won the Noble Prize for the treatment of diphtheria by specific antiserum. Richet did receive the Noble Prize later, in 1913, but for his work on anaphylaxis.

Herricourt and Richet macerated an osteogenic sarcoma with a little water, filtered the liquid portion through fine silk and injected it into three animals. They used one ass and two dogs. After 6, 7 and 12 days the animals were bled and the serum used in the treatment of tumors in other individuals. In the one case they report, they started treatment by injecting 3 ml of the antiserum into the soft tissue around the tumor and continued the injections daily for forty days using a total of 120 ml of serum. Following an initial inflammatory reaction, they observed that the pain was not only ameliorated and the size of the tumor diminished, but the patient appeared cured.

Within three months they made another report with the results of the treatment of about 50 patients by several of their colleagues and claimed that in general the following results were obtained: the pain was diminished, the ulcerations were diminished, there was a decrease in volume of the tumor, the course of the disease was slowed down and there was a general improvement in the patient. However it was not possible to cure the patients completely. They raised the question of specificity of the antiserum because normal serum itself had some effect. Then there was not much more reported in connection with treatment.

Subsequently, through the years, other attempts have been reported of the use of humoral antibodies in the treatment of cancer. It occurred to investigator after investigator to use such serum therapy in the treatment of cancer but no real therapy was achieved. There have also been many studies of the use of antibodies in the treatment of cancer of mice where the assay can be carried out with genetically homogeneous mice showing the positive effect of such treatment.

As time went on other aspects of immunology have been applied to the study of cancer. For example efforts have been made to treat cancers by immunizing the patient with vaccines made from his own tumor or related human tumors or by stimulating the immune response of the patient by treatment with agents known to do so such as BCG, C-Parvum or levamisole.

These aspects of the application of immunology to tumors will be described in the papers at this symposium. In addition there is also one paper on the use of humoral antibodies raised against tumor. Such antibodies have been shown to become localized in the tumor in animals and in man and have been used to carry therapeutic as well as diagnostic amounts of radioisotopes to the tumor in the case of animals. They also show some promise in carrying cytotoxic agents to the tumor.

Active Specific Immunotherapy: Experimental Studies

R. W. Baldwin, M. V. Pimm and R. A. Robins

*Cancer Research Campaign Laboratories, University of Nottingham,
University Park, Nottingham NG7 2RD, England*

ABSTRACT

With a range of carcinogen-induced and spontaneous tumours, active specific immunotherapy with a vaccine of viable or irradiated cells in admixture with BCG successfully controlled only those tumours with substantial immunogenicity. Immunosuppression by whole body irradiation abrogated the response. With the most responsive tumour, the maximum challenge inoculum controlled was 1×10^6 cells, and treatment had to be given within 1-2 days to be predictably successful. Analysis of the cellular composition of tumours undergoing regression showed a preferential localisation of lymphocytes, but this was not seen in tumours uncontrolled as a result of delayed therapy or overchallenge. Lymphocytes from progressively growing tumours, and also those from tumours undergoing immunotherapy-induced regression, were cytotoxic for cultured tumour cells in a microcytotoxicity test. Moreover, these tumour derived lymphocytes afforded protection in an in vivo Winn assay at a ratio of six lymphocytes per tumour cell.

KEY WORDS

Immunotherapy; BCG; C. parvum; tumour derived lymphocytes; microcytotoxicity; Winn test.

INTRODUCTION

One approach to the immunotherapy of malignant disease is active specific immunostimulation with vaccines of viable or attenuated tumour cells as immunogen together with immunological adjuvants, particularly BCG or Corynebacterium parvum (Bartlett and Zbar, 1972; Bomford, 1975; Hanna and Peters, 1978). The objective of the present investigation was to examine the response of experimental rat tumours to this form of therapy, particularly to identify optimum components of the vaccines and the host responses involved. In addition the limitation of the effect, in relation to the intrinsic immunogenicity of the tumour and the tumour load controllable have been examined and this equated with failure of therapy to stimulate beneficial host responses.

RESPONSE OF IMMUNOGENIC AND NON IMMUNOGENIC TUMOURS

The first series of experiments was carried out with eight rat tumour lines, carcinogen-induced and of spontaneous origin, which had previously been characterised for immunogenicity, and minimum tumorigenic dose (Table 1). Animals received challenge inocula of single cell suspension, prepared by trypsin digestion of solid tumour tissue, subcutaneously in one flank, and simultaneously a vaccine of tumour cells and BCG in the contralateral flank. This vaccine contained tumour cells in direct admixture with 0.2-1.0 mg moist weight of BCG organisms (Glaxo Percutaneous Vaccine B.P., Glaxo Laboratories, Greenford, Middlesex, England). In most cases viable cells were incorporated into the vaccine, it having previously been established that the local host responses generated by BCG were sufficient to control outgrowth of cells in the vaccine (Pimm, Hopper and Baldwin, 1978). With two tumours, mammary carcinomas Sp15 and Sp22, irradiated cells were used in the vaccine, since viable cells were not prevented from growth following admixture with BCG. Challenge inocula of cells of immunogenic tumour lines, sarcomas Mc5 and Mc7, hepatoma D23 and mammary carcinoma Sp4, were controlled by this regime of active therapy. Other tumours, with little or no immunogenic potential, were not suppressed, tumour incidences being comparable in treated and control rats, even with challenge inocula at the minimum tumorigenic dose.

TABLE 1 Active Specific Immunotherapy of Subcutaneously Transplanted Rat Tumours

Tumour/designation[1]	Tumour Immunogenicity index[2]	Vaccine Inoculum No. cells	BCG (mg)	Challenge Inoculum[3] No. cells	Takes Treated rats	Takes Controls
Sarcoma Mc5	5×10^6	2×10^6	0.5	2×10^6	0/6	6/6
Sarcoma Mc7	5×10^6	2×10^6	0.5	1×10^6	0/5	5/5
Hepatoma D23	5×10^5	1×10^5	0.2	5×10^4	0/5	5/6
Sarcoma Sp7	1×10^5	1×10^6	0.2	2×10^5	4/5	4/5
Mammary carcinoma Sp4	2×10^4	1×10^4	1.0	1×10^4	2/22	22/22
Sp15	1×10^3	5×10^6 IRR[4]	0.5	1×10^3	7/12	5/12
Sp22	No resistance	5×10^6 IRR	0.5	1×10^3	7/12	9/14
AAF57	No resistance	1×10^4	0.2	1×10^3	4/5	5/5

[1] MC - Induced with methylcholanthrene. D - Induced with 4-dimethylaminoazobenzene. Sp - Spontaneous origin. AAF - Induced with 2-acetylaminofluorene. All tumours transplanted in syngeneic WAB/Not rats.

[2] Maximum challenge inoculum controlled in rats immunised by graft excision or implantation of irradiated tissue. No resistance indicates no immunity to minimum inoculum for growth in control rats (1×10^3 cells).

[3] Injected contralaterally to vaccine inoculum.

[4] 15,000 R ^{60}Co γ-irradiation.

One implication of these findings is that the response to active specific therapy requires systemic immunity to the tumour, and therefore tests were carried out with two responsive tumours (Mc7 and Sp4) to examine the effect of simple immunosuppression (Table 2). Whole body irradiation with 450 R abolished the response, so that with both sarcoma Mc7 and mammary carcinoma Sp4 tumour growth was identical in treated and untreated irradiated animals, although therapy was again successful in rats not exposed to radiation.

TABLE 2 Abrogation of Active Specific Immunotherapy by Immunosuppressive Whole Body Irradiation

Tumour	Whole Body Irradiation[1] (R)	Vaccine Inoculum No. cells	BCG (mg)	Challenge Inoculum[2] No. cells	Takes Treated	Controls
Sarcoma Mc7	-	1×10^6	0.5	1×10^6	2/7	6/6
	450	1×10^6	0.5	1×10^6	7/7	0/6
Sarcoma Mc7	-	1×10^6	0.5	1×10^6	1/5	6/6
	450	1×10^6	0.5	1×10^6	5/6	5/6
Mammary carcinoma Sp4	-	1×10^4	0.5	1×10^4	2/10	10/10
	450	1×10^4	0.5	1×10^4	7/7	7/7

[1] ^{60}Co γ-irradiation, 7 R/min, 24-48 hours before challenge and therapy.

[2] Injected subcutaneously, contralaterally to vaccine inoculum.

MAGNITUDE OF THE RESPONSE AND OPTIMUM COMPONENTS OF THE VACCINE

Figure 1 illustrates the results of a series of experiments to determine the maximum challenge inoculum of one tumour, the sarcoma Mc7, controlled by a standard immunotherapeutic inoculum of 2×10^6 viable Mc7 cells in admixture with 0.5 mg moist weight of BCG at the same time as tumour challenge. Challenge inocula of 2×10^5 to 1×10^6 cells were controlled in almost all treated animals but grew out in the majority of control rats. Beyond 1×10^6 cells, however, the treatment was ineffective, 2×10^6 or 5×10^6 cell challenge being uncontrollable.

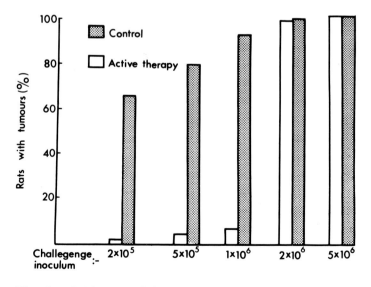

Fig. 1. Active specific immunotherapy of sarcoma Mc7 : comparison of therapeutic response against increasing challenge inocula.

A further series of tests was carried out to determine how long after challenge with 1×10^6 sarcoma Mc7 cells a therapeutic response was obtained (Fig. 2). In a group of 6 experiments when treatment was given on the day of challenge a mean of only 10% of animals were not cured by the therapy, although the response was variable with between 0% and 40% tumour takes in individual tests. When therapy was delayed the response became more variable, less predictable and generally poorer, so that by 7 days after challenge the majority of rats developed progressive growths in all tests.

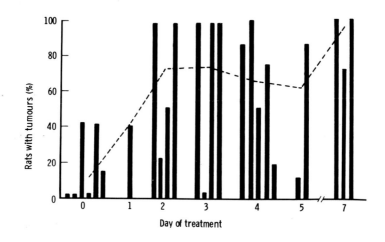

Fig. 2. Active specific immunotherapy of sarcoma Mc7 : effect of delaying treatment on therapeutic response

Having demonstrated therapeutic responses against a challenge inocula of 1×10^6 sarcoma Mc7 cells, with vaccines of viable cells in admixture with BCG, other tests were carried out to examine the response to BCG alone, irradiated cells alone (2×10^6) or vaccines of irradiated cells in admixture with BCG. Parallel tests were also carried out with Corynebacterium parvum (C.P.) (Wellcome CN6134, Wellcome Research Laboratories, Beckenham, Kent, England) at 0.7 mg dry weight of heat killed organisms/inoculum (Fig. 3). Essentially, BCG, C. parvum or irradiated tumour cells alone were ineffective in controlling the contralateral tumour challenge at the doses tested. Admixture of BCG or C. parvum with irradiated cells rendered them partially effective, with prevention of tumour growth in 40-60% of treated rats but the best response was seen with viable cells in admixture with either agent.

HOST RESPONSES

The studies described so far indicate that the immunotherapeutic effect observed with sarcoma Mc7 requires a radiosensitive response by the tumour host to a vaccine containing both tumour cells and adjuvant, and that this response is limited in its capacity to control tumour challenge with regard to both the number of tumour cells that may be controlled, and the timing of the treatment. In order to elucidate mechanisms involved in specific active immunotherapy-induced tumour regression, with a view to optimising and improving the responses obtained, the site of tumour challenge was chosen as an important area of investigation; clearly any cellular effector mechanism is likely to be evident at the site of tumour regression. Accordingly, the cellular composition of regressing tumours in immunotherapy

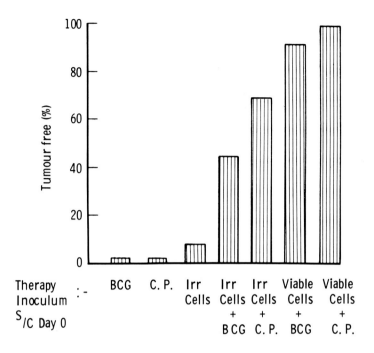

Fig. 3 Vaccine constituents required for therapeutic response against sarcoma Mc7.

treated rats was compared with that of progressively growing tumours in otherwise untreated rats. In addition, the cellular composition of tumours progressing in animals receiving challenge inocula too large to be controlled, or where therapy was given too late to be effective was also examined.

In order to identify and quantitate host cells present, challenge tumour inocula were removed, and disaggregated by mild enzyme digestion, and the constituent cell populations distinguished by acridine orange staining, Fc receptor rosette formation, and latex bead phagocytosis. In progressively growing tumours examined 7, 10 and 14 days after injection of tumour cells, cell digests consisted predominantly of sarcoma cells (60-80%), with considerable numbers of macrophages (20-30%) and a small number of lymphoid cells (less than 10%); these proportions did not change appreciably during the early phases of tumour growth examined. In contrast, tumours undergoing transient growth before regression in rats given active immunotherapy contained a marked lymphoid cell infiltrate (35-40%) during the early phases (7 and 10 days after injection). Only during later phases of regression (day 14) was the proportion of macrophages in the tumour increased to an average of 40%. Tumours growing progressively in rats given BCG only were found to have a cellular composition very similar to progressively growing tumours in normal rats. These findings indicated that a crucial early event in immunotherapy-induced tumour regression is lymphoid cell accumulation within the tumour. To examine this point further, the cellular content of tumours growing in rats given immunotherapy under suboptimal conditions was examined. Thus, when therapy is delayed until day 4 after tumour challenge, tumour growth is not controlled, and under these conditions the challenge tumours examined at day 10 did not show a marked lymphoid cell infiltration (Fig. 4). Similarly, in rats given immunotherapy but challenged with a high dose of tumour cells (4×10^6 cells), where again the challenge is not controlled, marked lymphoid cell accumulation does not occur (Fig. 5). With an intermediate challenge dose (2×10^6 cells), where tumour growth is retarded, but not controlled (see Fig. 2), there was some reduction in sarcoma cells present in the tumour, and a moderate increase in lymphoid cell accumulation.

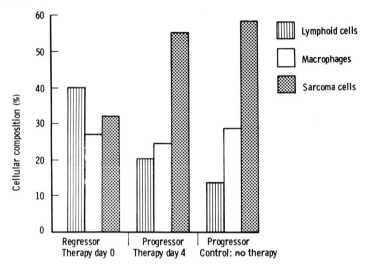

Fig. 4. Effect of delayed therapy on the cellular composition of challenge inocula during specific active immunotherapy of sarcoma Mc7.

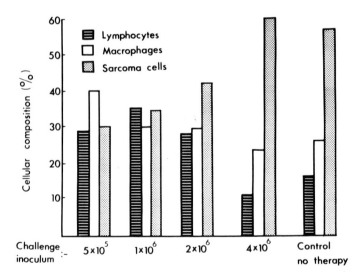

Fig. 5. Effect of increasing challenge inocula on the cellular composition during specific active immunotherapy of sarcoma Mc7.

These findings are consistent with the view that in this system at least, lymphoid cell infiltration and accumulation is an important early event in immunotherapy-induced tumour regression.

PROPERTIES OF TUMOUR DERIVED LYMPHOID CELLS

In order to characterise more fully the lymphoid component of progressively growing and regressor tumours, methods have been developed for the separation and purification of quantities of tumour derived lymphoid cells, and these have been tested against target tumour cells both in vitro and in vivo. Tumours were generally disaggregated by trypsin treatment, and lymphoid cells separated by nylon fibre column filtration. The in vitro test was a 48 hour microcytotoxicity assay in which the surviving target cells are quantitated by their uptake of ^{75}Se-selenomethionine (Brooks and others, 1978). Wherever possible, lymphoid cells were tested in parallel in Winn type assays, usually with a tumour cell to effector cell ratio of 1:6.

The results of two such experiments are summarised in Table 3. In these experiments spleen cells from normal rats were used as a control lymphoid cell population. As can be seen, these cells show prominent cytotoxic activity against Mc7 target cells; this activity is probably due to natural killer (NK) cells. However, when tested in vivo, spleen cells did not affect tumour growth.

TABLE 3 In Vivo and In Vitro Effects of Tumour Derived Lymphoid Cells

Source of Lymphoid Cells	Microcytotoxicity Test[1]		Winn Test[2]	
	Effector: target ratio	Percentage cytotoxicity	Effector: target ratio	Tumour takes
Normal spleen	100:1	84	6:1	6/6
Mc7 tumour (progressor)	100:1	72	6:1	0/6
Medium control	-	-	-	6/7
Normal spleen	100:1	43	6:1	6/6
Mc7 tumour (regressor)	100:1	52	6:1	1/6
Medium control	-	-	-	6/7

[1] Mc7 target cells surviving 48 hr incubation with lymphoid cells quantitated by uptake of ^{75}Se-selenomethionine. Cytotoxicity calculated with respect to medium control.

[2] Lymphoid cells and Mc7 tumour cells mixed and inoculated subcutaneously.

Lymphoid cells from 10 day old Mc7 progressor and regressor tumours showed in vitro cytotoxicity comparable with spleen cells tested on the same occasion. In the Winn assay, however, the tumour derived lymphoid cells exerted a marked antitumour effect, clearly reducing tumour takes in each test.

Thus lymphoid cells from within Mc7 tumours are capable of controlling tumour growth when present in admixture with sarcoma cells at an appropriate ratio. These experiments also show that the results of in vitro microcytotoxicity tests do not necessarily reflect the ability of lymphoid cells to control tumour growth in vivo.

DISCUSSION

This work demonstrates that active specific immunotherapy can control tumour growth, although the response is limited and certain criteria must be satisfied for a clear cut response to be obtained. Firstly, the tumour must be intrinsically immunogenic; therapy with mixed inocula of cells and BCG is ineffective against challenge inocula not also controllable in conventionally pre-immunised rats. Even with immunogenic tumours, treatment has to be initiated within a few days of challenge, and moderate immunosuppression by simple whole body irradiation abrogates the response. With the standard model employed, viable cells in admixture with BCG or C. parvum were more effective than vaccines incorporating radiation attenuated cells, and irradiated cells alone were virtually ineffective.

The present investigation is comparable to the extensive series of studies by Hanna and Peters (1978), Bartlett and Zbar (1972) and Bartlett and others (1978) on active specific immunotherapy with their transplanted line-10 guinea pig hepatoma. Here too a vaccine of irradiated cells alone was ineffective in controlling intradermal, subcutaneous or intravenous tumour challenge; although addition of BCG rendered the vaccine effective. Here, however, vaccines containing radiation attenuated cells were as effective as those containing viable cells, and indeed were more suitable, since mixed inocula of viable cells and BCG were occasionally tumorigenic. Although in the present work no attempt was made to optimise vaccine constituents in terms of BCG : tumour cell ratios, the studies by Hanna and Peters (1978) with the guinea pig hepatoma model, indicate that vaccines with high BCG : tumour cell ratios may be the most effective, and, furthermore, a second administration of cells and BCG, or irradiated cells alone, may improve the response. Hanna and Peters (1978) also investigated the therapeutic response in guinea pigs pre-sensitised to BCG, but observed no difference between sensitised and non-sensitised animals. Current work with the sarcoma Mc7 therapy model described in the present paper, also indicates that pre-sensitisation to BCG does not improve the therapeutic response, and may even exert a deleterious effect (Embleton and El-Sharkawy, unpublished).

In this form of active specific immunotherapy, it is likely that BCG or C. parvum act locally, and on the lymph node draining the site of the vaccine, to stimulate the generation of specific tumour immune responses, the effector cells of which may traffic to distant tumour foci. In an examination of lymph nodes draining the site of BCG and a more defined antigen (sheep red blood cells - SRBCs), Miller, Mackaness and Lagrange (1973) and Mackaness, Auclair and Lagrange (1973) found proliferative responses to SRBCs in the draining node were amplified by previous local injection of BCG. Furthermore the degree of immunopotentiation was quantitatively related to the level of cellular activity stimulated by the BCG. BCG injected to stimulate a distant lymph node did not influence the response to SRBCs, implying that to potentiate a response with BCG, both must impinge upon the same lymph node.

The effectiveness of tumour derived lymphoid cells in controlling tumour growth in vivo, and the accumulation of increased numbers of these cells in tumours undergoing immunotherapy-induced regression, are consistent with the concept, outlined above, that sensitised effector cells, generated in response to tumour vaccine inoculation, localise in the tumour undergoing therapy. The failure of lymphocytes to localise in the progressively growing tumours could be due firstly to insufficient induction of immunity, i.e., abrogation of the induction phase, or secondly to defects in the extravasation or accumulation of sensitised host cells at the tumour site, i.e., abrogation of the effector phase. Although it is not yet known which of these mechanisms is primarily responsible for failure of therapy, both potential effects could be due to tumour products, probably tumour antigen, and direct support for this hypothesis is provided by studies showing that active immunotherapy of the sarcoma Mc7, extensively used in the present work, can be abrogated, specifically, by systemic administration of tumour antigen in the form of intact irradiated cells or 3M KCl cell extracts (Baldwin, 1976). Furthermore, where active therapy is abrogated by soluble antigen, lymphocytic infiltration into the progressing tumour is markedly reduced (Hopper and Baldwin, unpublished observation).

In summary, these studies have delineated some of the requirements for control of tumour growth by active specific therapy and indicate that further evaluation of the host responses involved may provide a more rational, rather than empirical, approach to this form of treatment.

ACKNOWLEDGEMENTS

This work was supported by a grant from the Cancer Research Campaign, England, and by contract number NO1-CB-64042 with the Tumor Immunology Program, National Cancer Institute, National Institutes of Health, Bethesda, U.S.A.

REFERENCES

Baldwin, R. W. (1976). Role of immunosurveillance against chemically induced rat tumors. *Transplant. Rev.*, 28, 62-74.

Bartlett, G. L., and B. Zbar (1972). Tumor-specific vaccine containing Mycobacterium bovis and tumor cells: safety and efficacy. *J. Natl. Cancer Inst.*, 48, 1709-1726.

Bartlett, G. L., J. W. Kreider, D. M. Purnell, and A. J. Hockley (1978). Treatment of visceral tumor with BCG-tumor cell vaccine. *Cancer Immunol. Immunother.*, 4, 15-20.

Bomford, R. (1975). Active specific immunotherapy of mouse methylcholanthrene induced tumours with Corynebacterium parvum and irradiated tumour cells. *Br. J. Cancer*, 32, 551-557.

Brooks, C. G., R. C. Rees, and R. A. Robins (1978). Studies on the microcytotoxicity test. II. The uptake of amino acids ($[^3H]$ leucine or $[^{75}Se]$ methionine) but not nucleosides ($[^3H]$ thymidine or $[^{125}I]$ IUdR) or $^{51}CrO_4^{2-}$ provides direct and quantitative measure of target cell survival in the presence of lymphoid cells. *J. Immunol. Meth.*, 21, 111-124.

Hanna, M. G., and L. C. Peters (1978). Immunotherapy of established micrometastases with Bacillus Calmette Guerin tumor cell vaccine. *Cancer Res.*, 38, 204-209.

Mackaness, G. B., D. J. Auclair, and P. M. Lagrange (1973). Immunopotentiation with BCG. I. Immune responses to different strains and preparations. *J. Natl. Cancer Inst.*, 51, 1655-1667.

Miller, T. E., G. B. Mackaness, and P. M. Lagrange (1973). Immunopotentiation with BCG. II. Modulation of the response to sheep red blood cells. *J. Natl. Cancer Inst.*, 51, 1669-1676.

Pimm, M. V., D. G. Hopper, and R. W. Baldwin (1978). Host responses in adjuvant contact suppression of experimental rat tumours. *Develop. Biol. Stand.*, 38, 349-354.

Immunotherapy of Lung Cancer

E. Carmack Holmes

The Division of Oncology, Department of Surgery, UCLA School of Medicine, University of California, Los Angeles, CA 90024, U.S.A.

ABSTRACT

Lung cancer is a highly immunosuppressive disease. The extent of this immunosuppression correlates directly with the stage of disease and survival. Several studies have indicated that BCG immunotherapy is effective in patients with limited disease (stage I); but BCG does not appear to be effective in the more advanced stages. More purified products of BCG are being developed and are promising new agents. Levamisole has some immunopotentiation effects, and clinical trials in resected lung cancer patients using this agent are encouraging. Immunotherapy using specific lung tumor antigens is more difficult to perform, but at least two trials using these antigens are promising. Immunotherapy for lung cancer is clearly in its infancy. Therapeutic results to date have been modest. However the results are sufficiently encouraging to warrant further studies.

Keywords

BCG, C. parvum, immunotherapy, Levamisole, lung cancer, tumor antigens.

INTRODUCTION

The importance of immunologic factors in the progression of cancer has become increasingly apparent over the past several years. The relationship of host immunocompetence and prognosis has been emphasized, and immunodeficiency has been demonstrated to be associated with a poor prognosis in patients with lung cancer as well as in patients with other kinds of cancer. In addition, as with other human tumors, there is evidence which suggests that lung cancers contain antigens capable of evoking an immune response in the host under certain circumstances. Whether or not host immunological reactions directed towards these antigens play an important role in the relationship between the tumor and the host has yet to be determined. Therefore, while general immunosuppression correlates with a poor prognosis, it is not altogether clear what the role of immune reactions directed specifically toward tumor antigens is in the patient with cancer.

Patients with lung cancer are among the most profoundly immunosuppressed of all patients with solid neoplasms. They have impaired reactions of delayed cutaneous hypersensitivity. They are commonly unable to become sensitized to and to react to 2-4-dinitrochlorobenzene (DNCB). This inability to develop delayed cutaneous hypersensitivity to this topical antigen is closely related to prognosis and stage of disease. Patients with depressed delayed cutaneous hypersensitivity reactions to DNCB have a worse prognosis, are more likely to be resectable at the time of surgery, and of course tend to have more advanced disease (Holmes and Golub, 1976). In addition, lymphocytes from patients with lung cancer have been shown to have an impaired ability to undergo transformation *in vitro* upon stimulation with various antigens and mitogens (Holmes and Golub, 1976). This depression in lymphocyte transformation is associated also with a poor prognosis and more advanced disease. The mechanism of this immunosuppression in patients with lung cancer is not clear. It has been shown that serum from patients with lung cancer is capable of suppressing normal lymphocyte transformation (Giuliano and others, in press). This suggests the possibility that the tumor elaborates a humoral substance which is responsible for the suppression of cell-mediated immunity. However, more recent studies have indicated that lung cancer patients have suppressor lymphocytes and that these suppressor cells also may play a role in the immunosuppression of patients with lung cancer (Jerrell and others, 1978).

In view of these immunological findings, it is tempting to try to manipulate the immune response in an attempt to improve prognosis. Since our understanding of the mechanism of immunosuppression in patients with lung cancer is rudimentary, it is difficult to develop specific and effective therapy. There are several theoretical ways to attempt to manipulate the immune response. The immune response can be aroused by nonspecific stimulation with such agents as BCG and its subfractions. Corynebacterium parvum is also capable of nonspecifically stimulating the immune response. Theoretically, specific stimulation of immune response could occur when one uses tumor cells or their products and thereby specifically stimulates the immune response to tumor-associated antigens. Finally, a new class of agents, called immunopotentiators, recently has received considerable attention. Levamisole and Thymocin are among these agents. It is felt that Thymocin, and perhaps Levamisole, are capable of correcting certain deficiencies in cell-mediated immunity. These agents, however, do not appear to have an effect on an intact normal immune response. Since nonspecific stimulators such as BCG and C. parvum are not as effective in the immunocompromised host, it may be that immunopotentiators such as Thymocin and Levamisole will be more effective in patients with advanced disease, who are more likely to be immunosuppressed. However, the development of immunotherapeutic agents is very much in its infancy, and the mechanisms of action are poorly understood. In view of the limited number of immunotherapeutic agents available and our lack of understanding of their mechanism of action, and our equally poor understanding of the mechanism of immunosuppression in the patient with cancer, it is not surprising that the early trials with these immunotherapeutic agents have given rise to modest results. However, there have been some clearly detectable benefits of immunotherapy in patients with lung cancer. Clinical trials reported to date primarily have used nonspecific immunotherapy; a few studies have evaluated specific immunotherapy with tumor antigens or tumor cell vaccine.

Bacillus Calmette-Guérin (BCG)

BCG is an attenuated viable bovine tubercle bacillus and is prepared and supplied in a variety of ways (Bast and others, 1974). BCG has been evaluated extensively in animal tumor systems and also in clinical trials (Hanna, 1974; Nathanson, 1974). It is capable of nonspecifically stimulating the immune response in animals; however, there is no evidence that BCG is capable of correcting immunosuppression. Indeed, suppression of immune response by radiation therapy or corticosteroids abrogates the nonspecific immune stimulation occasioned by BCG. In animal models as well as in clinical studies the most effective way of using BCG to induce a systemic tumor resistance is by direct injection of BCG into the tumor (Hanna and others, 1976; Morton and others, 1974). Intralesional injection of BCG is capable of controlling cutaneous metastatic disease in a majority of patients so treated, and BCG currently is undergoing extensive evaluation as a nonspecific immune stimulator in a variety of solid tumors.

Several clinical trials suggest that BCG may be an effective adjunct to surgery in patients undergoing pulmonary resection for lung cancer. Foremost are those reported by McKneally and others (1976, 1978). In these studies patients were stratified on the basis of histological information and were randomized following complete resection to receive BCG injected intrapleurally or no intrapleural treatment. Both groups of patients received isoniazid (INH) therapy in order to minimize any risk of systemic BCG infection. Over the past five years 110 patients have been entered into this trial. Ten million colony-forming units of Tice strain BCG were injected. Intrapleural BCG appeared to have no effect on patients with stage II and III resected lung cancer. However, intrapleural BCG significantly improved the survival of patients with stage I disease. Of 66 patients with stage I disease entered into this trial, 30 received intrapleural BCG and INH therapy and 36 received INH therapy alone (Table 1). There was a highly significant difference in favor of the BCG-treated group in the stage I patients. However, as Table 1 indicates, survival in patients with stage II and stage III lung cancer was not affected by immunotherapy.

TABLE 1 Intrapleural BCG Immunotherapy in Resected Lung Cancer Patients

Therapy	Number of Patients[1]		
	Stage I	Stage II	Stage III
BCG + INH	3/30	8/13	7/12
INH	16/36	6/9	5/10
	$p=0.00066$	$p=0.3226$	$p=0.8587$

[1]Number of patients with recurrent cancer/total number in group.

The results of this carefully stratified and randomized trial are encouraging. The lack of effect in patients with stage II and stage III disease perhaps describes the limitation of BCG immunotherapy. Such patients are more immunosuppressed than patients with stage I lung cancer, and it is well known that BCG is more effective in the immunocompetent individual. A number of centers have initiated similar trials of intrapleural BCG in an attempt to corroborate these findings, but results of these trials are too preliminary at the present time to draw any meaningful conclusions.

Recent findings of Herberman and others (1978) suggest that BCG may not necessarily have to be administered into the intrapleural space in order to be effective in patients with surgically resected lung cancer. In these studies patients with stages I, II, and III lung cancer were randomized following pulmonary resection to receive BCG by the intradermal scarification technique or BCG plus whole tumor cells or no immunotherapy. The patients receiving immunotherapy with stage I disease had a significantly prolonged disease-free interval when compared to the nonimmunotherapy control group. These investigators also found that BCG immunotherapy was not as effective in the resectable stage II and III patients. These findings, therefore, corroborate the observations of McKneally that BCG immunotherapy is effective only in stage I resectable lung cancer. It is interesting to note that in these studies the addition of specific immunotherapy in the form of allogeneic irradiated tumor cells did not have any advantage over BCG alone.

Pouillart and others (1977) recently reported a clinical trial evaluating BCG in patients with resected lung cancer. Forty-three patients with resectable stage I and stage II lung cancer were randomized to receive intradermal BCG by scarification or no further treatment. These studies also indicate that BCG immunotherapy is effective in prolonging the disease-free interval in patients with stage I lung cancer. However, these investigators were unable to show an effect of BCG immunotherapy on the disease-free interval of patients with resectable stage II lung cancer.

Other forms of BCG immunotherapy are being evaluated in patients with lung cancer. In Japan a nonviable chemical extract that contains as its major component the cell-wall skeleton (CWS) of BCG has been evaluated extensively in patients with lung cancer. CWS was developed by Yamamura, and he and his coworkers have demonstrated its effectiveness in animal tumor models (Azuma and others, 1974). Early reports of randomized trials using CWS as an adjunct to surgery are quite promising (Azuma and others, 1974; Yasumoto and others, 1976). These more purified products of BCG are easier to quantitate and standardize, and for this reason may be more advantageous than BCG.

Since direct intratumor instillation of BCG appears to be the most effective way to induce systemic tumor resistance in animals, this technique has been evaluated in patients with resectable lung cancer. The BCG can be injected into the tumor percutaneously under fluoroscopic control or through the bronchoscope. The phase I evaluation of direct intratumor injection of BCG followed by pulmonary resection has indicated that the toxicity is quite acceptable. However, patients who have strong reactions to PPD prior to injection may have a significant febrile reaction for several days (Holmes and others, 1977; Holmes and others, in press). BCG injected in this fashion regularly results in the induction of a significant granulomatous inflammatory reaction and necrosis within the tumor. The histology of this reaction is very similar to that observed in the animal models when systemic antitumor immunity is established. This technique for immunotherapy of lung cancer has yet to be evaluated in a properly randomized phase III trial, and must be considered highly experimental. However, on a theoretical basis it has much attraction.

Corynebacterium Parvum

C. parvum is a killed, formalin-treated Corynebacterium. It has many features similar to BCG in animal tumor models. However, C. parvum has not been evaluated as extensively in patients with lung cancer as has been BCG. Israel (1973) has suggested that the use of C. parvum increases the rate of response to chemotherapy and the duration of response, and also diminishes the myelosuppressive effects of the chemotherapy. At the present time there is a large study in Western Europe evaluating the effects of C. parvum injected intrapleurally following pulmonary resection of patients with lung cancer in a protocol similar to the one using intrapleural BCG. While more than 200 patients have been randomized into this study, no therapeutic results are available at this time. C. parvum has certain theoretical advantages over BCG. C. parvum is a killed organism which cannot induce infection, whereas BCG on occasion has been known to cause infection in man. In addition, C. parvum can be administered intravenously with fewer toxic side effects than BCG. Intravenously administered C. parvum appears to localize in the pulmonary parenchyma, where a mild granulomatous process ensues with presumed stimulation of the regional mediastinal lymph nodes. For these reasons C. parvum has certain theoretical advantages over BCG. However, these advantages have yet to be demonstrated in appropriate clinical trials.

Levamisole

Levamisole (l-tetramisole) is a drug that has been widely used as an antihelmintic in animals and in man. In contrast to BCG and C. parvum, Levamisole is a chemical and can be taken orally in the form of a tablet. Levamisole is felt to be an immunopotentiator or an anti-anergic agent. Some studies suggest that Levamisole is capable of stimulating the immune response in patients who have been immunosuppressed, and in some patients with lung cancer Levamisole appears to be capable of reversing the suppression of lymphocyte function (Holmes and Golub, 1978). In view of the immunosuppression in lung cancer, Levamisole would appear to be an attractive agent for the treatment of this disease.

Several years ago a prospective randomized study was begun evaluating Levamisole as a surgical adjunct in patients with lung cancer (Israel, 1973). In this double-blind study patients who were candidates for thoracotomy were randomized to receive Levamisole 50 mg three times daily on the last three days prior to surgery, and a similar three-day course every second week for two years following surgery. The patients randomized to the control group received placebo in a similar fashion. Levamisole was given preoperatively in an attempt to prevent the immunosuppression caused by surgery. More than 200 patients were randomized into this study, and a significant follow-up period was obtained. When the two populations were considered as a whole, there was a slight trend in favor of the Levamisole-treated group in terms of disease mortality. However, results were not consistently statistically significant. When the patients were evaluated on the basis of body weight, there was a striking difference in favor of Levamisole in those who weighed less than 70 kg. The investigators felt that a fixed dose of 150 mg of Levamisole daily for three days every two weeks is an insufficient amount for patients weighing more than 70 kg, and that such patients were underdosed. They felt that the patients weighing less than 70 kg represented a population in which Levamisole therapy could be analyzed more accurately, and concluded that in such patients this dose is effective. Levamisole seemed to be especially effective in controlling disseminated disease, in that the first site of recurrence was most often intrathoracic rather than at distant sites (Amery, 1978).

Levamisole has very few side effects when given in the doses and the time schedules outlined in this study. Patients do complain occasionally of a bitter taste, with occasional gastric intolerance and on rare occasions increased nervousness. The current recommended dose is 2.5 mg/kg. While most investigators have not seen granulocytopenia in patients receiving this dose of Levamisole, severe granulocytopenia has been reported in patients with severe rheumatoid arthritis and other diseases who have received Levamisole therapy (Thornes, 1978). Levamisole is an attractive agent because the complications are rare, and theoretically it has the potential of reversing the immunosuppressive effects of lung cancer. However, the initial immunological investigations with this agent are somewhat controversial, and the clinical trials are in need of corroboration.

Specific Immunotherapy

The evaluation of specific immunotherapy in human cancer is not as extensive as that of immunotherapy with the nonspecific stimulators. Specific immunotherapy requires the use of tumor cells containing tumor-associated antigens or fractions of the tumor cell in various degrees of purification putatively containing tumor-associated antigens. Tumor cells are difficult to propagate in tissue culture, and the precise subfraction of the cell which contains the antigens that are presumably important in inducing an antitumor immune response in the host is not well delineated. For these reasons specific immunotherapy is more difficult and more expensive than nonspecific therapy.

Stewart and others (1978) have evaluated specific immunotherapy in patients with lung cancer. In this study, partially purified tumor antigens derived from lung cancer were mixed with complete Freund's Adjuvant and used as the specific immune stimulating agents. Patients who had undergone surgical resection for lung cancer were randomized to receive postoperative Methotrexate alone, tumor antigen with Freund's Adjuvant, or Methotrexate and tumor antigen with Freund's Adjuvant. The results of this phase II trial are relatively promising. The authors noted that the patients who received the immunotherapy with or without Methotrexate survived significantly longer than the patients treated with Methotrexate alone. As the investigators have suggested, more definitive stage III trials are indicated. Since these phase II trials did not include a control group receiving Freund's Adjuvant alone, it is difficult to separate the effect of Freund's Adjuvant from the effect of the tumor antigen. It is interesting to note that the patients who received the immunotherapy also developed delayed cutaneous hypersensitivity reactions to the immunizing lung cancer antigens. The patients who received Methotrexate alone did not develop cutaneous hypersensitivity to these antigens.

Another phase II trial using specific immunotherapy in patients with stage III lung carcinoma has been reported by Takita and others (1978). In this study patients with extensive stage III lung cancer who underwent complete resection of the tumor were randomized into a group receiving no further immunotherapy and a group receiving autologous tumor cell vaccine treated with Vibrio cholera neuraminidase. The tumor cell vaccine was given intradermally on the day of surgery and subsequently was repeated at two-week intervals. One month after surgery BCG was given intradermally every other week for four weeks and then once a month for the first year and every six weeks thereafter. Some of the patients in the control group and some of the patients in the immunotherapy group also received postoperative radiation therapy. Thirty resected patients were randomized, 15 into the control group and 15 into the immunotherapy group. The median survival in the control group was 12.1 months and the median survival in the treated group was 34.8 months. The difference between these two groups was statistically significant. Thus the results of this phase II study are encouraging. These investigators currently are involved in a phase III study to assess more accurately the effectiveness of this form of surgical adjuvant immunotherapy.

DISCUSSION

Only in the recent past has it been appreciated that lung cancer is a profoundly immunosuppressive disease. At the present time our knowledge of the mechanism of this immunosuppression is very superficial, and in the light of this ignorance it is difficult to manipulate accurately the immune response in favor of the patient. When we have a better understanding of the mechanisms of action of the immunotherapeutic agents and are better able to describe accurately the mechanisms of immunosuppression, we will be able to tailor the immunotherapy appropriately to the needs of the individual patient. For instance patients with stage I resectable lung cancer are not as profoundly immunosuppressed as patients with stage II and stage III lung cancer. Since it is known that certain immunotherapeutic agents such as BCG are not effective in the immunosuppressed patient, it is not surprising that the initial clinical trials using BCG indicate that this agent is effective only in early stage I resected lung cancer. To date the nonspecific immune stimulators have been consistently ineffective in more advanced immunosuppressed patients. Perhaps the immunorestorative agents such as Thymocin and Levamisole, and other agents yet to be described, will be most appropriate as immunotherapeutic agents in patients with more advanced lung cancer. The early results of the phase II trials suggest that perhaps a tumor antigen may be effective in the more advanced stages of disease.

It is clear, however, that immunotherapy as we now understand it is not a potent cytoreductive agent. All studies to date indicate that immunotherapy should be combined with surgery or radiation therapy or chemotherapy. Indeed, in the surgical adjuvant studies reported to date, immunotherapy has had a modest effect on survival and disease-free interval. However, it is encouraging that it has an effect at all. A number of clinical trials evaluating immunotherapy in the treatment of lung cancer are ongoing, and there is a continuing effort to develop more effective and less toxic agents, more effective routes of administration, and more effective and more highly purified tumor antigens.

ACKNOWLEDGMENT

These investigations have been supported in part by grant CA09010 awarded by the National Cancer Institute (DHEW).

REFERENCES

Amery, W. K. (1978). A placebo-controlled Levamisole study in resectable lung cancer. In W. D. Terry and D. Windhorst (Eds.), *Immunotherapy of Cancer: Present Status of Trials in Man*. Raven Press, New York. pp. 191-201.
Azuma, I., T. Taniyama, F. Hirao, and Y. Yamamura (1974). Antitumor activity of cell-wall skeleton of Mycobacteria and related organisms in mice and rabbits. *Gann*, 65, 493-505.
Bast, R. C., B. Zbar, T. Borsos, and H. J. Rapp (1974). BCG and cancer. *N. Engl. J. Med.*, 290, 1413-1420.
Giuliano, A. E., D. M. Rangel, S. H. Golub, E. C. Holmes, and D. L. Morton. Immunosuppression in lung cancer. *Cancer* (in press).
Hanna, M. G. Jr. (1974). Immunological aspects of BCG-mediated aggression of established tumors and metastases in guinea pigs. *Semin. Oncol.*, 1, 319-335.
Hanna, M. G. Jr., L. C. Peters, and I. J. Fidler (1976). The efficacy of BCG-induced tumor immunity in guinea pigs with regional and systemic malignancy. *Cancer Immunol. Immunother.*, 1, 171-177.
Herberman, R. B., E. Perlin, J. Reid, C. Miller, G. D. Bonnard, J. L. Weese, J. Glom, and R. J. Connor (1978). Immunotherapy of lung cancer with BCG and allogeneic tumor cells (abstract). *Proc. World Conf. Lung Cancer*, p. 28.
Holmes, E. C., and S. H. Golub (1976). Immunologic defects in lung cancer patients. *J. Thorac. Cardiovasc. Surg.*, 71, 161-168.
Holmes, E. C., K. P. Ramming, J. Mink, W. Coulson, and D. L. Morton (1977). New method of immunotherapy for lung cancer. *Lancet*, 2, 586-587.
Holmes, E. C., and Golub, S. H. (1978). Effect of Levamisole on cell-mediated immunity in patients with lung cancer. In M. A. Chirigos (Ed.), *Immune Modulation and Control of Neoplasia by Adjuvant Therapy*. Raven Press, New York. pp. 107-118.
Holmes, E. C., K. P. Ramming, M. E. Bein, W. F. Coulson, and C. D. Callery. Intralesional BCG immunotherapy of pulmonary tumors. *J. Thorac. Cardiovasc. Surg.* (in press).
Israel, L. (1973). Preliminary results of nonspecific immunotherapy for lung cancer. *Cancer Chemother. Rep. Part 3*, 4, 283-286.

Jerrells, T. R., J. H. Dean, G. L. Richardson, S. Vadlamudi, and R. B. Herberman (1978). Suppressor cell activity in immunodepressed lung and breast cancer patients (abstract). *Proc. Am. Assoc. Cancer Res., 19,* 73.

McKneally, M. F., C. Maver, H. W. Kausel, and R. D. Alley (1976). Regional immunotherapy with intrapleural BCG for lung cancer: Surgical considerations. *J. Thorac. Cardiovasc. Surg., 72,* 333-338.

McKneally, M. F., C. M. Maver, and H. W. Kausel (1978). Regional immunotherapy of lung cancer using postoperative intrapleural BCG. In W. D. Terry and D. Windhorst (Eds.), *Immunotherapy of Cancer: Present Status of Trials in Man.* Raven Press, New York. pp. 161-171.

Morton, D. L., F. R. Eilber, E. C. Holmes, J. S. Hunt, A. S. Ketcham, M. J. Silverstein, and F. C. Sparks (1974). BCG immunotherapy of malignant melanoma. *Ann. Surg., 180,* 635-643.

Nathanson, L. (1974). Use of BCG in the treatment of human neoplasms: A review. *Semin. Oncol. 1,* 337-350.

Pouillart, P., T. Palangie, P. Huguenin, P. Morin, H. Gautier, A. Lededente, A. Baron, and G. Mathe (1977). Attempt at immunotherapy with BCG of patients with bronchus carcinoma: Preliminary results. In S. E. Salmon and S. E. Jones, (Eds.), *Adjuvant Therapy of Cancer.* Elsevier/North Holland Biomedical Press, Amsterdam. pp. 225-235.

Stewart, T. H. M., A. C. Hollinshead, J. E. Harris, S. Raman, R. Belanger, A. Crepeau, A. F. Crook, W. E. Hirte, D. Hooper, D. J. Klaasen, R. F. Rapp, and H. J. Sachs (1978). Survival study of immunochemotherapy in lung cancer. In W. D. Terry and D. Windhorst (Eds.), *Immunotherapy of Cancer: Present Status of Trials in Man.* Raven Press, New York. pp. 203-216.

Takita, H., M. Takada, J. Minowada, T. Han, and F. Edgerton (1978). Adjuvant immunotherapy of stage III lung cancer. In W. D. Terry and D. Windhorst (Eds.), *Immunotherapy of Cancer: Present Status of Trials in Man.* Raven Press, New York. pp. 217-223.

Thornes, R. D. (1978). Interpretation of management of Levamisole-associated side effects. In M. A. Chirigos (Ed.), *Immune Modulation and Control of Neoplasia by Adjuvant Therapy.* Raven Press, New York. pp. 157-164.

Yasumoto, K., H. Manabe, M. Ueno, M. Ohta, U. Hidehiko, I. Akira, K. Nomoto, I. Azuma, and Y. Yamamura (1976). Immunotherapy of human lung cancer with BCG cell-wall skeleton. *Gann, 67,* 787-795.

Specific Active Immunotherapy and Specific Active Immunoprophylaxis in Lung Cancer

A. Hollinshead* and T. Stewart**

*Department of Medicine, George Washington University Medical Center,
Washington, D.C. 20037, U.S.A.
**Department of Medicine, University of Ottawa Medical Center,
Ottawa, Canada

ABSTRACT

Tumor-associated antigens (TAA) which are present on the cell in minute quantities have been isolated from the soluble products including inhibitory antigens (IA) obtained from separated membranes of each form of lung cancer. Other antigens which predominate on the lung cancer cell are not lung cancer-associated, but are found in many forms of cancer, and have been described previously as inhibitors (IA) which prevent the induction of the cell-mediated immune responses (CMI) by TAA. Current experiments suggest that IA are nucleoproteins which exist in plaques or patches at intervals along the lung cancer cell surface membranes. TAA of the main histologic types of lung cancer have been used successfully in specific active immunotherapy trials. These studies indicate that a strong cell-mediated immune reactivity is present for at least five years following immunization, and will probably last much longer. Real benefit might result from public health programs of prophylactic immunization in high-risk middle-age populations of heavy cigarette smokers.

INTRODUCTION

A definition of and description of the different types of tumor-associated antigens as well as the design of an active specific immunotherapy trial, with the example of lung cancer, have been presented (Hollinshead and Stewart, 1976; Hollinshead, 1978). The design of the lung cancer immunotherapy trial included the careful selection of TAA free of interfering, harmful or inhibitory material and combined with suitable adjuvant for slow delivery. An appropriately selected multimodality protocol permitted active specific excitation, both local and systemic, which resulted in an exclusive attack on the lung cancer through primary immunologic programing (Stewart and colleagues, 1976, 1977, 1978; Hollinshead, Stewart, and Takita, 1978a, 1978b; Takita, Hollinshead and Bjornsson, 1978). The patients in the Ottawa study (Fig. 1) were divided into two groups, those who were not immunized, some of whom received methotrexate and those who were immunized with TAA plus adjuvant, some of whom received methotrexate. Survival statistics as of October 1, 1978 are presented for those patients who had Stage I lung cancer; the classic life table analysis of survival shows a significant separation between the groups after 18 months (see Fig. 1). Up to the 24th month this is real survival, not based on life table methods. In contrast to the May 1978 evaluation at 95% confidence, the trial has now reached significance at the 1% level (see October 1978 confidence band, Fig. 1). Fifty-five patients with epidermoid carcinoma have thus far been entered into an ongoing specific active immunotherapy trial of TAA at the Roswell Park Memorial Institute, and at 2 1/2 years the immunotherapy arm appears to parallel the experience in Ottawa, with the group receiving Freund's adjuvant alone intermediate in survival between the control group and the immunotherapy group at this point (Hollinshead, Stewart, and Takita, 1978). The Georgetown University Medical Center and the George Washington University Medical Center in Washington, D.C. have established a protocol for chemoimmunotherapy, similar to one of the groups in the Ottawa trial, but, in addition, plan to study unresectable patients in order to decide whether or not chemoimmunotherapy has any usefulness in combination with radiotherapy.

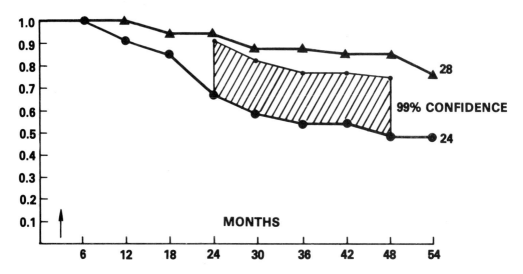

➤ indicates the 3 month point at which all therapy was finished
lower line – two Stage I groups in control, with or without MTX
upper line – two Stage I groups immunized, with or without MTX
99% confidence band – becomes significant after 18 months

Fig. 1. Specific active immunotherapy Stage I lung cancer (1 October 1978)

A multi center trial has begun in Canada, and will have three arms in a total of 300 patients with Stage I and II lung cancer following curative surgery. One hundred will serve as control. One hundred will receive Freund's complete adjuvant once per month for three months, and 100 will receive Freund's adjuvant homogenized with the appropriate allogeneic pooled tumor-associated antigens once per month for three months.

One of the more difficult problems is that of deciding which forms of immunologic testing should be done in a battery of tests to be used for monitoring various types of therapy. As summarized in Table 1, there are four different types of immunologic testing: nonspecific and specific cellular and humoral evaluation; these are discussed in detail elsewhere (Hollinshead, 1978). One of the more useful tests for monitoring specific active immunotherapy trials has been the use of skin tests with tumor-associated antigens for immunosurveillance during the trial (Hollinshead and Stewart, 1976). Since the vaccine is prepared from pooled separated tumor cell membranes of the same histologic type, which are solubilized by low frequency sonication, and the tumor-associated antigens further separated and purified, it is possible, at the same time, to prepare a number of skin tests for preservation at -70°C, to be used in monitoring the patients entered into the protocol. Although autologous TAA and allogeneic TAA probably do not share all sequences, most of the peptide chain appears to be shared in common, the terminal amino acid sequences appear to be similar, and the antigens from different tumors have lines of identity in immunoelectrophoresis. However, to insure that these reactivities were well defined, at the local as well as at the systemic level, we performed some autologous skin testing on a few patients. One example is shown in Fig. 2. In this instance, the patient was monitored with allogeneic TAA, and was skin test negative, converting to a positive reaction one month after the second course of immunotherapy. An enhanced reaction was seen to allogeneic TAA one month after the third course of immunotherapy. Thus, in this patient, the skin test permitted an assessment of conversion and of enhancement of the cell-mediated immune response to the vaccination. Thirteen months later, or 16 months after surgery, the autologous TAA, which had been purified and stored at -70°C, was titered, and a markedly sensitive response to 500 ng TAA was evident. Thirty-one months later, or 47 months after surgery, the patient was tested simultaneously with both the autologous and the allogeneic TAA. As shown in the legend for Fig. 2, a very strong, positive delayed hypersensitivity reaction at 48 hours, with 100% mononuclear cell infiltrate seen in the biopsy of the site, was produced both to the autologous and to the allogeneic TAA. Almost all patients on specific active immunotherapy, with long term survival, show a retention of strong delayed hypersensitive reactions to skin tests of lung cancer TAA, some examples of which are shown in Table 2.

TABLE 1 Immunologic Testing During Immunotherapy

Nonspecific
    Cellular: DHR-ST - recall antigens
                CBC, lymph. count, RBC sedrate
                 B, T cells - %, total
                 LMIT - inhibitory antigens
                       other tissue antigens
    Humoral: ELISA - cancer serum index
                RIA - IA, IA complexes
Specific
    Cellular: DHR-ST - TAA
                 Phase I booster test
                 Phase II & III conversion and levels
                DHR-ST of TAA + IA
                Titration of IA levels
                CF-ID (breast ca; squamous cell ca)
                LMIT - TAA (early stage melanoma +
                       early stage breast ±
                       early stage bladder ±)
    Humoral: ELISA - TAA, interfering serum components
                RIA - TAA, TAA complexes

We have described the abrogation of lung cancer patients' cell-mediated immune responses to TAA after mixture and incubation with IA (Hollinshead and Stewart, 1976; Hollinshead, Stewart, and Takita, 1978). There appeared to be a hundred times as many IA on the oat cell membranes as compared with the squamous cell membranes from lung tumors. Similar fractions and controls from normal lung are not inhibitory. These IA are present in the insoluble cell membrane material as well as from the heavy soluble membrane material which stays at the cathode or very top of the gel and does not separate on our standard polyacrylamide gel electrophoresis. Material at the top of the gel can be sliced, eluted, concentrated and remixed with the antigens for titrations as described elsewhere. Rosenberg (1975) felt that the platinum complexes disrupted antigenic masking on animal tumor cells, exposing what he

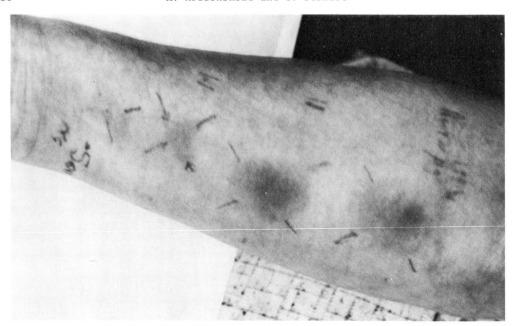

Autologous TAA skin tests of a lung cancer patient on TAA vaccine (allogeneic) plus methotrexate.

(N.B. Patients were monitored using 100 µg TAA protein of the same allogeneic material as the vaccine. Autologous TAA was not regularly used for monitoring patients in the trial, and, in the above patient was not tested until 16 months after surgery, 13 months after the last vaccination; see above picture.)

Months after surgery	Titration (TAA protein/0.1 ml skin test)	Delayed hypersensitivity reaction at 48 hours (millimeters induration)
16	112.28 micrograms	40x40
16	11.2 micrograms	36x35
16	1.1 micrograms	20x15
16	500 nanograms	13x11
16	50 nanograms	negative
47	112.28 micrograms	30x30

[Note reactivity to stored autologous antigen (4 years storage at -70°C)]

Allogeneic TAA Skin Tests of the Same Patient

Months after surgery:	one	two	three	four	forty-seven
MM induration at 48 hours: (100 µg TAA)	3	4	40	60	37x47

(Note conversion to positive reaction one month after second course of immunotherapy)

Fig. 2. Autologous and allogeneic skin testing in the same patient.

TABLE 2 Retention of Strong Delayed Hypersensitive Reactions to Skin Tests of Pooled Allogeneic Antigen in Patients on Specific Active Immunotherapy

Diagnosis	Type of allogeneic TAA tested	Months after surgery	mm Induration response at 48 hours to 100 µg TAA skin test
Epidermoid-Adenocarcinoma	Adenocarcinoma	56	27x22
	Epidermoid		37x32
Epidermoid	Epidermoid	36	45x31
Epidermoid	Epidermoid	36	47x54
Epidermoid-Adenocarcinoma	Epidermoid	20	20x19
	Adenocarcinoma		21x22
Adenocarcinoma	Adenocarcinoma	26	15x12

thought might be new antigens at the cell surface which generate a host immune reaction. He noted the appearance of densely stained patches associated only with the tumor cell membranes and not with normal cell membranes, and identified these as possibly masking antigens containing DNA. Most nucleic acids and nucleoproteins are weaker antigens and may escape host immune surveillance. In previous studies of the cell surface markers on leukemic and pre-leukemic cells (Hollinshead and colleagues, 1977), we observed patches on the cell membranes, which were not present on normal white blood cells (Fig. 3a). Lung cancer and melanoma cultured cells were obtained from Dr. G. Cannon and WI-38 fibroblasts were obtained from Mr. Feinmark. Because most alterations in the cell occur during fixation, we used a procedure of fixation and of staining with methyl green pyronin Y as previously described (Hollinshead, 1964). As shown in Fig. 3a, DNA containing patches were observed only on the tumor cell membranes. We also observed the effect of DNAse incubated with IA separated from primary lung and primary melanoma tumor cells. As shown in Table 3, with suitable controls, reactivity was seen, and, thus, it is possible that the inhibitory antigens consist of DNA nucleoproteins, as illustrated in the schema (Fig. 3b).

We have demonstrated that most patients receiving specific active immunotherapy overcome CMI interference by the inhibitory antigens whilst patients during cancer recurrences fail to overcome the inhibition (Hollinshead and Stewart, 1976; Hollinshead, Stewart, and Takita, 1978).

Some of our patients on immunotherapy showed a reactivation of the vaccine site four years or more after receiving immunotherapy. One such example is patient RW, who (1) showed suspicious bronchial cytology in 1968; (2) showed a left upper lobe shadow in 1974 with no symptoms; (3) had a thoracotomy in June of 1974, left upper lobectomy for squamous cell carcinoma, no hilar involvement, T_2 (4.5 cm), N_0M_0; on microscopy "tumor blood vessel invasion noted;" (4) issued immunotherapy (TAA + FCA) on June 28, August 8 and September 16, 1974; (5) hemoptysis in May 1976, suspicious 1 cm lesion on x-ray of left lung, regressed by time of tomograms 1 month later; (6) trachea and left main bronchus recurrent disease November 1977; (7) local irradiation to trachea plus adriamycin, BCNU, Mtx plus citrovorum rescue; (8) in the middle of 1978 vaccine site on the left thigh was active. This active site was biopsied, and showed granulation tissue only. (9) August 16, 1978, patient is free of disease on broncoscopy and biopsy. Feels well. An additional example: patient CMC had surgery March 19, 1974, and had T_2 N_0M_0 adenocarcinoma. At the time of this report, this patient, who was placed on immunotherapy, remains disease free. During May and June of 1978, this patient reported a reactivation of his vaccine site, and there was granulation tissue only. Thus, in this individual, there is evidence of 4 1/2 years of activity.

 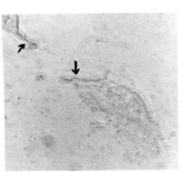 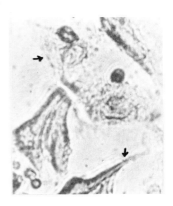

WI-38　　　　　　　6278　　　　　　　　400X　　CALU
LUNG CELL　　　　MELANOMA CELL CULTURE　　　LUNG CANCER CELL
CULTURE　　　　　　　　　　　　　　　　　　　　CULTURE

Approx. 1200X

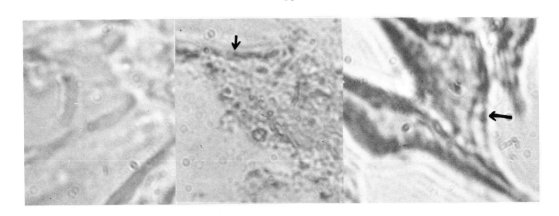

(photography courtesy of Jno Randall)

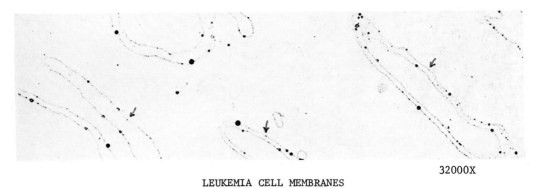

LEUKEMIA CELL MEMBRANES　　　　　　32000X

Fig. 3a. DNA-protein patches (possible inhibitory antigens)

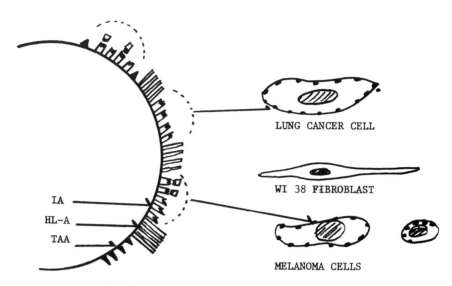

Fig. 3b. Nucleoprotein (DNA) inhibitory antigens (IA) in clusters or patches on the cancer cell plasma membranes.

TABLE 3 Effect of DNAse Treatment on Inhibitory Antigens on the Cancer Cell Plasma Membrane, after Separation of the Membranes and Release of the Antigens by Low Frequency Sonication

	DNAse Reactivity	PAGE BAND at cathode*
Sigma DNAse 104 units		Faint line
Worthington DNAse 104 units		Faint line
Calf thymus DNA 50 units		Faint line
DNA plus S. DNAse	+	Negative
DNA plus W. DNAse	+	Negative
Lung Cancer Inhibitory Antigens (IA) (100 micrograms)		Thick band
Lung Ca IA plus S. DNAse	+	Negative
Lung Ca IA plus W. DNAse	+	Faint band
Melanoma Inhibitory Antigens (100 micrograms)		Thick band plus dark area below
Melanoma IA plus S. DNAse	+	Faint band
Melanoma IA plus W. DNAse	+	Faint line

*(Sigma DNAse lot 46C-0132, electrophoretically purified, free of RNAse, 2085 Kunitz units per microgram.
 Worthington DNAse, bovine pancrease, free of RNAse, purified, 2979 Kunitz units per microgram.
 Calf thymus DNA, Calbiochem. lot 52497, no. 2618, A grade, histone free
 IA: partially soluble, contains nucleoprotein by spectrophotometry and only partially separable on polyacrylamide 3.5% gel at top.
 6.25 mM $MgSO_4$ used to activate DNAse; incubation with IA at 37°C/30'.

In conclusion, there is retention of strong activity to pooled antigen for over 5 years, as well as to autologous antigen, and this suggests that a strong delayed hypersensitivity reaction toward soluble tumor-associated antigens can be engineered, and that this dual systemic and local response will last at least for 5 years following immunization, and probably much longer.

This raises a real and very important practical question. In a high risk population of heavy cigarette smokers, male, would prophylactic immunization reduce the incidence of eventual lung cancer and would those who develop the disease have a better response to conventional therapy? As shown in Fig. 4a, the death rate of male smokers on 1 1/2 packs per day is 12 to 28 times greater than the death rate of males who never smoked in terms of lung cancer deaths alone. Occupations in uranium mining and in asbestos industry are associated with 3 to 5 times the death rate due to lung cancer by the average male, and it is unknown to what degree such occupational risks are increased in males who smoke. In the Ottawa area, 25 out of 500 heavy smokers in the age range of 51 to 55 years have cancer within 2 to 5 years, thus a risk of 1 in 20. It is possible to test this general population or a population at higher risk, in order to see whether or not prophylactic immunization would reduce the incidence of eventual lung cancer in these groups. As shown in Fig. 4b, lung cancer mortality is greatest in the age group from 55 to 74 years of age. Therefore, real benefit may accrue from such a public health program if it is initiated between the ages of 51 to 55, five years or more before the usual average time of diagnosis of clinical lung cancer. In Fig. 5, we present a possible immunoprophylaxis protocol, which includes another arm for study of the possible beneficial effect of vitamin A derivatives, to see if the effects are additive. In our experience, no loss of time from work needs to be anticipated from the morbidity of TAA immunization. Animal studies on immunoprophylaxis (virus-induced TSTA: Hollinshead, McCammon, and Yohn, 1972; carcinogen-induced TSTA: Prager and colleagues, 1973) have been successful. Many immune theories of surveillance have come under some doubt in recent years, but what is wrong in middle age high risk groups with immune surveillance by polyvalent cancer antigens?

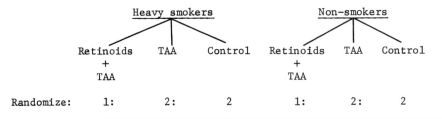

Fig. 5. Outline for lung cancer immunoprophylaxis protocol immunosurveillance in high risk groups.

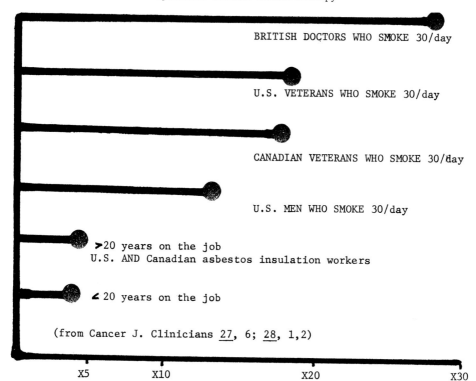
Fig. 4a. Mortality ratio: no. exposed/no. unexposed.

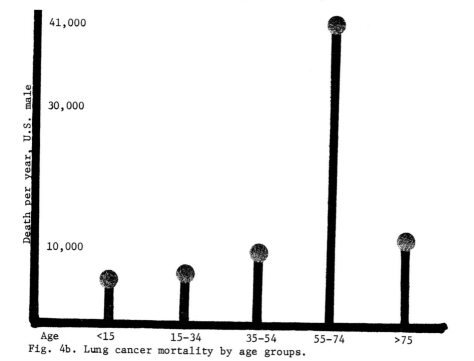
Fig. 4b. Lung cancer mortality by age groups.

REFERENCES

Hollinshead, A. (1964). Incorporation of 8-azaguanine and growth inhibition in mammalian spinner cultures. Exptl. Cell Res., 34, 144-154.

Hollinshead, A.C. (1978). Active-specific immunotherapy. In Immunotherapy of Human Cancer, Raven Press, New York. pp. 213-233.

Hollinshead, A., J.R. McCammon, and D.S. Yohn. (1972). Immunogenicity of a soluble transplantation antigen from adenovirus 12-induced tumor cells demonstrated in inbred hamsters (PD-4). Can. J. Microbiol., 18, 1365-1369.

Hollinshead, A.C. and Stewart, T.H.M. (1976). Lung tumor antigens: specific active immunotherapy trials. In Proc. III Internatl. Symp. Detection and Prevention of Cancer, Vol. IV, part 2, Marcel Dekker, New York. p. 52.

Hollinshead, A., T.H.M. Stewart, and H. Takita. (1978). Phase II trials of specific active immunotherapy for lung cancer patients. In S. Weinhouse (Ed.), Proc. Am. Soc. Clinical Oncol., Vol. 19, Abstract C-76, Waverly Press, Inc., Maryland. p. 325.

Hollinshead, A.C., T.H.M. Stewart, and H. Takita. (1978). Tumor-associated antigens: their usefulness as biological drugs. In F. Muggia and M. Rozencweig (Eds.), Lung Cancer: Progress in Therapeutic Research, Raven Press, New York. pp. 501-518.

Hollinshead, A.C., C.S. White, C.T. Ladoulis, L.D. Ellis, R.O. Gilcher, and L.S. Lessin. (1977). Cell surface markers on pre-leukemic cells I. Soluble proteins. Comp. Leukemia Res., Elsevier/North-Holland Biomedical Press. pp. 256-259.

Prager, M.D., A.C. Hollinshead, R.J. Ribble, and I. Derr. (1973). Immunity induction by multiple methods, including soluble membrane fractions to a mouse lymphoma. J. Natl. Cancer Inst., 51, 1603-1607.

Rosenberg, B. (1975). Possible mechanisms for the anti-tumor activity of platinum coordination complexes. Cancer Chemother. Rep., Part 1, 59, 589-598.

Stewart, T.H.M., A.C. Hollinshead, J.E. Harris, R. Belanger, A. Crepeau, G.D. Hooper, H.J. Sachs, D.J. Klassen, W. Hirte, E. Rapp, A.F. Crook, M. Orizaga, P.S. Sengar, and S. Raman. (1976). Immunochemotherapy of lung cancer. Ann. N.Y. Acad. Sci., 277, 436-466.

Stewart, T.H.M., A.C. Hollinshead, J.E. Harris, R. Sankaranarayanan, R. Belanger, A. Crepeau, A.F. Crook, W.E. Hirte, D. Hooper, D.J. Klassen, E.F. Rapp, and H.J. Sachs. (1977). A survival study of specific active immunochemotherapy in lung cancer. In R.G. Crispen (Ed.), Neoplasm Immunity: Solid Tumor Therapy, Franklin Institute Press, Philadelphia. pp. 37-48.

Stewart, T.H.M., A.C. Hollinshead, J.E. Harris, R. Sankaranarayanan, R. Belanger, A. Crepeau, A.F. Crook, W.E. Hirte, D. Hooper, D.J. Klassen, E.F. Rapp, and H.J. Sachs. (1978). Specific active immunochemotherapy in lung cancer: A survival study. In Proc. EORTC Plenary Session on Adjuvant Therapies and Markers of Post-surgical, Minimal Disease, Springer Verlag, Paris. in press.

Takita, H., A.C. Hollinshead, and S. Bjornsson. (1978). Chemotherapy, surgery and immunotherapy of inoperable lung cancer, In S. Weinhouse (Ed.), Proc. Am. Soc. Clinical Oncol., Vol. 19, Abstract C-50, Waverly Press, Inc., Maryland. p. 319.

Specific Antibodies as Carriers of Diagnostic or Therapeutic Doses of Radioisotopes (or Drugs)

David Pressman

*Department of Immunology Research, Roswell Park Memorial Institute,
666 Elm Street, Buffalo, New York 14263, U.S.A.*

ABSTRACT

This paper demonstrates the use of specific antitumor antibodies to carry diagnostic or therapeutic amounts of radioactivity to a tumor and its metastases and contains some comments about the possible use of antibody to carry cytotoxic agents to a tumor for therapy.

KEYWORDS

antibody
antitumor antibody
radioiodine
localizing antibody
labeled antibody

Historically Hericourt and Richet in 1895 were the first to report an attempt to achieve a therapy of cancer by use of antitumor antibodies. Although they initially reported very positive results the therapy was not continued. Subsequently others reported similar attempts but no lasting therapy was developed. Our approach to the problem was to see if antitumor antibody can be prepared which actually goes to the tumor when injected intravenously. This we proposed to do by attaching a radioactive isotope to the antibody raised against a tumor and see if the antibody carries the radioactivity to the tumor. If it does, then antibody might be used to carry cytotoxic or diagnostically useful amounts of a radioisotope to the tumor.

In our work starting 30 years ago (Pressman and Keighley, 1948) we showed that antibodies can be iodinated without destroying the antibody specificity and that radiolabelled anti-organ antibody preparations could be made which would localize in the specific organ. Thus we showed that anti-rat-kidney antibody which had been shown by Masugi (1933) to be cytotoxic to kidney did localize in the kidney primarily in the glomerular tuft (Pressman, Hill and Foote, 1949). We also showed that antibodies could be prepared which localized in other organs for example anti-rat-lung antibodies could be raised which would localize in the lung (Tsuzuku, Yagi and Pressman, 1967). Organ specific antibodies could be purified to remove antibodies which cross localized in other organs. Our experiments showed therefore that there are differences in the vascular beds of the blood vessels passing through the different organs and that organ specific antibodies could recognize these differences so that specific localization could be obtained (Pressman and Yagi, 1973).

In 1953 Dr. Leonhard Korngold and I (Pressman and Korngold, 1953) published the first demonstration that antibody raised against a tumor becomes fixed in the tumor. This was for a mouse tumor, the Wagner osteogenic sarcoma. In 1958, Blau, Day and I (Pressman, Day and Blau, 1958) showed that radioiodinated antitumor antibodies could be used to locate a rat tumor, the Murphy lymphosarcoma, by external scanning.

In 1960, Bale, Spar and Goodland were the first to show a therapeutic effect of radioactive iodine carried to tumor by an antibody, i.e., anti-fibrin antibody. This was done with the Murphy lymphosarcoma tumor in rats. Fibrin can be considered to be a tumor antigen since it is present in many rapidly growing tumors. We had previously shown that anti-fibrin antibodies localize in rapidly growing tumors (Day, Planinsek and Pressman, 1959). Recently Ghose and Guclu (1974) showed cure of a mouse lymphoma with radioiodinated antibody.

We also developed the use of the paired label technique for control of the localizing antibody studies (Pressman, Day and Blau, 1957). Antitumor antibody labeled with one radioisotope of iodine is compared in localizing properties with normal globulin labeled with another isotope of iodine. Both are injected into the tumor bearing animal simultaneously and the distribution of the two radioisotopes determined in the tumor and in various organs and tissues. The paired label technic has the advantage in that it demonstrates just how much antibody and just how much control globulin is localized in each portion of the tumor. This is particularly important when the tumor is not uniform but contains portions in various states of necrosis and vascularization. Thus a value for nonspecific localization is available as a control for the antibody localization value in each piece of tumor assayed.

The problem has now progressed to where positive results have been obtained by others in investigations involving human subjects primarily where the tumor from different individuals contain a common tumor antigen. Thus, Ghose and his colleagues have recently shown by radioactive scanning that radioiodinated goat antibodies raised against a human renal cell tumor can be rendered specific by suitable absorption and, when injected into a patient with a renal cell tumor, carries radioactive iodine to the tumor and its metastases (Belitsky, Ghose and colleagues, 1978a, 1978b). His control was Technicium labeled sulfur colloid. Goldenberg and his colleagues (1978) have more recently shown in a paper published a few months ago, that antibodies prepared in goats against carcinoembryonic antigen when radioiodinated and injected into patients bearing CEA producing tumors do accumulate in the tumor so that it can be located by radioscanning. Again the control was Technicium and Technicium labeled albumin rather than iodinated normal goat globulin.

Previously Mahaley, Day and colleagues (1965) showed that anti-human brain tumor antibodies localized in human brain tumors by injecting anti-brain tumor antibodies labeled with ^{125}I and normal goat globulin labeled with ^{131}I into a patient and measuring the amount of each isotope in samples of the tumor removed at surgery. Their calculations needed correction for antibodies which localized elsewhere than the tumor. However localization was definitely demonstrable by radioautography.

Since anti-fibrin antibodies have been known to be fixed rapidly in certain animal tumors (rapidly growing tumors) in which fibrin is being deposited (Day,

Planinsek and Pressman, 1959a, 1959b; Spar, Goodland and Bale, 1959). McCardle and colleagues (1966), Marrack and colleagues (1967) and Spar and colleagues (1967) have studied the situation in humans and found that the anti-fibrin antibodies are fixed in many human tumors.

Several considerations are involved in getting localization of antibody in a tumor. One is that the antibody must be directed to specific surface components of the tumor cells since localization must first take place on the cell surface. Subsequently the radioantibody may be interiorized. Lewis, Pegrum and Evans (1974) showed that radioiodinated anti-lymphocyte antibodies localized on lymphocyte are interiorized in the capping reaction. Guclu, Tai and Ghose (1975) also showed interiorization with antitumor antibody as did Robert and Revillard (1976).

It is easy for antibody to come into contact with single cells in culture or in the circulation or in ascites fluid and much *in vitro* work has been done with cell cultures demonstrating the cytotoxicity of cytotoxic agents coupled with antibody (Ghose and Blair, 1978). Localization on a solid tumor is much more difficult because then localization has to take the place as the blood or lymph carrying the antibody passes through the tumor. The proportion of the total antibody injected which can contact the tumor in unit time is determined by what proportion of the cardiac output reaches the tumor. This can be very small and thus a long time may be required for localization of antibody in tumor even if the blood is completely cleared of localizing antibody as it passes through the tumor which seems to be the case.

That the blood is cleared essentially completely of localizing antibody as it passes through the specific organ has been shown to be the case for kidney, lung, and adrenal[1] and may be just as efficiently cleared by tumor as it passes through the tumor. However the small fraction of the total circulation passing through the tumor in unit time may limit the amount of localization which can take place before the radioantibody is eliminated from the circulation by ordinary metabolic processes. In an attempt to aid in the fixation of antitumor antibodies in a tumor, Order (1976) has used a catheter to lead the administered radiolabel directly to the blood vessels supplying the tumor.

This problem might be overcome if it were possible to use antitumor antibodies produced in humans. In order to avoid the ethical problem of producing antibodies in humans, Dr. Grossberg and I (Pressman, Chu and Grossberg, 1978) have just found that it might be possible to carry out the procedure without any raised antibody by separating tumor-binding globulin from normal human serum and using it in place of raised antibody. CEA binding globulin was isolated from normal human serum by affinity chromatography using CEA coupled to Sepharose as a solid adsorbant. Those globulins from normal human serum which were absorbed on the CEA Sepharose were eluted and showed specific binding for CEA in a radioimmune assay. The theoretical basis for this experiment is that since normal globulin is a mixture of immunoglobulins produced by different lines of immunoglobulin producing cells, there can be isolated from the mixture those immunoglobulins which interact with particular antigen, in this case the CEA antigen. The immunoglobulin has the advantage over animal antibody in that it is a product of humans and thus may stay in the circulation for a longer time than xeno-anti-

---

[1] See the following references: Pressman, Eisen and Fitzgerald, 1949; Blau, Day and Pressman, 1957; Tamanoi, Yagi and Pressman, 1961; Hiramoto, Yagi and Pressman, 1958.

bodies. Immunoglobulin acting like antibodies with distinct binding specificities had previously been isolated from normal sera of other species by the use of an affinity absorbent. We showed it first in our laboratory (Pressman and Korngold, 1952) and subsequently others showed it. Thus, in 1952, we (Pressman and Korngold, 1952) showed that it is possible to isolate globulins from normal rabbit serum capable of localizing in kidney, lung or liver of rats. Thus, specific tissue binding globulins were isolated by treating radioiodinated normal serum globulin with the insoluble portion of kidney tissue and then eluting the adsorbed globulins by use of alkali or heat and these globulins showed a preferential localization in kidney. When the absorbant was the insoluble part of lung tissue or of liver tissue, globulins capable of localizing in lung or in liver, respectively, were obtained.

Subsequently, Winkler, Adetugbo and Lehrer (1972) reported that globulins isolated from normal rabbit serum by absorption on insoluble adsorbents to which azobenzenesulfonate groups (or azobenzoate groups) had been coupled, showed specific binding to proteins containing the azobenzenesulfonate group (or containing the azobenzoate group) depending on the adsorbent originally used. Indeed they were able to obtain separately each of the hapten binding globulins from the same batch of serum.

More recently, Sela, Wang and Edelman (1975) and Sela and Edelman (1977) reported the isolation of cell binding antibodies from normal globulins of several species. Other laboratories have reported the presence of "natural" antibodies in sera but did not isolate them. For example, Hager and Tompkins (1976) found "antibodies" in normal rabbit serum that react with human adenocarcinoma cells but did not isolate them. Irie, Irie and Morton (1974) detected a natural antibody in human sera reacting with antigen on human cultured cells grown in fetal bovine serum.

Another possible problem in the use of antibodies to carry radioactivity to the tumor is the effect of tumor antigen or antitumor antibody in the circulation of the tumor bearing host on the localization of administered antitumor antibody. Goldenberg and colleagues (1978) has shown that the presence of a relatively high CEA concentration in the circulation, 250 ng/ml, did not prevent localization of iodinated anti-CEA antibody on a CEA producing tumor nor does the presence of the large amount of fibrinogen in the circulation interfere with the fixation of antifibrin antibodies on fibrin in a tumor.

If the presence of antibody in the circulation becomes a problem, it may be possible to remove it from the circulation, and as it is formed, by thoracic duct canulation as described by Dr. Sam Rose (Rose, 1973).

Investigations on the use of antibodies to carry cytotoxic agents to tumors are getting more attention. The work in this area has been recently reviewed by Ghose and Blair (1978). Thus besides radioactive isotopes, diphtheria toxin, methyltrexate, chlorambacil, trenimon, daunorubicin, adriamycin, phenylenediamine mustard, phospholipase C, glucoseoxidase and boron[2] have been coupled to antitumor antibodies and in several cases have shown increased tumoricidal activity when tested with tumor cells *in vitro*. Some positive cytotoxic activity has been reported *in vivo* in certain cases (Ghose and Blair, 1978). It would be very

---

[2]See Hawthorne, Wiersema and Takasugi (1972). Boron becomes cytotoxic when irradiated with neutrons since each boron atom which absorbs a neutron splits to one α particle and one lithium nucleus.

worthwhile to radiolabel the antibodies carrying drugs so that the localization of the drug antibody complex can be determined in order to see if the drug, when attached to the antibody, localizes the antibody where the drug tends to localize rather than where the antibody itself would otherwise localize.

REFERENCES

Bale, W. F., I. L. Spar, and R. L. Goodland. (1960). Experimental radiation therapy of tumors with I^{131}-carrying antibodies to fibrin. Cancer Res. 20, 1488-1494.

Belitsky, P., T. Ghose, J. Aquino, J. Tai, and A. S. MacDonald. (1978a). Radionuclide imaging of metastasis in renal cell carcinoma patients by ^{131}I-labeled anti-tumor antibody. Radiology 126, 515-517.

Belitsky, P., T. Ghose, J. Aquino, S. T. Norvell, and A. H. Blair. (1978b). Radionuclide imaging of primary renal cell carcinoma patients by ^{131}I-labeled antitumor antibody. J. Nucl. Med. 19, 427-430.

Blau, M., E. D. Day, and D. Pressman. (1957). The rate of localization of antirat kidney antibodies. J. Immunol. 79, 330-333.

Day, E. D., J. A. Planinsek, and D. Pressman. (1959a). Localization *in vivo* of radioiodinated anti-rat-fibrin antibodies and radioiodinated rat fibrinogen in the Murphy rat lymphosarcoma and in other transplantable rat tumors. J. Natl. Cancer Inst. 22, 413-426.

Day, E. D., J. A. Planinsek, and D. Pressman. (1959b). Localization of radioiodinated rat fibrinogen in transplanted rat tumors. J. Natl. Cancer Inst. 23, 799-812.

Ghose, T., and A. H. Blair. (1978). Antibody-linked cytotoxic agents in the treatment of cancer: Current status and future prospects. J. Natl. Cancer Inst., in press (Sept. issue).

Ghose, T., and A. Guclu. (1974). Cure of a mouse lymphoma with radio-iodinated antibody. Europ. J. Cancer 10, 787-792.

Goldenberg, D. M., F. DeLand, E. Kim, S. Bennett, F. J. Primus, J. R. van Nagel, Jr., N. Estes, P. DeSimone, and P. Rayburn. (1978). Use of radiolabeled antibodies to carcinoembryonic antigen for the detection and localization of diverse cancers by external photoscanning. New Engl. J. Med. 298, 1384-1388.

Guclu, A., J. Tai, and T. Ghose. (1975). Endocytosis of chlorambucilbound antitumor globulin following "capping" in EL4 lymphoma cells. Immunol. Commun. 4, 229-242.

Hager, J. C., and W. A. F. Tompkins. (1976). Antibodies in normal rabbit serum that reacts with tissue-specific antigens on the plasma membranes of human adenocarcinoma cells. J. Natl. Cancer Inst. 56, 339-344.

Hawthorne, F., R. J. Wiersema, and M. Takasugi. (1972). Preparation of tumor-specific boron compounds. I. *In vitro* studies using boron-labeled antibodies and elemental boron as neutron targets. J. Med. Chem. 15, 449-452.

Hericourt, J., and C. Richet. (1895). Traitement d'un cas de Sarcome par la Serotherapie. Comptes Rendus Acad. Sci. 120, 948-950.

Hiramoto, R., Y. Yagi, and D. Pressman. (1958). *In vivo* fixation of antibodies in the adrenal. Proc. Soc. Exp. Biol. Med. 98, 870-874.

Irie, R. F., K. Irie, and D. L. Morton. (1974). Natural antibody in human serum to a neogantigen in human culture cells grown in fetal bovine serum. J. Natl. Cancer Inst. 52, 1051-1057.

Lewis, C. M., G. D. Pegrum, and C. A. Evans. (1974). Intracellular location of specific antibodies reacting with human lymphocytes. Nature 247, 463-465.

Mahaley, M. S., Jr., J. L. Mahaley, and E. D. Day. (1965). The localization of radioantibodies in human brain tumors. II. Radioautography. Cancer Res. 25, 779-793.

Marrack, D., M. Kubala, P. Corry, M. Leavens, J. Howze, W. Dewey, W. F. Bale, and I. L. Spar. (1967). Localization of intracranial tumors. Comparative study with ^{131}I-labeled antibody to human fibrinogen and neohydrin-^{203}Hg. Cancer 20, 751-755.

Masugi, M. (1933). Uber das weden der spezifiscen varanderungene der niere und der leber durch das nephrotoxin bzw. das hepatotoxin. Beitr. Pathol. 91, 82.

McCardle, R. J., P. V. Harper, I. L. Spar, W. F. Bale, G. Andros, and F. Jiminez. (1966). Studies with iodine-131-labeled antibody to human fibrinogen for diagnosis and therapy of tumors. J. Nucl. Med. 7, 837-847.

Order, S. E. (1976). The history and progress of serologic immunotherapy and radiodiagnosis. Radiology 118, 219-223.

Pressman, D., T. M. Chu, and A. L. Grossberg. (1978). A CEA-binding immunoglobulin isolated from normal human serum by affinity chromatography. J. Natl. Cancer Inst., submitted for publication.

Pressman, D., E. D. Day, and M. Blau. (1957). The use of paired labeling in the determination of tumor-localizing antibodies. Cancer Res. 17, 845-850.

Pressman, D., E. D. Day, and M. Blau. (1958). Radioactive anti-tumor antibodies. Proc. 2nd UN Intl. Conf. on Peaceful Uses of Atomic Energy, Geneva 24, 236.

Pressman, D., H. N. Eisen, and P. J. Fitzgerald. (1949). The zone of localization of antibodies. VI. The rate of localization of anti-mouse-kidney serum. J. Immunol. 64, 281-287.

Pressman, D., R. F. Hill, and F. W. Foote. (1949). The zone of localization of anti-mouse-kidney serum as determined by radioautographs. Science 109, 65-66.

Pressman, D., and G. Keighley. (1948). The zone of activity of antibodies as determined by the use of radioactive tracers; the zone of activity of nephritoxic antikidney serum. J. Immunol. 59, 141-146.

Pressman, D., and L. Korngold. (1952). Experimental hypersensitivity. Science 116, 433.

Pressman, D., and L. Korngold. (1953). The *in vivo* localization of anti-Wagner osteogenic sarcoma antibodies. Cancer 6, 619-623.

Pressman, D., and Y. Yagi. (1973). Radioisotopes in immunology. In N. R. Rose, F. Milgrom, and C. J. van Oss (Eds.), Principles of Immunology, Chapter 8, Macmillan Publ. Co., New York. pp. 103-109.

Robert, M., and J. P. Revillard. (1976). Fate of antibodies bound to lymphocyte surface. I. Study with complement-dependent cytotoxicity, indirect immunofluorescence and radiolabelled antibodies. Ann. Immunol. (Inst. Pasteur) 127C, 129-144.

Rose, S. (1973). Augmentation of immune activity of elimination of antibody and its implications in cancer. J. Surg. Oncol. 5, 137-166.

Sela, B. A., and G. Edelman. (1977). Isolation by cell-column chromatography of immunoglobulins specific for cell surface carbohydrates. J. Exp. Med. 145, 443-449.

Sela, B. A., J. L. Wang, and G. J. Edelman. (1975). Antibodies reactive with cell surface carbohydrates. Proc. Natl. Acad. Sci. 72, 1127-1131.

Spar, I. L., W. F. Bale, D. Marrack, W. C. Dewey, R. J. McCardle, and P. V. Harper. (1967). ^{131}I-Labeled antibodies to human fibrinogen. Diagnostic studies and therapeutic trials. Cancer 20, 865-870.

Spar, I. L., R. L. Goodland, and W. F. Bale. (1959). Localization of I^{131} labeled antibody to rat fibrin in a transplantable rat lymphosarcoma. Proc. Soc. Exp. Biol. Med. 100, 259-262.

Tamanoi, I., Y. Yagi, and D. Pressman. (1961). Rate of localization of anti-rat lung antibody. Proc. Soc. Exp. Biol. Med. 106, 769-772.

Tsuzuku, O., Y. Yagi, and D. Pressman. (1967). Preparative purification of lung-localizing rabbit anti-rat lung antibodies *in vitro*. J. Immunol. 98, 1004-1010.

Winkler, M. H., K. Adetugbo, and G. M. Lehrer. (1972). Specific hapten binding activity in normal sera. Immunol. Commun. 1, 51-68.

Overview of Controlled Prognostic Evaluations of Levamisole Immunotherapy in Clinical Cancer

Willem K. Amery

Janssen Pharmaceutica N.V., B-2340 Beerse, Belgium

ABSTRACT

Up to May 1978, results from 26 controlled clinical studies, evaluating the prognostic impact of giving levamisole to cancer patients treated with surgery, irradiation and/or chemotherapy, were available. The analysis of these data has led to a series of guidelines concerning the use of levamisole in clinical cancer: these guidelines are summarized. Levamisole is well tolerated by most oncological patients. Therefore, this new treatment modality appears to hold great promise.

KEYWORDS

Levamisole, adjuvant treatment, immunotherapy, clinical effects.

INTRODUCTION

As in several other diseases, a trend seems to emerge in clinical cancer therapy, indicating that a distinction ought to be made between a more aggressive, remission-inducing therapeutic approach on the one hand, and measures that are aimed to maintain what one has achieved by the previous treatment modalities on the other. This dichotomy is well illustrated by the application of surgery and irradiation, for the first, aggressive approach, and of immunotherapy for the maintenance type of treatment.

This presentation aims at reviewing the controlled experience with levamisole as an adjuvant immunotherapy in clinical cancer.

MATERIAL AND METHODS

All clinical studies, available up to May 1978, have been considered. Only those trials that had evaluated the potential effect of adjuvant treatment with levamisole on the prognosis of clinical cancer patients, have been retained for the analysis.

There were 26 such studies, but nine of these were still in a very early stage as judged from the number of patients included and/or the duration of follow-up (Brincker, Thorling & Jensen, 1976; Cabanillas and colleagues, 1977; Chahinian and colleagues, 1978; Chang, Wiernik & Lichtenfeld, 1978; Grandval and colleagues, 1976; Smith & de Kernion, 1978; Valdivieso and colleagues, 1977;

Wanebo and colleagues, 1978; Wright and colleagues, 1978). Therefore, the conclusions have mainly been based upon the 17 more advanced evaluations (Amery, 1978; Debois, 1978; D'Souza, Daly & Thornes, 1978; Gonzales & Spitler, 1978; Hall and colleagues, 1978; Klefström, 1978; Miwa, Orita & Tanaka, 1977; Mussche & Kluyskens, 1978; Pavlovsky and colleagues, 1978; Pines, 1977; Rojas and colleagues, 1978; Stephens, 1978; Verhaegen, 1978; Vuopio, 1978; Yap and colleagues, 1979). In all, 1 474 levamisole-treated patients and about 1 600 controls, most of them untreated or treated with a placebo, were followed in the 26 studies evaluated here. A break-down of these studies by disease category and by the type of primary treatment is given in Fig. 1 and 2.

FINDINGS AND COMMENTS

I. Efficacy

The results provided by these studies clearly suggest several tentative conclusions that are relevant to further studies of this drug and to its clinical usefulness. These conclusions, which further substantiate and expand earlier preliminary conclusions (Amery and colleagues, 1977), are listed below.

A. Dosage regimen

Intermittent treatment (two consecutive days every week or three consecutive days every fortnight) is probably at least as effective as continuous treatment. Also, the weekly and the fortnightly intermittent schedules seem to be equally effective. Two consecutive days every week may, therefore, be the preferable regimen since it is practical for the patient and since the total amount of drug given is only two sevenths of the amount that the patient would take if he followed the continuous treatment schedule.

The daily dose should be adapted to the bodyweight of the patient or to his body surface area. A daily target dose of 2.5 mg/kg or 85 mg/m^2, divided over two or three intakes, seems appropriate.

Levamisole treatment should be started as early as possible, but synchronous treatment with cytotoxic therapies is to be avoided. The following guidelines may be given:
- with surgery: treatment to be started during the last two or three days preceding the operation;
- with radiotherapy and with continuous cytoreductive chemotherapy: treatment with levamisole is started one to two days after termination of the cytotoxic therapy;
- with cyclic chemotherapy: levamisole is given in-between the cycles.

B. Patient eligibility

All histological types of tumor may be eligible in certain stages (see further).

Patients with more advanced, but still potentially curable malignancies seem to profit most from levamisole treatment.

Immunodeficient patients may be the outstanding group of candidates for levamisole treatment, but reliable criteria to establish such a deficiency are still to be developed. There is some evidence that the pre-treatment absolute lymphocyte count may be useful for that purpose in certain conditions.

C. What results can be hoped for ?

No reduction of the tumor load: levamisole should not be used as a monotherapy.

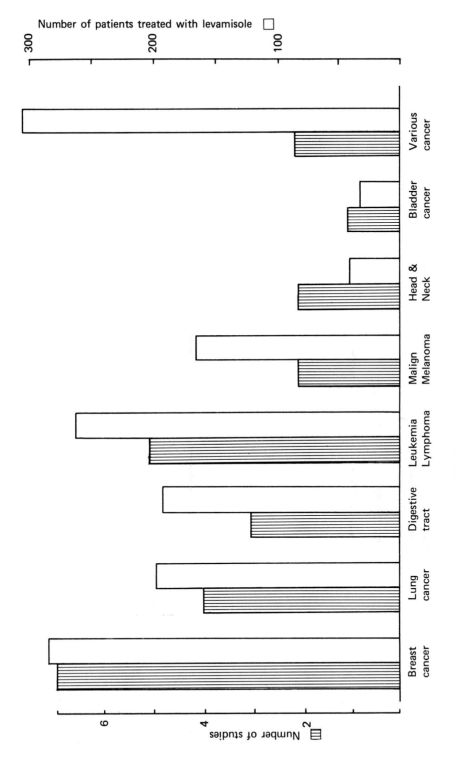

Fig. 1. Break-down by disease group.

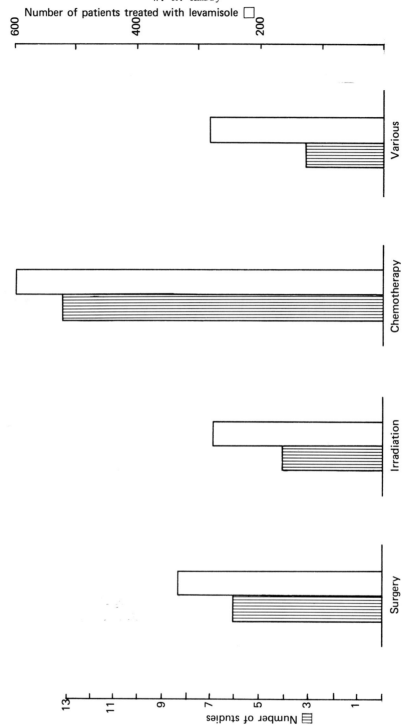

Fig. 2. Break-down by type of primary treatment.

Levamisole consolidates anti-cancer effects obtained by other treatment modalities. Whithin this context it may be particularly effective against metastasis formation.

The addition of levamisole to cyclic chemotherapy has resulted in some instances, but not in other, in an increased remission rate. The presence or absence of such an effect may, amongst other factors, be related to the marrow toxicity induced by the cytostatic agent since levamisole treatment has been reported to enhance bone marrow reconstitution in patients treated with cytostatic chemotherapy.

II. Acceptability

In general, levamisole has been very well tolerated (only very few patients have stopped their treatment because of side-effects): it may, therefore, be considered as a very suitable agent for adjuvant treatment especially in patients who are clinically free of disease.

The only major side-effect reported is a reversible, allergic destruction of peripheral granulocytes leading to clinical agranulocytosis. This complication is rare: it was found in only 0.4 % of the patients treated with levamisole in the studies reviewed here. The condition was always reversible in these patients on stopping levamisole treatment and never fatal.

CONCLUSION

If used appropriately, levamisole holds great promise as an adjunct to classical anti-cancer treatment since it is well tolerated and since it appears to stabilize the effects obtained by other anti-cancer therapies.

REFERENCES

Amery, W.K., F. Spreafico, A.F. Rojas, E. Denissen and M.A. Chirigos (1977). Adjuvant treatment with levamisole in cancer. A review of experimental and clinical data. Cancer Treatm. Rev., 4, 167-194.

Amery, W.K. (1978). Final results of a multicenter placebo-controlled levamisole study of resectable lung cancer. Cancer Treatm. Rep. (in press).

Brincker, H., K. Thorling and K.B. Jensen (1976). Prolongation of the duration of remission in acute myeloid leukemia (AML) with levamisole. Paper presented at the Spring Meeting of Scandinavian Hematologists, Aarhus, Denmark, June.

Cabanillas, F., V. Rodriguez, E.M. Hersh, G. Mavligit, G.P. Bodey and E.L. Middleman (1977). Chemoimmunotherapy of advanced non-Hodgkin's lymphoma (NHL) with CHOP-BLEO + levamisole. Abstract for the Meeting of the American Society of Clinical Oncology, Denver, CO, May.

Chahinian, A.P., E.A. Mandel, I.S. Jaffrey, A.S. Teirstein and J.F. Holland 1978). Randomized trial of chemotherapy with or without immunotherapy in advanced lung cancer. Janssen Research Products Information Service, Clinical Research Report No. R 12 564/87, July.

Chang, P., P.H. Wiernik and J.L. Lichtenfeld (1978). Levamisole (L), cytosine arabinoside (Ara-C), and daunorubicin (DNR) induction therapy of adult acute nonlymphocytic leukemia (ANLL). Abstract for the Meeting of the American Society of Clinical Oncology, Washington, DC, April.

Debois, J.M. (1978). Five-year experience with levamisole in cancer patients. Third interim report. Janssen Research Products Information Service, Clinical Research Report No. R 12 564/69, February.

D'Souza, D.P., L. Daly and R.D. Thornes (1978). Levamisole, BCG and warfarin as adjuvants to chemotherapy for increased survival in advanced breast cancer. Unpublished report.

Gonzalez, R. and L. Spitler (1978). Effect of levamisole as a surgical adjuvant therapy on malignant melanoma. Cancer Treatm. Rep. (in press).

Grandval, C.M., E. Bugnard, E. Cardama, R. Estevez, G. Paraskevas, P. Angelakis and R.D. Thornes (1976). Interim analysis of patient data from protocol No. 12 564/066. Janssen Research Products Information Service, Clinical Research Report No. R 12 564/47, September.

Hall, S.W., R.S. Benjamin, L. Heilbrun, U. Lewinski, J.U. Gutterman and G. Mavligit (1978). Chemoimmunotherapy of refractory malignant melanoma with actinomycin D and levamisole. In M.A. Chirigos (Ed.), Immune Modulation and Control of Neoplasia by Adjuvant Therapy, Raven Press, New York. pp. 131-140.

Klefström, P. (1978). Levamisole in addition to chemotherapy in advanced breast cancer. In H. Rainer (Ed.), Immunotherapy of Malignant Diseases, F.K. Schattauer Verlag, Stuttgart, New York. pp. 102-111.

Miwa, H., K. Orita and S. Tanaka (1977). Cancer immunotherapy with levamisole. Paper presented at the Convention of the Japanese Society of Cancer Therapy, Fukuoka, Japan, October.

Mussche, R.A. and P. Kluyskens (1978). Prognosis of primarily treated localized laryngeal carcinoma ameliorated through levamisole treatment. (in manuscript)

Pavlovsky, S., G. Garay, F. Sackmann-Muriel, A. Hayes, E. Svarch, M. Eppinger-Helft and R. Failace (1978). Chemoimmunotherapy with levamisole in acute lymphoblastic leukemia. (in manuscript)

Pines, A. (1977). BCG with and without levamisole in the treatment of patients with advanced squamous lung cancer following radical radiotherapy. Cancer Immunol. Immunother., 3 (Suppl.), 34 (Abstract No. 85).

Rojas, A.F., J.N. Feierstein, H.M. Glait and A.J. Olivari (1978). Levamisole action in breast cancer Stage III. In W.D. Terry and D. Windhorst (Eds.), Immunotherapy of Cancer: Present Status of Trials in Man, Raven Press, New York. pp. 635-645.

Smith, R. and J. de Kernion (1978). Preliminary report on the use of levamisole in bladder cancer. Cancer Treatm. Rep. (in press).

Stephens, E. (1978). Levamisole and FAC treatment in disseminated breast cancer. Cancer Treatm. Rep. (in press).

Valdivieso, M., A. Bedikian, M.A. Burgess, V. Rodriguez, E.M. Hersh, G.P. Bodey and G.M. Mavligit (1977). Chemoimmunotherapy of metastatic large bowel cancer. Nonspecific stimulation with BCG and levamisole. Cancer, 40, 2 731-2 739.

Verhaegen, H. (1978). Postoperative levamisole in colo-rectal cancer. In H. Rainer (Ed.), Immunotherapy of Malignant Diseases, F.K. Schattauer Verlag, Stuttgart, New York. pp. 94-101.

Vuopio, P. (1978). A randomized controlled study of levamisole as an adjunct to maintenance cytostatic chemotherapy in adult acute leukemia. Janssen Research Products Information Service, Clinical Research Report No. R 12 564/71, March.

Wanebo, H.J., E. Hilal, E.W. Strong, H.F. Oettgen and C.M. Pinsky (1978). Randomized trial of levamisole in patients with squamous cell carcinoma of the head and neck. Cancer Treatm. Rep. (in press).

Wright, P.W., L.D. Hill, A.V. Peterson, R. Pinkham, L. Johnson, T. Ivey, I. Bernstein, C. bagley and R. Anderson (1978). Preliminary results of combined surgery and adjuvant BCG and levamisole treatment of resectable lung cancer. Cancer Treatm. Rep. (in press).

Yap, Hwee-Yong, G.R. Blumenschein, G.N. Hortobagyi, C.K. Tashima, A.U. Buzdar, J.U. Gutterman, E.M. Hersh and G.P. Bodey (1979). Chemoimmunotherapy for advanced breast cancer. Cancer (in press).

Problems of Nonspecific Immunotherapy in Cancer

R. L. Ikonopisov

Department of Clinical Dermatooncology and Immunotherapeutic Oncology, Research Institute, Medical Academy, Sofia-Darvenitza 1156, Bulgaria

ABSTRACT

Nonspecific immunotherapy of cancer in man by means of a variety of bacterial, polysaccharide and synthetic preparations has become more or less a conditio sine qua non. Amongst immunostimulants BCG vaccine has gained widest acceptance in the clinical practice dealing with a variety of neoplastic diseases in man. The author's broad experience with BCG immunostimulation in malignant melanoma, breast, lung, gastrointestinal, head and neck cancer, etc., is the basis on which he discusses a few controversial aspects of clinical practice with BCG vaccine. Evidence for the overall benefit for cancer patients from BCG comes from randomized trials of a number of investigators among whom only Cunningham's, Pinsky's and Constanzi's groups reach negative conclusions. Follows a discussion on the immunogenicity and virulence of different BCG strains with a comparison between Pasteur and Sofia 1 strains, the preferable route of administration, some side effects and hazards of BCG in clinical use, the phenomenon of tumour growth enhancement, the relationship between immunosuppressive drugs and immunopotentiating agents. Finally the author's personal experience within a limited period of BCG administration impresses with the observed complete annihilation of its beneficial effects within a 5-year period. This raises the question of how long should BCG vaccine be administered to the patient treated for cancer. Temporarily accepting the view for a 'nonstop' BCG treatment a scheme for BCG vaccination is offered, which is characterized by a gradual increase of the intervals in between BCG manipulations or a decrease of the overall number of mycobacterias introduced into the host organism.

INTRODUCTION

Experimental and clinical data on the beneficial effect of immunostimulants, adjunct to classical modalities of anticancer treatment, render immunotherapy increasingly important in the complex approach to the cure of neoplastic disease in man. Immunotherapy has already become a well defined medico-biological discipline with its own specific principles and modalities and with the use of biological and chemically synthesized reagents of identifiable site and nature of activity. The broad spectrum of nonspecific weapons for active immunotherapy includes the live attenuated BCG vaccine (as well as its subcellular fractions MER and CWS in mineral oil), corynebacterium parvum, levamisole, thymosin, the synthetic double stranded RNA – Poly I: Poly C, Poly A:Poly U, vitamin A and other bacterial, polysaccharide and synthetic preparations.

Recently it has been demonstrated that immunotherapy with C. parvum, for instance, prolongs remission and survival in disseminated forms of breast cancer, lung cancer and sarcoma (Israel and Edelstein, 1976). Spitler et al. (1978) found no difference between levamisole and placebo treated patients with malignant melanoma in contrast to the striking effect of levamisole in resectable lung cancer (Amery, 1978) and stage III breast cancer (Rojas et al., 1978).

However, most of the investigators seem to have focussed their interest on the immunotherapeutic role of the live attenuated BCG vaccine and its subcellular fractions.

Presently clinical experience with BCG in the treatment of leukaemia (Mathé and others, 1968, 1978; Mathé, 1976; Powles and others, 1973, 1978, Gutterman and others, 1974, 1978), in malignant melanoma (Morton and others, 1970, 1974, 1976, 1978; Nathanson, 1972; Krementz and others, 1971; Bluming and others, 1972; Ikonopisov, 1972, 1973, 1975; Pinsky and others, 1973, 1978, Gutterman and others, 1973, 1974, 1975, 1978), in lung cancer (Hadjiev and Kavaklieva-Dimitrova, 1969; Takita and Brugarolas, 1973, Yamamura, 1978), in breast cancer (Gutterman and others, 1975, Hortobagyi and others, 1978), in Hodgkin's disease (Sokal and Aungst, 1969; Bakemeir and others, 1978), in non-Hodgkin's lymphoma (Jones and others, 1978), in cancer of the head and neck (Donaldson, 1972), in bladder cancer (Morales and others, 1978), in colo-rectal cancer (Moertel and others, 1978; Engstroem and others, 1978; Mavligit and others, 1978), in osteogenic sarcoma (Eilber and others, 1978) has established a permanent role for BCG nonspecific immunostimulation in the combined therapeutic approach to malignant melanoma in man.

Despite the broad spectre of tumours in whose treatment BCG has been administered, studies on the therapeutic value of BCG and its precise mechanism of action have in fact raised more questions than they have answered.

Personal Observations and Comments

During the period 1970-1976 we have administered BCG immunostimulation to more than 500 patients suffering from a variety of solid tumours, such as malignant melanoma, breast cancer, lung cancer, gastrointestinal cancer, head and neck cancer, etc. In the course of this work and from a review of the clinical material and results (in preparation for publication) some problems of practical and scientific importance occurred. The first problem arising from our own and other investigators' experience concerns the actual benefit for cancer patients from BCG nonspecific immunotherapy in terms of remission and survival.

Data on the following two tables (Table 1 and 2) furnishes some information on results from BCG treatment of malignant melanoma, one of the most appropriate models for immunotherapy.

TABLE 1 Results of Immunotherapy Adjunct to Surgery of Malignant Melanoma*

Extent of disease	Therapy	Results	Investigator(s)
Minimal Residual Disease (MRD)	BCG	Prolonged remission and survival	Bluming and others (1972)
MRD	BCG	"	Ikonopisov (1972)
MRD	BCG	"	Gutterman and others (1973)
MRD	BCG	"	Morton and others (1974)
MRD	BCG	Increased recurrences	Cunningham and others (1978)
MRD	BCG	No prevention or delay in recurrences	Pinsky and others (1978)
MRD	BCG + tumour cells	Decreased recurrences	Morton and others (1978)

*Compiled from J.U. Gutterman and others (1976), Med. Clin. N. Amer., 60, 3, 441. In: W. D. Terry and D. Windhorst (Eds). Immunotherapy of Cancer: Present Status of Trials in Man, Raven Press, New York, 1978.

TABLE 2 Results of Chemoimmunotherapy in Malignant Melanoma in Man*

Extent of disease	Therapy	Results	Investigator(s)
Disseminated	DTIC + BCG	Prolonged survival	Ikonopisov (1973, 1975)
Disseminated	DTIC + BCG	Prolonged remissions and survival	Gutterman and others (1974, 1975, 1978)
Disseminated	DTIC+VCR+BCG +tumour cells	Prolonged remission	Currie and McElwain (1975)
MRD	DTIC + BCG	Increased recurrences	Cunningham and others (1978)
MRD	DTIC + BCG	Decreased recurrences	Beretta and others (1978)
Disseminated	BCNU+Hydroxyurea+DTIC(BHD)+BCG	No increase in response rate and survival	Constanzi (1978)

*Compiled from J.U. Gutterman and others (1976), Med. Clin. N. Amer., 60, 3, 441, and Progr. Cancer Res. Ther., vol.6, In: W. D. Terry and D. Windhorst (Eds), Immunotherapy of Cancer: Present Status of Trials in Man, Raven Press, New York, 1978.

Altogether the results from randomized trials on immunotherapy and chemotherapy in malignant disease indicate that there is some negative experience and conflicting results from the clinical application of BCG vaccine, C.Parvum and levamisole. It is evident that broadest is the experience with BCG in malignant melanoma, where 3 out of 13 studies appear to be discouraging (Pinsky and others, 1978; Cunningham and others, 1978; Constanzi, 1978). It should be remembered that although positive results prevail in trials in malignant melanoma these effects are on the whole rather modest, but as it has been previously pointed out, they are clinically and scientifically important.

The fact that work with BCG vaccine reveals a favourable direction towards an increase in remission rates, prolongation of remissions and overall survival warrants the continuation of intensified studies. No doubt controversies of present results in immunotherapy with BCG lie within the different setting of the studies, the differences in the strain, lot and form of the BCG preparation, the dose-effect relationship, the route of administration, etc., etc.

There is full agreement that information on the biological and biochemical properties of a BCG strain and its forms (liquid, lyophilized, fresh-frozen, fresh culture harvested) is of great importance in comparative studies conducted on broad national or international scales. Maximal standardization of BCG preparations is being attempted with regard to immunogenicity, allergogenicity and virulence of the BCG vaccine. Bluming and others (1972) as well as Mackaness and others (1973) have demonstrated that the lyophilized Pasteur BCG vaccine is presently the optimal one in this respect. Our own studies show that Pasteur BCG results in more marked skin reactions of the delayed type hypersensitivity (DTH) as compared to Sofia BCG strain (Table 3) (Fig.1)

TABLE 3 <u>Comparison Between the Strength of Skin Reactions of Delayed Type Hypersensitivity (DTH) Induced by Pasteur BCG and Sofia BCG Strains</u>

Patient	Sex	Age	Clinical Stage*	Skin Reactions of DTH	
				BCG (S)	BCG (P)
I.V.G.	F	76	I	+	++
A.I.G.	F	68	I	+	+++
Z.T.B.	F	38	III	±	+++
P.S.P.	F	67	III	±	++
V.C.M.	M	28	III	+	++
H.S.D.	M	36	III	±	+++

*Clinical stage I   - local disease
 Clinical stage II - regional lymph node metastases
 Clinical stage III - disseminated disease

Whether these differences reflect only the fact of a higher number of viable organisms in the Pasteur BCG ($1,6 \pm 0,8.10^8$ versus 3 to 17.10^6 per ml of the Sofia strain) is open to question. Sofia BCG strain possesses marked immunogenicity and virulence. The readjustment of the dose of BCG delivered at a given site according to the strength of the skin reactions (Pinsky and others, 1978) is certainly within variable limits of desirable approximation. The strength of the skin

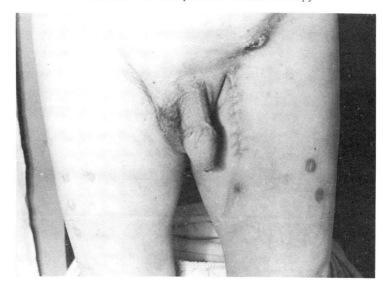

Fig. 1 A comparison between the strength of skin reactions of DTH induced by Pasteur BCG (left limb) and Sofia BCG strain (right limb)

reactions at the site of BCG vaccination may be associated with the allergogenic properties of the vaccine and the local destruction of a number of mycobacterial bodies at a certain level of antituberculin allergy. Dose of BCG viable organisms now expressed in colony forming units (CFU) is probably the most important factor in BCG immunotherapy. It should be of such a magnitude so as to provide for a truly systemic administration and spread of BCG within the host organism. The tendency for achieving a direct dose-effect relationship (Bartlett and others, 1972; Hersh, 1978) based on Mathe's postulate for maintaining a permanent mycobacterial septicaemia should not be regarded as irrevocably orthodoxal since high doses may enhance T-suppressor activity (Geffard and Orbach-Arbouys, 1976). Weiss (1978) has also claimed that undertreatment in immunotherapy in animal systems is more effective than overtreatment.

It has been postulated that certain bacteria can increase immunogenicity of antigenic substances but to do so they must be very close to the target tissue. It follows that BCG should be administered in the area of the primary lesion (Baldwin and Pimm, 1973; Cascinelli and others, 1977), so that organisms should be in the most immediate contact with the target cells or should drain through the same lymphatic pathways as the tumour cells from the original primary site.

Moreover, a previously negative site injected with a weaker BCG preparation is subjected to immunologic conversion after administering a more potent BCG preparation at a distal point as regards the regional lymph node basin.

However, antigenic information is evidently received and processed not only by peripheral lymph node tissue but also by central compartments of the immune system (Fig. 2).

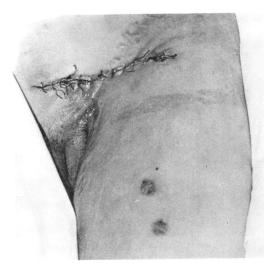

Fig. 2 Marked skin reactions of delayed type hypersensitivity of BCG challenge over the skin of a limb whose regional (inguinal) lymph node basin has been dissected twice

The route of administration of the BCG vaccine poses another problem. Which route is most effective? Oral, respiratory, intradermal, subdermal, intravenous, intraarterial, intralymphatic? By subcutaneous injection, skin scarification or percutaneous multipuncture? Although there is strong opinion in favour of the intradermal route, providing for a more accurate quantitation of BCG in CFU, its side effects - intradermal cold abscessus and shallow ulcers (Sparks, 1976), etc., etc. (Fig. 3) renders the method rather unattractive. Moreover, Rosenthal (1973) claims that multiple puncture provides for a fourfold higher uptake of BCG living mycobacterial bodies via the skin lymphatics and the small blood vessels as compared to the intradermal route. Hersh (1978) reports that Mackaness has recently found that more organisms were introduced into the skin and more delivered to the lymph nodes by scarification than by any other technique. On the basis of our own experience the percutaneous multipuncture with the Heaf gun has been quite satisfactory, delivering enough viable mycobacterial bodies so as to induce skin reactions of DTH of different strength and displaying no local or side effects.

Another problem comprises the recognized hazard of potentially inducing by BCG an immunologic enhancement of the tumour growth (Sparks, 1976). Although immunologic enhancement is a particular issue in the tumour-host relationship, one is worried by the possibility that it may occur as the result of an innate property of the BCG strain itself. In this respect the findings of Turcotte and Quevillon (1976) are highly intriguing as these authors have found that BCG strains Montreal, Pasteur, Tice, Moreau and possibly many others possess three distinct morphological phenotypes - smooth (I), filamentous (II) and rough (III) in superficial 14-day cultures - displaying characteristic biochemical and biological properties. It has been demonstrated that while phenotype I enhances, phenotype III inhibits tumour growth. One can hardly doubt the exceptional importance of such findings in terms of the possibility for differential harvesting, subculturing and genetic selection of phenotypes profitable in immunotherapy of cancer.

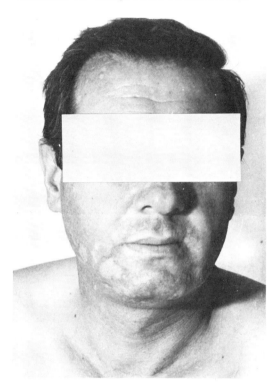

Fig. 3 Vitiligo occurring in a patient following
BCG immunostimulation

A further question of importance is the relationship between immunosuppressive drugs and immunopotentiating agents. The well-known scheme of Currie and Bagshawe (1970) in administering successfully C.parvum precisely on the 12th day following cyclophosphamide administration in murine fibrosarcomata is presumably valid for the particular combination between tumour strain, chemotherapeutic and immunotherapeutic agents. Our studies (unpublished) on the adequate timing between immunostimulants and immunosuppressive drugs suggest that starting with immunopotentiation and delivering the immunosuppressive drugs in the period of the expected peak of immune response could be an alternative approach since a temporary and moderate identation is observed in the curve of the overall boosted level of humoral and/or cellular immunity.

Finally an important question from a practical point of view is the duration of nonspecific immunostimulation once it has been initiated and included in combined therapeutic approach to cancer treatment. We did the following study:

During the period 1970-1972 a total of 215 melanoma patients were subjected to BCG immunostimulation in addition to the treatment required by the clinical stage of the disease (Table 4). Results in survival rates were compared in a historical trial to the survival of a control group of another 215 melanoma patients treated during the period 1967-1969, corresponding in numbers per year and in the clinical

TABLE 4 Melanoma Patients Subjected to BCG Nonspecific Immunotherapy According to Clinical State (TNM Classification) and Number per Year (1970-1972)

Clinical Stage*	TNM Classification	1970	1971	1972	Total
I	$T_{1-4}N_0M_0$	26	22	26	74
II	$T_{1-4}N_{1-3}M_0$	23	30	15	68
III	$T_{1-4}N_{1-3}M_1$	28	28	17	73
	Total	77	80	58	215

*Clinical stage I   - local disease
Clinical stage II - regional lymph node metastatic involvement
Clinical stage III - disseminated disease

stage of the disease. The BCG vaccine (Sofia strain) was given by percutaneous multipuncture at two ports of entry with a Heaf gun in a clockwise direction at each session for a period of one year: the first three months weekly and monthly thereafter. The following were the results (Table 5):

TABLE 5 Survival of Patients with Malignant Melanoma at the Third, Fourth and Fifth Year after Ending ± BCG Treatment

Without BCG vaccine			With BCG vaccine		
Survival	Year	Percentage	Percentage	Year	Survival
3 yrs	1969	58.3	75.3	1972	3 yrs
4 yrs	1968	46.7	62.5	1971	4 yrs
5 yrs	1967	36.2	35.8	1970	5 yrs

It is evident that patients treated with BCG survive in much higher percentages as compared to controls in the third and fourth year, whereas survival of the group treated with BCG drops in the fifth year following immunization to the level of survival of the control group. The effect of BCG administered to the studied group of patients for a period of one year has been gradually exhausted and in fact annihilated by the end of the fifth year of follow up, therefore proving that BCG immunostimulation should be carried out continuously.

This is a problem which has already confronted clinical immunotherapists of cancer. N. N. Trapeznikov and others (1977) report that in an 18 months follow up of patients treated for primary melanoma, recurrence of the disease was recorded in 44.4% of the patients subjected to surgery only and in 18.2% of the patients receiving surgery + BCG. At a 24 months follow up these percentages are already approximately identical with 45.4% for the first group of patients and 41.7% for the latter. Mathé (1976) has insisted on indefinitely prolonging the course of immunotherapy with BCG vaccinating claiming that 'the effect of BCG on the immune system function persists as long as does the BCG-itis and the latter should be maintained by repeated applications of BCG - over 5 years' and concluding that we should adapt ourselves to the fact that the cancer patient is to 'survive in the company of BCG.'

We have already embarked on a study of a 'nonstop' BCG immunostimulation as an adjunct to combined treatment of cancer patients at a regimen of gradually prolonging the inbetween intervals of BCG applications after the following scheme:

The first year - weekly for three months and monthly thereafter,
The second year - at monthly intervals.
The third year and thereafter - at quarterly intervals.

Whether such a prolonged BCG regimen should display the same phenomenon of gradual annulation of the immunostimulatory effects because, in the phrase of Woodruff (1973), 'something surely gets tired of being stimulated and ceases to react', remains to be seen. Meanwhile this and the already discussed problems await their rational solution.

REFERENCES

Amery, W. K. (1978). A placebo-controlled levamisole study in resectable lung cancer. In: W. D. Terry and D. Windhorst (Eds), Progr. Cancer Res. Ther., vol. 6. Immunotherapy of Cancer: Present Status of Trials in Man, Raven Press, N. Y., pp. 191-203.

Bakemeir, R. F., W. Costello, J. Horton and V. T. DeVitta (1978). BCG immunotherapy following chemotherapy-induced remission in stage III and IV Hodgkin's disease. In: W. D. Terry and D. Windhorst (Eds), Progr. Cancer Res. Ther., vol. 6. Immunotherapy of Cancer: Present Status of Trials in Man, Raven Press, N. Y., pp. 513-519.

Baldwin, R. W., M. V. Pimm (1973). BCG immunotherapy of local spontaneous growths and postsurgical pulmonary metastases of a transplanted rat epithelioma of spontaneous origin. Int. J. Cancer, 12, 420-427.

Bartlett, G. L., B. Zbar and H. J. Rapp (1972). Suppression of murine tumour growth by immune reaction to the Bacillus Calmette-Guerin strain of Mycobacetrium bovis. J. Natl. Cancer Inst., 48, 245-257.

Blumming, A. Z., C. L. Vogel, J. L. Ziegler and others (1972). Immunological effects of BCG in patients with malignant melanoma. A comparison of two modes of administration. Ann. Int. Med., 76, 405-411.

Cascinelli, N., G. P. Balzarini, V. Fontana, S. Orefice, U. Veronesi (1977). Intralymphatic administration of BCG in melanoma patients. Cancer Immunol. Immunother., 2, 157-161.

Constanzi, J. J. (1978). Chemotherapy and BCG in the treatment of disseminated malignant melanoma. In: W. D. Terry and D. Windhorst (Eds), Progr. Cancer Res. Ther., vol. 6, Immunotherapy of Cancer: Present Status of Trials in Man, Raven Press, N. Y., pp. 87-95.

Cunningham, T. J., D. Schoenfeld, L. Nathanson and others (1978). A controlled study of adjuvant therapy in patients with stage I and II malignant melanoma. In: W. D. Terry and D. Windhorst (Eds), Progr. Cancer Res. Ther., vol. 6, Immunotherapy of Cancer: Present Status of Trials in Man, Raven Press, N. Y., pp. 19-27.

Currie, G. A. and K. D. Bagshawe (1970). Active immunotherapy with Coryne-bacterium parvum and chemotherapy in murine fibrosarcoma. Brit. med. J., 2, 541-544.

Currie, G. A. and T. J. McElwain (1975). Active immunotherapy as an adjunct to chemotherapy in the treatment of disseminated malignant melanoma. A pilot study. Brit. J. Cancer, 31, 143-156.
Donaldson, R. C. (1972). Methotrexate plus Bacillus Calmette-Guerin (BCG) and isoniazid in the treatment of cancer of the head and neck. Am. J. Surg., 124, 527-534.
Eilber, F. R., C. M. Townsend and D. L. Morton (1978). Adjuvant immunotherapy of osteosarcoma with BCG and allogeneic tumour cells. In: W. D. Terry and D. Windhorst (Eds), Progr. Cancer Res. Ther., vol. 6, Immunotherapy of Cancer: Present Status of Trials in Man, Raven Press, N. Y., pp. 299-304.
Engstroem, P. F., A. R. Paul, R. B. Catalano and others (1978). Fluorouracil versus Fluorouracil + BCG in colorectal adenocarcinoma. In: W. D. Terry and D. Windhorst (Eds). Progr. Cancer Res. Ther., vol. 6, Immunotherapy of Cancer: Present Status of Trials in Man, Raven Press, N. Y. pp. 587-597.
Geffard, M. and S. Orbach-Arbouys (1976). Enhancement of T-suppressor activity in mice by high doses of BCG. Cancer Immunol. Immunother., 1, 1/2, 41-45.
Gutterman, J. U., G. Mavligit, C. M. McBride and others (1973). Active immunotherapy of recurrent malignant melanoma by systemic BCG, Lancet, 1, 1208-1212.
Gutterman, J. U., G. Mavligit, J. A. Gottlieb and others (1974). Chemoimmunotherapy of disseminated malignant melanoma with Dimethyl-triazeno-imidazole carboxamide and Bacillus Calmette-Guerin. N. Engl. J. Med., 291, 592-597.
Gutterman, J. U., G. M. Mavligit, G. M. Burgess and others (1976). Immunotherapy of breast cancer, malignant melanoma and acute leukaemia with BCG: prolongation of disease-free interval and survival. Cancer Immunol. Immunother., 1, 1/2, 99-109.
Gutterman, J. U., G. M. Mavligit, C. M. McBride and others (1978). Postoperative immunotherapy for recurrent malignant melanoma. In: W. D. Terry and D. Windhorst (Eds), Progr. Cancer Res. Ther., vol. 6, Immunotherapy of Cancer: Present Status of Trials in Man, Raven Press, N. Y., pp. 35-57.
Hadjiev, S. and Ya. Kavaklieva-Dimitrova (1969). Application du BCG dans le cancer chez l'homme. Folia med., 11, 8-14.
Hersh, E. M. (1978). Discussion on disseminated melanoma. In: W. D. Terry and D. Windhorst (Eds), Progr. Cancer Res. Ther., vol. 6, Immunotherapy of Cancer: Present Status of Trials in Man, Raven Press, N. Y., pp. 655-664.
Hortobagyi, G. N., J. U. Gutterman, G. R. Blumenschein and others (1978). Chemoimmunotherapy of advanced breast cancer with BCG. In: W. D. Terry and D. Windhorst (Eds), Progr. Cancer Res. Ther., vol. 6, Immunotherapy of Cancer: Present Status of Trials in Man, Raven Press, N. Y., pp. 655-664.
Ikonopisov, R. L. (1972). The rational of immunostimulation procedures in the therapeutic approach to malignant melanoma of the skin. Tumori, 58, 2, 121-127.
Ikonopisov, R. L. (1973). Nonspecific immunostimulation with BCG in malignant tumours. Proc. Bulg.-Soviet Symposium on Cancer Immunol., Moscow, pp. 73-74 (in Russian).
Ikonopisov, R. L. (1975). The use of BCG in the combined treatment of malignant melanoma. Behring Inst. Mitt., 56, 206-214.
Israel, L. and R. Edelstein (1976). Nonspecific immunostimulation with Corynebacterium parvum in human cancer. Immunologic agents of neoplasia. 26th Symposium, Baltimore, Maryland, Williams & Williams, pp. 485-505.

Krementz, E. T., M. S. Samuels, J. H. Wallace and others (1971). Clinical experience in immunotherapy of cancer, Surg. Gynec. Obstetr., 133, 209-217.
Jones, S. E., S. E. Salmon, T. E. Moon and J. J. Butler (1978). Chemotherapy of non-Hdgkin's lymphoma with BCG: a preliminary report. In: W. D. Terry and D. Windhorst (Eds), Progr. Cancer Res. Ther., Vol.6, Immunotherapy of Cancer: Present Status of Trials in Man, Raven Press, N. Y., pp. 519-529.
Mackaness, G. B., D. J. Auclair and P. H. Lagrance (1973). Immunopotentiation with BCG. I. Immune response to different strains and preparations. J. Natl. Cancer Inst., 51, 1655-1667.
Mathé, G., J. L. Amiel, L. Schwarzenberg and others (1968). Demonstration de l'efficacité de l'immunothérapie active dans le leucémie aigue lymphoblastique humaine. Rev. Franç. Etudes Clin. et Biol., 13, 454-459.
Mathé, G. (1976). Surviving in the company of BCG. Cancer Immunol. Immunother., 1, 1/2, 3-7.
Mathé, G., L. Schwarzenberg, F. de Vassal and others (1978). Chemotherapy followed by active immunotherapy in the treatment of acute lymphoid leukaemias for patients of all ages: results of ICIG protocols, 1, 9 and 10, prognostic factors and therapeutic indications. In: W. D. Terry and D. Windhorst (Eds), Progr. Cancer Res. Ther., vol.6, Immunotherapy of Cancer: Present Status of Trials in Man, Raven Press, N. Y., pp.451-471.
Mavligit, G. M., J. U. Gutterman, M. A. Malahey and others (1978). Systemic adjuvant immunotherapy and chemoimmunotherapy in patients with colorectal cancer (Dukes' C classification): prolongation of disease-free interval and survival. In: W. D. Terry and D. Windhorst (Eds), Progr. Cancer Res. Ther., vol.6, Immunotherapy of Cancer: Present Status of Trials in Man, Raven Press, N. Y., pp. 597-604.
Moertel, C. G., M. J. O'Connell, R. E. Ritts, Jr. and others (1978). A controlled evaluation of combined immunotherapy (MER-BCG) and chemotherapy for advanced colorectal cancer. In: W. D. Terry and D. Windhorst (Eds), Progr. Cancer Res. Ther., vol.6, Immunotherapy of Cancer: Present Status of Trials in Man, Raven Press, N. Y. pp. 573-587.
Morales, A., D. Eidinger and A. W. Bruce (1978). Adjuvant BCG immunotherapy in recurrent superficial bladder cancer. In: W. D. Terry and D. Windhorst (Eds), Progr. Cancer Res. Ther., vol.6, Immunotherapy of Cancer: Present Status of Trials in Man, Raven Press, N. Y., pp. 225-231.
Morton, D. L., F. R. Eilber, R. A. Malmgren and others (1970). Immunological factors which influence response to immunotherapy in malignant melanoma. Surgery, St. Louis, 68, 158-162.
Morton, D. L., F. R. Eilber, E. C. Holmes and others (1974). BCG immunotherapy of malignant melanoma: summary of a seven-year experience. Ann. Surg., 180, 635-643.
Morton, D. L., F. R. Eilber and E. C. Holmes (1976). BCG immunotherapy as a systemic adjuvant to surgery in malignant melanoma. Med. Clin. N. Amer., 60, 3, 431-439.
Morton, D. L., E. C. Holmes, F. R. Eilber and others (1978). Adjuvant immunotherapy of malignant melanoma: preliminary results of a randomized trial in patients with lymph node metastases. In: W. D. Terry and D. Windhorst (Eds), Progr. Cancer Res. Ther., vol. 6, Immunotherapy of Cancer: Present Status of Trials in Man, Raven Press, N. Y. pp. 57-65.
Nathanson, L. (1971). Experience with BCG in malignant melanoma. Cancer Res., 12, 99-103.

Pinsky, C., Y. Hirshaut and H. Oettgen (1972. Treatment of malignant melanoma by intralesional injection of BCG. Proc. Amer. Assoc. Cancer Res., 13, 21-25.

Pinsky, C., Y. Hirshaut, H. J. Wanebo and others (1978). Surgical adjuvant immunotherapy with BCG in patients with malignant melanoma. In: W. D. Terry and D. Windhorst (Eds), Progr. Cancer Res. Ther., vol. 6, Immunotherapy of Cancer: Present Status of Trials in Man, Raven Press, N. Y., pp. 27-35.

Powles, R. L. (1973). Immunotherapy for acute myelogenous leukaemia. Brit. J. Cancer, 28 (Suppl.1), 262-284.

Powles, R. L., J. Russell, T. A. Lister and others (1978). Immunotherapy for acute myelogenous leukaemia: analysis of a controlled clinical study 2 1/2 years after entry of the last patient. In: W. D. Terry and D. Windhorst (Eds), Progr. Cancer Res. Ther., vol.6, Immunotherapy of Cancer: Present Status of Trials in Man, Raven Press, N. Y., pp. 315-329.

Rojas, A. F., J. N. Feierstein, H. M. Glait and J. Olivari (1978). Levamisole action in breast cancer stage III. In: W.D. Terry and D. Windhorst (Eds), Progr. Cancer Res. Ther., vol.6, Immunotherapy of Cancer: Present Status of Trials in Man, Raven Press, N. Y. pp. 635-647.

Rosenthal, S. R. (1973). BCG vaccination and leukaemia mortality. Natl. Cancer Inst. Monogr., 39, 189-192.

Sokal, J. E. and C. W. Aungst (1969). Response to BCG vaccination and survival in advanced Hodgkin's disease. Cancer, 24, 128-134.

Sparks, F. C. (1976). Hazards and complications of BCG immunotherapy. Med. Clin. N. Amer., 60, 3, 499-509.

Spitler, L. E., R. W. Sagebiel, R. G. Glogan and others (1978). A randomized double-blind trial of adjuvant therapy with levamisole versus placebo in patients with malignant melanoma. In: W. D. Terry and D. Windhorst (Eds), Progr. Cancer Res. Ther., vol.6, Immunotherapy of Cancer: Present Status of Trials in Man, Raven Press, N. Y. 73-79.

Takita, H. and A. Brugarolas (1973). Adjuvant immunotherapy for bronchogenic carcinoma. Cancer Chemother. Rev., 4, 293-298.

Trapeznikov, N. N. (1977). In: Clinical Immunotherapy of Cancer (with R. L. Ikonopisov and V. V. Yavorski), S. Med. i fizk., p. 304, (in Bulgarian).

Turcotte, R. and M. Quevillon (1976). Antitumour activity and their biological properties of two phenotypes isolated from BCG. Cancer Immunol. Immunother., 1, 1/2, 25-30.

Weiss, D. W. (1976). MER and other mycobacterial fractions in the immunotherapy of cancer. Med. Clin. N. Amer., 60, 3, 473-498.

Woodruff, M. F. A. (1973). Discussion of G. Mathé's paper on 'Attempts at using systemic immunity adjuvants in experimental and human cancer therapy'. In: Immunopotentiation, Ciba Foundation Symposium 18 (new series), Elsevier, Excerpta Medica, North Holland, Associated Scientific Publishers, Amsterdam, London, New York, pp.328.

Yamamura, Y. (1978). Immunotherapy of lung cancer with oil-attached cell-wall skeleton of BCG. In: W. D. Terry and D. Windhorst (Eds), Progr. Cancer Res. Ther., vol.6, Immunotherapy of Cancer: Present Status of Trials in Man, Raven Press, N. Y., pp. 173-181.

Chemoimmunotherapy: Basic Principles and Clinical Examples*

J. G. Sinkovics**, C. Plager*** and M. J. McMurtrey****

The University of Texas System Cancer Center,
M.D. Anderson Hospital and Tumor Institute, Houston, Texas 77030, U.S.A.
*Supported by the Kelsey-Leary Foundation, Houston, Texas;
Don and Sybil Harrington and Baker and Taylor Drilling Company,
Amarillo, Texas, U.S.A.
**Professor of Medicine and Chief, Section of Clinical Tumor Virology and
Immunology and Solid Tumor Clinics (Service), Department of Medicine
***Assistant Professor of Medicine, Department of Medicine
****Associate Professor of Surgery, Department of Surgery

ABSTRACT

The addition of viral oncolysates to regimens of combination chemotherapy and BCG did not provide further benefits for patients with malignant melanoma. Patients at high risk of relapse remained tumor-free after receiving adequate adjuvant chemotherapy. The additional benefits of BCG are estimated to be marginal only. Response rate of patients with stage III disease receiving chemotherapy, BCG and melanoma lysates was not better than that of patients receiving chemotherapy and BCG but length of survival appears to be slightly better for the former group. Patients with metastatic sarcomas experienced prolonged stabilization of disease, i.e. retardation of disease progression, on chemotherapy, BCG and sarcoma viral oncolysates, but updated reanalysis of this patient population is overdue.

INTRODUCTION

Chemoimmunotherapy may be counterproductive. If chemotherapy is highly immunosuppressive and immunotherapy is given at the time of profound immunosuppression, immunotherapy may become ineffective. If chemotherapy is given immediately after immunotherapy, the immunoreactive clone(s) of lymphoid cells proliferating in response to the immune stimuli may be annihilated, thus creating the state of immunological unresponsiveness or immune tolerance. Chemotherapy therefore should not produce protracted immunosuppression and immunotherapy should be given at the time when immune faculties are recovering from short pulses of chemotherapy (Fig. 1).

Potential advantages of chemoimmunotherapy include 1. reduction of tumor mass; 2. selective immunosuppression, i.e. less harm to cytotoxic macrophages and lymphoid cells than to B lineage lymphoid cells producing serum factors capable of blocking lymphocyte mediated-cytotoxicity; and 3. expression of tumor-associated antigens in increased density during brief periods of immunosuppression following each course of chemotherapy, thus increasing the vulnerability of tumor cells to

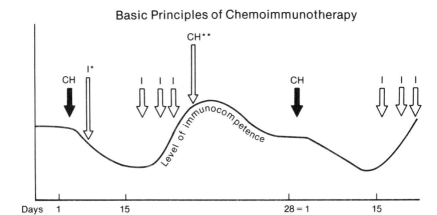

Fig. 1   CH   = chemotherapy in short cycles permits recovery ("overshoot")
         I   = immunotherapy: right timing is at recovery phase of immunocompetence
         I*   = immunotherapy given immediately after chemotherapy may be ineffective
         CH** = chemotherapy given immediately after immunotherapy may annihilate immunoreactive clones

immune attack (for review, see Harris and Sinkovics, 1976).

Major modalities of immunotherapy are 1. nonspecific immunostimulation operating through activation of macrophages and nonselective killer lymphocytes; 2. tumor-specific immunity operating through selectively cytotoxic lymphocytes and specific antibodies which a. potentiate lymphocyte mediated cytotoxicity; b. arm macrophages; c. reset the balance of complexes formed by soluble tumor antigens and antibodies so that their blocking effect on lymphocyte-mediated cytotoxicity will be abrogated; and d. lyse tumor cells in the presence of complement. Further modalities of immunotherapy include products of lymphoid cells, such as molecular mediators of the delayed hypersensitivity reaction, immune RNA and transfer factor; and interferon. This latter substance not only inhibits viral replication; it also functions as a regulator of cell growth. While interferon may inhibit the growth of certain tumor cells (Burkitt's lymphoma; osteosarcoma; melanoma), tumor cells may gain interferon-resistance (for review, see Harris and Sinkovics, 1976) and tumor cell populations inhibited by interferon may lose their sensitivity to radio- and chemotherapy due to their reduced growth fraction.

A new idea concerns the possible elimination of suppressor cells. If the physiologic function of embryonic antigens is the activation of suppressor cells in order to provide protection for the allogeneic fetus against maternal rejection and if carcinoembryonic antigen-producing tumors by imitating fetal growth survive in the immunoreactive host because carcinoembryonic antigens activate similar clones of suppressor cells (Sinkovics, 1976), the critical suppressor cell clone could be intentionally activated in a given host by the injection of purified embryonic or carcinoembryonic antigens and this clone, replicating in response to the embryonic antigens, could then be annihilated by an immediate dose of chemotherapy (cyclophosphamide; methotrexate). A previous, unpublished, single experiment by one of us (JGS) with a transplantable mouse lymphoma suggests this sequence of events, i.e. accelerated growth of tumor in mice injected with embryonic cells and retarded growth and even rejection of tumor in mice injected with embryonic cells first and then with cyclophosphamide (Fig. 2).

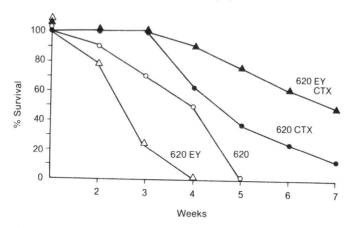

Fig. 2

Among the unsolved problems of human tumor immunotherapy is the question of clinical efficacy of human tumor antigen preparations, in particular when administered in combination with nonspecific immunostimulation. An outstanding clinical trial in Canada claims benefit for patients with bronchogenic carcinoma receiving nonimmunosuppressive chemotherapy with high dose methotrexate and leucovorin rescue and immunotherapy with bronchogenic carcinoma cell wall antigens incorporated in complete Freund's adjuvant (Stewart and others, 1976). In contrast, two other well conducted trials failed to show benefit when nonspecific immunostimulation with BCG was combined with attempted tumor-specific immunization in the case of acute myelogenous leukemia (Powles and others, 1977) and malignant melanoma (McIllmurray and others, 1977; Hedley and others 1978). In the latter case, tumor enhancement appears to have occurred.

This communication will briefly review our experience with the chemoimmunotherapy of oat cell carcinoma of the lung, malignant melanoma and sarcomas.

RESULTS

Small (Oat) Cell Carcinoma

This clinical trial has recently been published in detail (Holoye and others, 1978). Table 1 indicates no benefit from the addition of BCG to combination chemotherapy (vincristine, doxorubicin and cyclophosphamide) in the treatment of this tumor. However, patients with partial response to chemotherapy experienced longer median survival when BCG was added to the regimen. The difference in median survival in these two groups of partial responders (chemotherapy versus chemotherapy plus BCG) was of borderline significance.

TABLE 1 Chemoimmunotherapy for Small (Oat)
Cell Bronchogenic Carcinoma

Treatment & results	Number of patients	Approximate median survival (weeks)	P value
CR VAC	18	50	0.68
CR VAC BCG	11	48	
PR VAC	10	37	0.04
PR VAC BCG	12	52	
NoR VAC	10	9	0.13
NoR VAC BCG	6	19	

Slight prolongation of median survival is at lowest level
of significance for patients in partial remission (PR)
CR = complete remission
VAC = vincristine, adriamycin, cyclophosphamide
BCG = Bacille Calmette Guerin (Chicago)

Malignant Melanoma

This clinical trial is now being published in detail (Sinkovics and others, 1979). The major aim of this trial was to determine the clinical efficacy of chemotherapy plus BCG versus chemotherapy, BCG and melanoma viral oncolysates. Stage I (primary tumor) and stage III (regional lymph node metastases) patients were treated in an adjuvant fashion, i.e. after rendered surgically tumor-free. Most patients with stage I disease had deep lesions located on the trunk. Clark's levels were in the range of 3.5 and Breslow's thickness ranged from 2.06 mm to 3.34 mm. All patients with stage III disease had large palpable tumors (stage III B); in most cases the lymph nodes were matted, tumor cells broke through the capsule of lymph nodes and were observed in fibroadipose tissue between lymph nodes. Two major modalities of chemotherapy were used: semustine (methylCCNU) and dacarbazine; or vincristine, actinomycin D and dacarbazine. Five-day courses of chemotherapy were spaced at 28 day intervals. There is evidence that neither dacarbazine (Bruckner and others, 1974) nor semustine (Berd and others, 1978) is overtly immunosuppressive. No significant difference in the two chemothrapeutic regimens is apparent, but these results will be analyzed in another publication. The two major modalities of immunotherapy were Chicago BCG applied by the scarification technique and lysates of allogeneic cultured melanoma cells referred to as "viral oncolysates". Melanoma cell viral oncolysates were prepared from established cell line #2124 (Sinkovics and others, 1978) after inoculation of the tumor cells with the weakly cytopathogenic PR8 human influenza A virus. Immunotherapy was given on days 17 and 24 in between courses of chemotherapy. Tables 2 and 3 show that relapses, i.e. progressive disease, was apparently retarded in patients with bad prognosis stages I, II and III B disease, but the addition of melanoma viral oncolysates to the chemotherapy plus BCG regimen did not increase the effect. The two patients who from stage I disease advanced to death in the chemotherapy, BCG and melanoma viral oncolysate group had unusually deep (4.5 and 6 mm) primary tumors on the trunk or in the head and neck region. There are no patients in the chemotherapy plus BCG group with comparable tumors; therefore the two groups of patients with stage I disease (chemotherapy plus BCG versus chemotherapy, BCG plus melanoma lysates) were not comparable. In the chemotherapy, BCG plus melanoma lysate group, several other patients with deep (Clark's level 4 and 5) and ulcerated primary melanomas remain tumor-free at two years (Sinkovics and others, 1979).

TABLE 2 Patients with Stages I-II Malignant Melanoma

Treatment*	Stage	Avg Depth	Number of Pts	Patients Alive NED	Progressing	Patients Dead with Tumor
Ch BCG	I	level 3.4 2.06 mm	20	19	1	0
Ch BCG MVO	I	level 3.5 3.34 mm	8	6	0	2
Other**	I	level 3 1.48 mm	16	14	1	1
Ch BCG	II	-	3	1	1	1
Ch BCG MVO	II	-	1	0	0	1

*After wide excision   **Surgical excision only; dacarbazine only; BCG only

Ch = chemotherapy   BCG = Bacille Calmette-Guerin   MVO = Melanoma viral oncolysate

TABLE 3 Patients with Stage III B Malignant Melanoma

Treatment*	Number of Pts		Time Intervals (Months)			Patients Alive NED(%)	Progressing	Patients Dead	Relapsed Adequately Treated	(%)
			Stage I-III	Treatm-Failure	NED					
Ch BCG	F	13	29.5	7	>16.4					
	M	21	35.8	8.4	>17.3					
	Total	34	33.4	8	>15.9	17 (48.5)	2	15	12/24	(50)
Ch BCG MVO	F	14	44.9	9.75	>22.3					
	M	16	24.3	7.3	>29.1					
	Total	30	35	8.4	>26	13 (43.3)	2	15	8/19	(42)

F = female   M = male   NED = no evidence of disease
*After surgical dissection of regional lymph nodes

A comparison of patients with stage III B disease receiving either chemotherapy and BCG or chemotherapy, BCG and melanoma lysates also failed to show any benefit for those patients who also received melanoma lysates (Sinkovics and others, 1979). Table 3 shows tumor-free survival for 46% of patients in these two treatment groups at median follow-up time from beginning of treatment of 12 months. Delay of disease progression was observed especially in women receiving chemotherapy, BCG and melanoma lysates, but in this particular group of patients progression from stage I to stage III disease was also the longest, thus slower progression of disease could be attributed rather to unknown factors of host-tumor relationship than to the effect of melanoma lysates (Sinkovics and others, 1979). Fig. 3 depicts the survival curves of patients with stage III B disease.

There are several long-term (over 2 years) tumor-free suvivors in the chemotherapy, BCG plus melanoma lysate treatment group and this is reflected in the curve showing slightly but not significantly better survival for this group. Practically all patients with stage III B disease (with melanoma cells in the fibroadipose tissue in between involved lymph nodes; and with high likelihood of subclinical stage IV disease) were expected to relapse. While longer observation is warranted, it is evident at this point that patients with very poor prognosis may survive on chemotherapy plus BCG tumor-free and that the addition of melanoma lysates did not improve the efficacy of the regimen (Sinkovics and others, 1979).

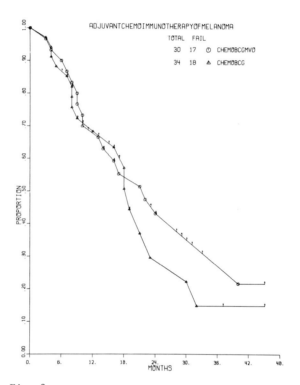

Fig. 3

Patients with stage IV (metastatic) disease were not treated in an adjuvant (prophylactic) fashion. Time intervals from treatment to death were 6.2; 12.8; and 5.1 months for patients receiving chemotherapy and BCG; chemotherapy, BCG and melanoma lysates; or chemotherapy alone (Table 4). The average survival of those few patients who achieved tumor-free status exceeds 17 months in both the chemotherapy plus BCG and chemotherapy, BCG and melanoma lysate treatment groups (Sinkovics and others, 1979). Table 4 shows that 73% of adequately treated patients eventually progressed on chemoimmunotherapy (either BCG or BCG plus melanoma lysates) versus 86% of patients progressing on chemotherapy only. Again, no benefit from adding melanoma lysates to chemotherapy and BCG was evident.

TABLE 4 Patients with Stage IV Malignant Melanoma

Treatment	Number of Pts	Patients Alive			Dead with Tumor	Progressing or Dead Adequately Treated	(%)	Time Intervals (Months)			
		NED	PR/St	Progr				from Treatment to Death	Duration of Progressive Disease	Duration of PR/St Disease	Duration of NED State
Ch BCG -	24	2	3	3	16	13/18	19/26 (73)	6.25	>15.3	>9	>17
Ch BCG MVO	11	1	1	0	9	6/8		12.8	-	>14	>17
Ch - -	19	0	3	5	11	12/14	(86)	5.1	> 4.2	>14.6	-

PR/St = Partial remission and/or stable

Fig. 4 summarizes all patients in the 2 major treatment groups (chemotherapy plus BCG versus chemotherapy, BCG and melanoma lysates) as analyzed at 3 points in time. Retardation of disease progression was evident early for patients receiving chemotherapy, BCG and melanoma lysates. This effect is not evident upon entering increasing numbers of patients into treatment and extending the observation period to 2 years. All patients included in this report entered treatment for 6 months or longer.

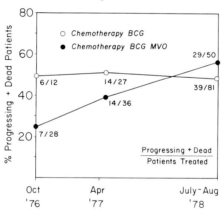

Fig. 4

MVO = Melanoma Viral Oncolysates (Allogeneic)

Sarcomas

All patients had metastatic disease. The chemotherapy consisted of vincristine, cyclophosphamide, doxorubicin (Adriamycin) and dacarbazine. After 500 mg/m^2 doxorubicin was given, it was replaced by actinomycin D. Immunotherapy consisted of either BCG or BCG plus allogeneic viral oncolysates of established human sarcoma cell lines (Sinkovics and others, 1977a). Early analyses indicate retardation of disease progression in patients receiving chemotherapy, BCG and sarcoma lysates (Sinkovics and others, 1977b), but reanalysis of data is now long overdue. A report will soon follow with up-dated analysis of this trial.

DISCUSSION

The median disease-free interval for patients with stage III (regional lymph node metastases) malignant melanoma after surgical dissection is approximately 8 months and median survival is less than 12 months (Paterson and others, 1978). However, these patient populations include also those who were found to have microscopic regional lymph node metastases at the time of elective regional lymph node dissection for stages I or II (local recurrence and satellitosis) disease. Patients with stage III B disease in the clinical trial reported herein all had large, palpable, invasive regional lymph node metastases and highly probable subclinical stage IV disease at the beginning of treatment. Surgical tumor removal seldom if ever cures these patients (Geelhoed and others, 1977). It is estimated that at 2 years less than 20% of these patients would be tumor-free, if treated surgically only; an analysis of surgically treated comparable stage III B patients is underway and will be reported elsewhere in comparison with the patients described in this report. Thus, in this preliminary report, it appears that chemotherapy plus BCG was effective in preventing relapses in patients at high risk. It is not possible to express the magnitude of the benefits, if any, that BCG added to the chemotherapy regimens. Based on the difference of disease progression in adequately treated patients with stage IV melanoma, a 13% margin in favor of BCG exists. This is a trend at best, which is clearly not significant. The addition of melanoma lysates, i.e. melanoma specific immunization, failed to provide any clinically evident benefits. This trial (with newer patients entered within the past 6 months) will be analyzed again at 3 and 5 years.

REFERENCES

Berd, C., Wilson, E., Bellet, R. E., and Mastrangelo, M. J. (1978). MethylCCNU adjuvant chemotherapy of malignant melanoma is not immunosuppressive. Proc. 14th Ann. Meet. Am. Assoc. Clin. Oncol., 19, 349 (abstr. C-171).
Bruckner, H. W., Mokyr, M. B., and Mitchell, M. S. (1974). Effect of imidazole-4-carboxamide 5-(3, 3-dimethyl-1-triazeno) on immunity in patients with melanoma. Cancer Research, 34, 181-183.
Geelhoed, G. W., Breslow, A., and McCune, W. S. (1977). Malignant melanoma: Correlation of long-term follow-up with clinical staging, level of invasion and thickness of primary tumor. Am. Surg., 43, 77-85.
Harris, J. E., and Sinkovics, J. G. (1976). The Immunology of Malignant Disease. 2nd ed. Mosby, St. Louis.
Hedley, D. W., McElwain, T. J., and Currie, G. A. (1978). Specific active immunotherapy does not prolong survival in surgically treated patients with stage II B malignant melanoma and may promote early recurrence. Brit. J. Cancer, 37, 491-496.
Holoye, P. Y., Samuels, M. L., Smith, T., and Sinkovics, J. G. (1978). Chemoimmunotherapy of small cell bronchogenic carcinoma. Cancer, 42, 34-40.
McIllmurray, M. B., Embleton, M. J., Reeves, W. G., Langman, M. J. S., and Deane, M. (1977). Controlled trial of active immunotherapy in management of stage II B malignant melanoma. Brit. Med. J., 1, 540-542.
Paterson, A. H. G., McPherson, T. A., and Williams, D. J. (1978). Malignant melanoma (stage III B): A pilot study of adjuvant chemo-immunotherapy. Cancer Treatm. Rep., 62, 571-573.
Powles, R. I., Russell, J., Lister, T. A., Oliver, T., Whitehouse, J. M. A., Malpas, J., Chapuis, B., Crowther, D., and Alexander, P. (1977). Immunotherapy for acute myelogenous leukemia: A controlled clinical study 2 1/2 years after entry of the last patients. Brit. J. Cancer, 35, 265-272.

Sinkovics, J. G. (1976). Suppressor cells in human malignant disease. Brit. Med. J., 1, 1072-1073.

Sinkovics, J. G., Gyorkey, F., Kusyk, C., and Siciliano, M. J. (1978). Growth of human tumor cells in established cultures. In H. Busch (Ed.), Methods in Cancer Research, Vol. 14, Academic Press, New York. pp. 243-322.

Sinkovics, J. G., Plager, C., McMurtrey, M. J., Papadopoulos, N. E., Waldinger, R., Combs, S., Romero, J. J., and Romsdahl, M. M. (1979, in press). Adjuvant chemoimmunotherapy for malignant melanoma. In R. Crispen (Ed.), Cancer Immunology: Experimental and Clinical, The Fifth Chicago Symposium, Sept. 13-15, 1978, Franklin Institute Press, Philadelphia, Penn.

Sinkovics, J. G., Plager, C., McMurtrey, M., Romero, J. J., and Romsdahl, M. M. (1977a). Immunotherapy of human sarcomas. In Management of Primary Bone and Soft Tissue Tumors, Year Book Medical Publishers, Chicago, Illinois, pp. 361-410.

Sinkovics, J. G., Plager, C., and Romero, J. J. (1977b). Immunology and immunotherapy of patients with sarcomas. In R. Crispen (Ed.), Neoplasm Immunity: Solid Tumor Therapy, Franklin Institute Press, Philadelphia, Penn. pp. 211-219.

Stewart, T. H. M., Hollinshead, A. C., Harris, J. E., Belanger, R., Crepeau, A., Hooper, G. D., Sachs, H. J., Klaassen, D. J., Hirte, W., Rapp, E., Crook, A. F., Orizaga, M., Sengar, D. P. S., and Raman, D. (1976). Immunochemotherapy of lung cancer. Ann. New York Acad. Sci., 277, 436-466.

Intrapleural Corynebacterium Parvum as Adjuvant Therapy in Operable (Stage I and II) Bronchial Non-small Cell Carcinoma; Preliminary Report

The Ludwig Lung Cancer Study Group

The study group is : J. Stjernswärd, M. Kaufmann (Ludwig Institute for Cancer Research, Switzerland), M. Zelen, K. Stanley (Frontier Science and Technology Research Foundation, Inc., Boston. Mass., USA), D.S. Freestone, R. Bomford, M.T. Scott, T. Priestman (The Wellcome Research Laboratory, Beckenham, England), C. Mouritzen and U.W. Henriques (Dept. of Thoracic and Cardiovascular Surgery and Institute of Pathology, Aarhus Kommunehospital, Aarhus, Denmark), N. Konietzko, D. Greschuchna, v. Melchner, W. Maassen and W. Hartung, W. Wierich (Ruhrland Klinik and Institute of Pathology, Essen, Germany), J. Vogt-Moykopf, D. Zeidler, H. Toomes and K. Wurster (Thoraxchirurgische Spezial-Klinik and Institute of Pathology, Heidelberg, Germany), F. Krause, R. Rios and R. Spanel (Thoraxchirurgische Abteilung Fachkrankenhaus für Lungen- und Bronchialerkrankungen and Institute of Pathology, Löwenstein, Germany), J. Orel, M. Benedik, B. Hrabar and D. Ferluga and T. Rott (Clinical Center, Dept. of Thoracic Surgery and Institute of Pathology, Ljubljana, Yugoslavia), S. Plesnicar (The Institute of Oncology, Ljubljana, Yugoslavia), H.A. Rostad, J.R. Vale (Rikshospital, Oslo, Norway), S. Hagen, S. Birkeland (Ulleval Hospital, Oslo, Norway), T. Harbitz, R. Nissen-Meyer (Aker Hospital, Oslo, Norway), P. Lexow (Institute of Pathology, Oslo, Norway), L. Rodriguez, V.O. Björk, K. Böök and J. Willens (Karolinska Sjukhuset, Thoracic Clinic and Institute of Pathology, Stockholm, Sweden), E. Grädel, J. Hasse, P. Holbro and P. Dalquen (Kantonsspital, Thoraxchirurgische Klinik and Institute of Pathology, Basel, Switzerland), L. Eckmann (Tiefenauspital, Chir. Univ.-Klinik, Bern, Switzerland), B. Nachbur, T. Liechti (Inselspital, Dept. of Thoracic and Cardiovascular Surgery, Bern, Switzerland), H. Cottier and A. Zimmermann (Institute of Pathology, Inselspital, Bern, Switzerland), W. Maurer, M. Kaufmann, P. Froelicher (Bürgerspital, Surgical Dept., Solothurn, Switzerland), H. Denck, N. Pridun (Krankenhaus der Stadt Wien-Lainz, Chir. Abt., Vienna, Austria), K. Karrer (Institute for Cancer Research, University of Vienna, Austria), H. Holzner (Institute of Pathology, Vienna, Austria).

ABSTRACT

The Ludwig Lung Cancer Study Group aims to investigate the role of immunotherapy as adjuvant treatment modality in operable (Stage I and II) bronchial non-small cell carcinoma. The participants are 12 european clinics and institutes. With a proven accrual of 350 patients per year the group offers a sharp tool in clinical oncology with regards to bronchial carcinoma. The accrual phase of the first trial will close in January 1979, starting a new protocol in February 1979.

The ongoing randomized clinical trial aims to determine if postoperative intrapleural administration of Corynebacterium parvum in patients with "radical" resected bronchial non-small cell carcinoma can (a) increase the tumor recurrence-free interval or (b) increase survival. Furthermore the study aims to identify high and low risk patient subgroups after biological and immunological investigations.

The possibility of giving C. parvum intrapleurally in humans was investigated in a Phase-I-toxicity study. A dose of 7 mg has been adopted for the clinical trial since this dose combines a measurable systemic effect (increase of neutrophil and monocyte counts) with acceptable toxicity. The main morbidity was fever, flu-like symptoms and chest pain.

As of October 1st, 1978, 400 patients have entered into the trial. It is too early to make a definite comparisons of the treatments with regards to disease-free interval or survival. The average follow-up time for the analyzed cases is approximately 6 months. So far, 13% of the patients have proven tumor recurrence and 8% have expired. Preliminary indications are that surgical stage, histology, and type of resection are prognostic factors.

INTRODUCTION

The overall prognosis of patients with bronchogenic carcinoma is poor. The 5-year survival including all stages varies between 5 and 15% (Selawry and Hansen, 1973).

For curative treatment, surgical resection remains the most effective therapy. But even after so-called radical resection, the results are difficult to accept. In spite of radical removal of the primary tumor, there is in reality a high number of patients with loco-regional incomplete resection and/or microdisseminations at time of surgery (Matthews et al, 1973).

The search for an adjuvant therapy which may be active both locoregionally and systematically is an ethical obligation. Postoperative as well as preoperative radiotherapy has not improved survival in operable cases, nor has chemotherapy so far fulfilled the hopes in non-small cell cancers (Second Nat. Cancer Inst. Conf. on Lung Cancer).

The role of immunotherapy in itself or as an adjuvant treatment modality appears from the result of two randomized studies (McKneally et al, 1976) (Amery, 1976) promising but needs much further investigation. The large number of patients with bronchogenic carcinoma, the maximal tumor reduction after intended radical resection, the frequent tumor recurrence in spite of radical resection within the first two postoperative years and the inefficiency of present adjuvant treatments offers a realistic opportunity to investigate conclusively whether prognosis could be improved by immunosupportive approaches.

In 1977 we created a study group which in an international, cooperative, randomized, clinical trial, aims to determine if postoperative intrapleural administration of Corynebacterium parvum in patients with "radical" resected bronchogenic non-small cell carcinoma (classified as Stage I and II postsurgically) can :

> increase the tumor recurrence-free interval or
>
> increase survival.

Furthermore, the study aims to identify high and low risk patients for Stage I and II operable bronchogenic non-small cell carcinoma after biological and immunomorphological investigations. The trial was activated in July 1977.

During the study various additional investigations are carried out in a search for gaining new information on the biology of the disease :

1) Immunomorphological analysis of cellular infiltrates within or in the neighborhood of the primary tumor, and of metastases; analysis of cellular composition of tumor draining lymph nodes (Kaufmann et al, 1977 a and b) and correlation of the findings with the clinical development of the patient.

2) Centralized pathological review, not only standardizing the histopathological evaluation but also trying to identify subgroups with varying prognosis within operable non-small cell lung cancer patients. Classification according to criterias found in the primary tumor as well as sites and numbers of lymphnodes found to be tumor involved.

3) Distribution of intrapleural and intravenous pre- and postoperative administered isotopically labeled C. parvum in humans as a base for a future protocol.

In order to balance the treatment groups with respect to prognosis, the patients are stratified at entry according to the type of surgery (lobectomy or pneumonectomy). Within each stratification group the patient is assigned to the randomized treatment regimen surgery plus placebo or surgery plus C. parvum.

With regards to accrual of patients for clinical trials in cancer, it should not take longer than 2 years and the number must be large enough to obtain a conclusive answer either positive or negative, within a relatively short period of time. The trial should be able to determine if the treatment modality improves the prognosis or not, even in biological different subgroups of patients. Single small and non-randomized trials in adjuvant cancer therapy are a waste of both time and resources. They interject a great deal of noise into the literature and hamper true medical progress. It has been determined that 400 eligible patients (200 per treatment arm) are required for the present study. If disease-free interval would in fact be increased by 50% (that is a change from 3 to 4.5 years) this increase will then be detected with a probability of about 90% using a statistical test with $p = .05$. To the investigator this means that negative or positive, the results of the study will be conclusive.

At the present 12 surgical university clinics, the majority from Europe, and their local institutes of pathology are participating in the trial. The statistical center is the Frontier Science and Technology Research Foundation, Boston. The Coordinating Center is the Ludwig Institute for Cancer Research, Switzerland. C. parvum is provided by the Wellcome Research Laboratories.

All patients with a proven bronchogenic non-small cell carcinoma and classified post surgical and histologically as Stage I and II are considered for the trial. Patients are followed every third month during the first three years and then every six months during the first five years postoperatively or until death.

The possibility of giving C. parvum intrapleurally in man was investigated in a Phase-I-toxicity study prior to the randomized trial. C. parvum was given in a dose varying from 0.1 - 14 mg postoperatively in 63 patients operable for bronchial carcinoma. A single dose of 7 mg C. parvum intrapleurally 6 - 10 days postoperatively has been found as optimal since this dose combines a measurable systemic effect (increase of neutrophil and monocyte counts) with acceptable toxicity. The main morbidity was fever, flu-like symptoms and chest pain (Ludwig Lung Cancer Cooperative Group 1978 a and b).

RESULTS

As of October 1st, 1978, 400 patients have been entered into the study. Entry onto the current study will terminate and a new protocol will begin in January 1979. The rate of accrual is approximately 25 patients per month.

The distribution of patients by age, sex, tumor stage, histology and type of resection appear to be well-balanced between the two groups of treatment, as can be expected by the large number of patients. 80% of the patients represent the tumor Stage I, 20% the Stage II. The distribution of various histological tumor types is within known proportions with more squamous cell carcinomas than adenocarcinomas, and large cell carcinomas. Lobectomies have been performed in 54% of the patients, pneumonectomies in 30%, bilobectomies in 8% and other types of resections, e.g. sleeve resections, in another 8%.

Complications of the treatments were recorded subjectively as either none, mild, moderate, severe, life-threatening, or lethal, according to a prearranged standardization. No death in connection with the adjuvant therapy has been observed. Significant more complications have been found for the patients receiving C. parvum. A fever of greater than 38° was observed in 39% of the patients assigned to C. parvum as opposed to 7% of the patients assigned to placebo.

In concordance with the Phase I Study, C. parvum intrapleurally appears to have also a significant systemic effect, as revealed by an increase in the total number of circulating WBC, mainly due to neutro- and monocytes. 63% of the patients assigned to C. parvum had a greater than 50% increase in monocytes as opposed to 37% of the patients assigned to placebo, who had a similar increase. It is too early to make any definitive comparisons of the treatments with regard to disease-free interval or survival. At this time, the average follow-up time for an analyzed case is approximately six months. So far, 13% of the patients have proven tumor recurrence and 8% have expired. However, it is possible to identify tentatively a few high risk groups so that we may improve the stratification of the next study.

It appears as if the bilobectomy patients form a high risk group, see figure 1. It has not been analysed so far, whether the poor prognosis of the bilobectomy-patients is connected with the resection of upper and middle lobe or with the resection of middle and lower lobe. From the knowledge of lymphatic tumor spread it could be expected that surgical treatment of lower lobe cancers by bilobectomy may result in early recurrence.

The present Ludwig Lung Cancer Group with a proven accrual of more than 300 eligible patients per year offers a sharp tool for analysing different treatment modalities with the aim to improve the prognosis of cancer patients.

It is obvious, that besides the results of recurrence-free interval and survival, such trials automatically will contribute important biological information such as relapse pa-tern of bronchogenic carcinomas after radical resection, and identification of high and low risk subgroups of patients.

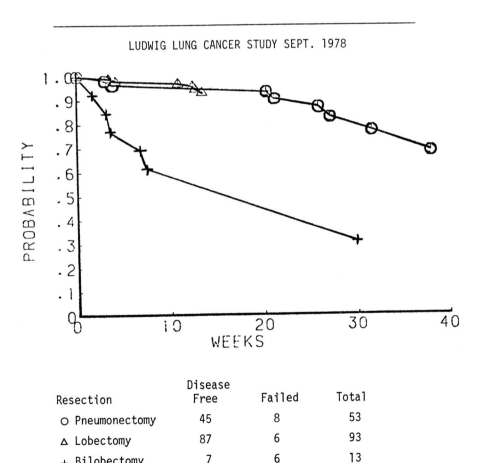

Fig. 1. Disease-Free Interval by Type of Resection

REFERENCES

Amery, W. K. (1976). Double blind levamisole trial in resectable lung cancer. Ann. N.Y. Acad. Sci. 277, 260-268.

Kaufmann, M., Wirth, K., Scheurer, J., Zimmermann, A., Luscieti, P., Stjernswärd, J. (1977). Immunomorphological lymph node changes in patients with operable bronchogenic squamous cell carcinoma. Cancer 39, 2371-2377.

Ludwig Lung Cancer Study Group. (1978). Search for the possible role of "immunotherapy" in operable bronchial non-small cell carcinoma (Stage I and II): A Phase I study with Corynebacterium parvum intrapleurally. Cancer Immunol. and Immunotherapy 4, 69-75.

Ludwig Lung Cancer Study Group. (1979). Search for the role of immunotherapy with Corynebacterium parvum intrapleurally in operable non-small cell carcinoma. Proceeding to Workshop in "Recent trends in immunodiagnosis and immunotherapy of malignant tumors - relevance for surgery." Eds. Herfarth, Flad et al., Springer Verlag, 210-213 (in press)

Matthews, M. J., Kanhouwa, S., Pickren, J., Robinette, D. (1973) Frequency of residual and metastatic tumor in patients undergoing curative surgical resection for lung cancer. Cancer Chemotherapy Reports, Part 3, Vol. 4, No. 2, 63-67.

McKneally, M. F., Maver, C., Kausel, H.W. (1976). Regional immunotherapy of lung cancer with intrapleural BCG. Lancet I, 377.

Second Nat. Cancer Inst. Conf. on Lung Cancer, Airlie House, Virginia, USA (in press).

Selawry, O. S., Hansen, H. H. (1973). Lung Cancer in : Cancer Medicine. Ed. J.F. Holland and E. Frei, Lea and Febinger, Philadelphia, 1473-1518.

Reprint request : Ludwig Institute for Cancer Research, CH 1066 Epalinges, Switzerland or to any of the participants listed.

Symposium No. 19 — Chairman's Summary

David Pressman

The first report of the use of immunotherapy in the treatment of cancer was that of Herricourt and Richet in 1895. They prepared antisera against a human osteogenic sarcoma and reported remarkable effects obtained in altering the course of the disease when the serum was injected around and into tumors in human patients. Their second report dealt with 50 patients and was still more optimistic. Nevertheless they ceased pursuing this line of endeavor although others have studied it subsequently.

At the current time the use of antibodies as therapy is still being pursued but there are additional therapies where direct immunization with a tumor vaccine is used and where nonspecific stimulation is used.

All of these fields were covered in Symposium No. 19.

Tumor antibodies have the potential of being used therapeutically or diagnostically when used to carry radioisotopes or chemotherapeutic agents to a tumor. At the symposium it was shown that in the case of carcinoembryonic antigen tumor binding immunoglobulins can be isolated from normal human serum and thus provide a source of tumor specific immunoglobulins as carriers of radioactive or cytotoxic agents.

In the area of specific active immunotherapy in humans an expanding program was described using antigens isolated from each type of lung cancer. Results from a 5 year study indicated significant extension of the disease free interval in those patients which had been immunized.

Reports of nonspecific immune stimulation concerned both general systemic immune stimulation and adjuvant contact stimulation where the stimulating agent is injected directly into tumor. The stimulating agents used were: Bacillus Calmette Guerin (BCG), Corynebacterium Parvum (CP) or Levamisole. One report concerned the use of BCG cell wall skeleton in man. This has the added advantage of avoiding any infection with BCG besides appearing to contain active stimulating principal. Extension of life expectancy in lung cancer and leukemia without serious side reaction was observed. Patients with malignant pleurisy were particularly aided when intrapleural injection was used.

Another report was that the administration of BCG intrapleurally following pulmonary resection resulted in a statistically significant increase in survival in patients so treated when compared to a concominant randomized controlled group. Patients with Stage I resected lung cancer showed a significant disease free interval. However the same intrapleural treatment with BCG postoperatively did not result in a significant increase in survival over controls in patients with Stage II and Stage III lung cancer.

Other reports involved the use of BCG in melanoma and breast cancer as well as in lung cancer and leukemia and indicated BCG caused prolonged remissions when used in addition to local surgery and in addition to chemotherapy in disseminated melanoma. Benefits of BCG immunotherapy were reported to be lost within 5 years of cessation of therapy and thus indicates requirement of continued BCG treatment.

Levamisole, a chemical which can be taken orally, is felt to augment a depressed immune response. Studies have shown that Levamisole is capable of improving immunocompetence in patients with lung cancer. A trial comparing Levamisole to Placebo in patients with resected lung cancer has indicated some benefit of Levamisole therapy.

Experiments in animals involving the effect of BCG and CP intravenously and intralesionally were reported to parallel the results obtained in clinical studies on humans. Moreover it was frequently possible to suppress subcutaneous tumour growth by contact with BCG or CP even with tumours displaying little or no immunogenicity as defined by the capacity to induce tumour immunity in preimmunized rats.

Lymphocyte Subpopulations in Tumor Immunity
An Overview of the Workshop Immunology: Lymphocytes

Alberto J. L. Macario

Division of Laboratories and Research, New York State Department of Health,
Albany, New York 12201, U.S.A.

Over 30 contributions were selected by the Scientific Committee of the XII International Cancer Congress for the Workshop entitled Immunology: Lymphocytes. These contributions deal with various aspects of tumor immunology, including quantitation, characterization, generation, regulation and function of lymphocyte subpopulations involved in the immune reaction elicited by malignant tumors. The properties of the cell populations which mediate, suppress and amplify anti-tumor immunity are described, as well as antibodies and other factors which interact with cells and modify them. T-, B-, K- and NK-cells and macrophages are studied to understand how they are generated, activated against tumors, regulated, and how they ultimately exercise their functions, either protecting from, or enhancing tumor growth. Along these lines, thymic factors, immune RNA, tumor antigens, alpha-fetoprotein, soluble serum and tissue-culture molecules released by lymphoid and tumor cells, are examined. Cell-surface markers, enzymes and cAMP levels, and response to mitogens such as PHA and Con A, are determined. Experiments are carried out mostly in mice, in vivo and in vitro, but also human materials are used in some instances. Thus the relevance of studies in animals and under artificial in vitro conditions to the human in vivo situation can be evaluated. The effect of disease stage, drugs, radiation and therapeutic intervention upon lymphocyte number and function, including migration, is also investigated.

Although most reports deal with more than one problem, they cluster in distinct areas according to their major topic of study. This trend has served to arrange the papers under the following headings.

I. QUANTITATION OF LYMPHOCYTE SUBPOPULATIONS IN CIRCULATION, LYMPHOID ORGANS AND TUMOR TISSUES.

Berghaus, Heilmann and Sasse report that the absolute number of T cells (which form rosettes with sheep erythrocytes) is diminished in 50- to 60-year old cancer patients, as compared to healthy individuals within the same age bracket. B-cell (immunoglobulin positive) number is not modified but in the cases of pancreatic carcinoma examined by the authors, where they found it increased. Nakamura and coworkers find in humans that T-lymphocytes are not abundant within tumor tissues but they are selectively located in contact with cancer cells. Very few T-lymphocytes are found in the stroma. Mc Credie, MacDonald and Wood observe a reduction of the cells that bring about ADCC (antibody-dependent cell-mediated cytotoxicity) in the peripheral blood of humans after the 65th year of age, patients with advanced cancer and/or following radiotherapy. In the latter case, the values return to normal after 12 weeks unless the therapy is unsuccessful.

II. PHA RESPONSIVENESS AS A TOOL FOR STUDYING CANCER

Kreutzmann and co-workers report that PHA stimulation of normal human peripheral blood lymphocytes decreases fluorescence polarization in a test which utilizes fluorescein diacetate. Such a decrease in fluorescence polarization does not occur if lymphocytes from cancer patients are stimulated with PHA. Wetter, Linder and Brandhorst communicate that plasma-cell leukemia cells respond in vitro to PHA by increasing DNA and immunoglobulin light-chain synthesis, whereas the cells from non-leukemic plasmacytomas do not.

III. CYCLIC ADENOSINE MONOPHOSPHATE (cAMP)

Olkowski and McLaren find low levels of cAMP in peripheral blood lymphocytes of cancer patients which are mitogen-hyporesponders and have metastases. Successful therapy is accompanied by normalization of mitogen responsiveness, and levels of anti-tumor cytotoxic T-cells and cAMP in peripheral lymphocytes.

IV. BINDING OF LYMPHOCYTES TO TUMOR AND OTHER CELL TYPES

Rivenson, Madden and Nowakowski show that mouse Leydig cells, maintained in vitro for years, bind rat lymphocytes as well as Leydig cells freshly obtained from the animal. Rivero and co-workers have previously shown that in pleural and ascitic fluids from cancer patients, lymphocytes tend to adhere to macrophages, forming rosettes. They now report that the neoplastic cells in these fluids never bind lymphocytes.

V. EFFECT OF IMMUNOPOTENTIATORS, IMMUNOSUPPRESSORS AND CHEMOTHERAPEUTIC AGENTS UPON LYMPHOCYTE SUBPOPULATIONS

Taniguchi and co-workers study the effect of Toxohormon, Propionibacterium acnes and BCG upon ATP-ase and acid phosphatase levels in spleen lymphocytes of rats bearing Yoshida's ascitis and Sato's lung tumor. The enzyme levels diminish when the tumors spread. Toxohormon lowers the enzyme levels, but the other agents increase them while augmenting anti-tumor resistance. Anaclerio and co-workers find that cyclophosphamide, azathioprine and daumomycin abrogate generation of mouse-spleen suppressor cells whereas adriamycin, Corynebacterium parvum and BCG have the opposite effect. Dumitrescu and co-workers examine the effect of pyrithiamine upon DNA synthesis and migration of mouse thymocytes. Their results suggest that the drug promotes DNA synthesis and migration of thymocytes to the spleen. Nabih modifies the surface of Raji cells with 1-2-cyclohexeno, 4-(B-diethyl-aminoethyl) thioxanthone and then injects these cells in mice. These mice develop the capacity to phagocytize Plasmodium-bearing erythrocytes.

VI. THYMIC FACTORS

Ueda, Yamagata and Sugawa report that human fetal thymic extracts increase in vitro the responsiveness to Con A of peripheral blood lymphocytes obtained at early, but not of those obtained at late, stages (III and IV) of uterine cancer. Bena and Mordoh report a similar finding. Fetal thymic extracts in vitro increase PHA-responsiveness of peripheral blood lymphocytes from cancer patients, but not of lymphocytes from normal individuals.

VII. IMMUNE RNA

Fukushima and co-workers obtain RNA from lymphoid cells of rabbits immunized with human rectal and lung cancer-cell antigens. This RNA confers the capacity to kill tumor cells in vitro to human lymphocytes from healthy donors of the same blood group as the cancer patient.

VIII. SOLUBLE SUPPRESSOR FACTORS

Ting and co-workers find that sera from cancer patients suppress in vitro immune T-cell mediated cytotoxicity against the murine lymphoma FBL-3, and reduce the response of non-immune mouse lymphocytes to Con A and PHA, but not to LPS. Kieler and co-workers report that several factors, which inhibit DNA synthesis, are released by mouse spleen cells in vitro, under various experimental conditions. These factors are released by non-presensitized cells when cultivated in serum-free medium in the absence of any stimulation and after stimulation with PHA. They are also released by tumor pre-sensitized cells if stimulated with the relevant tumor antigen. In each case, the DNA-synthesis inhibitory factor produced has a distinctive molecular weight. Braun and co-workers also find that cytoplasmic extracts from mouse spleen cells inhibit DNA synthesis in PHA-stimulated syngeneic and allogeneic lymphocytes, as well as in two human lymphoblastoid cell lines. These extracts also inhibit in vivo induction of cytotoxic T-lymphocytes and enhance tumor incidence in BALB mice inoculated with an AKR lymphoma within a glass cylinder.

IX. SUPPRESSOR CELLS AND ALPHA-FETOPROTEIN

Albala, Flinton and Chester report that the BW5147 mouse (AKR) lymphoma cells inhibit PHA-induced blastogenesis in cultures of AKR spleen cells. The degree of inhibition is proportional to the number of lymphoma cells added to the spleen cell cultures. On these grounds the authors calibrate a procedure for the disclosure of lymphoma cells in circulation. They can detect by this method fewer cells than it is possible by any other usual procedure. Macario and Mizejewski find that C57L/J mice bearing a syngeneic subcutaneous hepatoma develop splenomegaly and high serum alpha-fetoprotein (AFP) levels, which parallel tumor and spleen size progression. The authors postulate that AFP causes proliferation and activation of spleen suppressor cells. To test this working hypothesis they attempt at transferring suppression with spleen cells from tumor-bearing, splenomegalic, mice to syngeneic recipients. Spleen cells obtained on day 35 post-tumor implantation and given intraperitoneally increase tumor incidence, while cells obtained on day 14 and 21 reduce it. Tumor incidence is also increased when the spleen cells are inoculated subcutaneously along with the challenging tumor cells. AFP potentiates this effect. Padarathsingh and co-workers report that spleen cells from BALB mice bearing the plasmacytoma ADJ-PC5 show poor blastogenic response to T- and B-cell mitogens. However, these spleen cells can recognize tumor antigens, as shown by the migration inhibition test and are able to reject tumor cells in the Winn's assay.

X. ANTITUMOR ANTIBODIES

Braun and co-workers report that BALB mice bearing an AKR lymphoma (L15) in a glass-cylinder, as well as mice that have rejected the tumor, show antibodies which recognize surface antigens in L15. These antibodies are demonstrable by immunoprecipitation and cytolytic assays. Braun, de Kohan and Pasqualini use the same tumor-mouse system and demonstrate antibodies capable of inducing antibody-dependent cell-mediated cytotoxicity in normal BALB spleen cells. The cells involved adhere to nylon, and are Thy 1.2 negative and Fc-receptor positive. Ioachim, Paluch and Dorsett study the cellular infiltrates in lung tumors and presence of antibodies in these infiltrates. Cellular infiltrates are more pronounced in squamous than in oat-cell carcinoma. Within each of these two types, the infiltrates are more abundant in the more differentiated forms. Plasma cells are predominantly associated with well differentiated squamous cell carcinoma. From these, anti-tumor antibodies are eluted and shown to react specifically with the tumor cells, but not with nonpulmonary tumors or cells from normal adult and fetal lung.

XI. NATURAL KILLER (NK) CELLS

Linna and Lam discuss natural resistance to Marek's disease. It can be transferred from adult, resistant chickens, to newly hatched birds, which have not yet developed resistance, with spleen cells. T- and B- cells and macrophages can be removed from the spleen-cell inoculum, without altering its ability to confer adoptive-resistance. Thus the cells involved in the latter resemble the NK cells of mammals. Djeu and co-workers find that interferon and interferon inducers (poly-I:C) increase NK cells in normal mouse spleens. This effect is obtained giving the agents in vivo, and also in vitro, to spleen cells in culture. Poly-I:C induction of NK reactivity requires macrophages while interferon acts directly on NK cells. Chang and Log describe three transplantable SJL/J reticulum cell neoplasms which display killer activity against two different targets among the several tested. This killer effect is observed in vitro and it can also be conferred to proper F_1 recipients by the tumor cells. In the latter case, host cells seem to be involved in the generation of NK activity, which is mediated by a cell sensitive to UV, 10,000 r and freeze-thawing, and resistant to treatment with anti-Thy 1.2 serum and complement.

XII. LYMPHOCYTE MALIGNANCIES AND AUTOIMMUNITY

The preceding work (by Chang and Log) and that of Murphy and Roths describe situations in which malignant lymphoid proliferations are accompanied by immunological phenomena which might injure the tumor host. The latter two authors describe a new lymphoproliferative disorder associated with autoimmune complex glomerulonephritis in mice (spontaneous mutation). This animal model is considered suitable for elucidating the pathogenic relationship between lymphoproliferation, autoimmunity and lymphoma development.

XIII. CELL-MEDIATED IMMUNITY

Schirrmacher and Shantz study two variants of a DBA/2 mouse lymphoma, one which metastasizes and another which does not. While the latter generates specific cytotoxic T-cells, the metastasizing variant does not do so. Thus a correlation is found between inability to generate cytotoxic lymphocytes and metastatic potential. Cudkowicz, Nakano and Nakamura describe a murine model system in which parental cytotoxic T-lymphocytes (CTL) are generated in vitro by co-cultivation of F_1 spleen and parental lymphoid cells containing cortisone resistant, Thy-1 positive, non-nylon adherent T-cells, and macrophages. F_1 cells, syngeneic to the CTL, or sharing H-2D determinants, are not lysed by CTL, but they can block lysis of the parental target by CTL. Thus CTL are autoreactive. Amagai, Katsura and Kishida describe the induction in vitro of cytotoxic cells against the mastocytoma P815 by co-cultivation of DBA/2 spleen and x-irradiated tumor cells. Generation of these killer cells depends upon the presence of nylon wool non-adherent, Thy 1.2, positive cells and macrophages. Pretreatment of spleen donors with cyclophosphamide or hydrocortisone acetate abolishes subsequent generation of the killer cells in vitro. Collavo and co-workers perform studies with adult mice, thymectomized, lethally x-irradiated and bone-marrow reconstituted, which develop tumors if inoculated with Moloney murine sarcoma virus (M-MSV). Reconstitution of these mice with normal lymphoid cells confers them the ability to reject M-MSV-induced tumors by means of cytotoxic T-cells. Sugimoto and co-workers describe the in vitro response of spleen cells from A/J mice bearing the C1300 neuroblastoma to soluble and insoluble tumor antigens. This response is mediated by T-cells which increase with time after tumor implantation up to two weeks, but they decline afterwards as the tumor continues to grow.

Search for Immunosuppressive Effects of Suppressor Cells and Alpha-Fetoprotein in a Mouse-Hepatoma Syngeneic System

Alberto J. L. Macario and Gerald J. Mizejewski

*Division of Laboratories and Research, New York State Department of Health,
Albany, New York 12201, U.S.A.*

ABSTRACT

Growth of hepatoma-BW7756 in syngeneic C57L/J mice accompanies a rise in serum alpha-fetoprotein (AFP) concentration in the first 3 weeks of tumor growth, after which high levels are maintained. Spleen size and cell number increase as does AFP.

Since AFP reportedly stimulates suppressor cells causing immunosuppression, we hypothesized that the observed increase in spleen cell number is due to hyperplasia of suppressor cells under the constant stimulation of tumor-derived AFP thereby explaining the failure of tumor rejection. We tested this hypothesis by transferring, intraperitoneally, spleen cells obtained at weekly intervals post-transplantation, to syngeneic recipients inoculated with a tumor-cell dose producing 50% tumor incidence. Thus far, incidence was not increased in over 200 spleen-cell recipients examined. Spleen cells obtained on days 14 and 21 post-transplantation reduced tumor incidence from 56% (35/63 normal mice) to 12% (4/32 spleen-cell recipients) ($p<0.0001$). This protective effect could not be abrogated by AFP administered with the spleen cells over 7 to 10 days after tumor challenge. Spleen cells inoculated subcutaneously along with the tumor cells did, however, enhance tumor growth.

KEY WORDS

Suppressor cells; alpha-fetoprotein; mouse hepatoma BW7756; lymphocytes, anti-hepatoma; lymphocyte suppression; immunosuppression, mediated by alpha-fetoprotein.

INTRODUCTION

We describe here the first steps of our search for suppressor cells and for a role for alpha-fetoprotein (AFP) in the regulation of the immune response to a mouse hepatoma which produces this fetal protein.

The role and mechanism of action of AFP in the regulation of immune and lymphocyte reactivities are controversial matters. Data from different laboratories do not agree (Murgita and Tomasi, 1975, 1975a; Sell, Sheppard and Poler, 1977; Sheppard and co-workers, 1977). One element that obscures the results and aids the controversy is that AFP from different sources are tested on cells or animals which are not syngeneic with the AFP donors. Moreover, AFP effects are sought upon reactions which are irrelevant to the response against antigens or molecules related in any known way to the fetal protein. Although we do not deny that these studies might be of some interest, we believe that a more direct and specific approach is cogent to elucidate the role of AFP in immunity. A system should preferably be utilized whereby cellular and humoral responses against cells producing AFP can be examined in a syngeneic combination of AFP and target and effector cells. Furthermore, the system should provide a tool for studying phenomena that are likely to occur in nature and thus bear some biopathological significance. Accordingly, we utilized a mouse-hepatoma syngeneic model, in which the tumor produces AFP proportional to its growth (Mizejewski and Dillon, 1978). We recently found that this is accompanied by a parallel increase in the spleen cell number. In view of these findings and since it was claimed that AFP activates suppressor cells (Alpert and co-workers, 1978; Murgita and co-workers, 1977), we hypothesized that AFP in our experimental model causes hyperplasia of the spleen suppressor-cell compartment thereby leading to failure of tumor rejection. We report here our experiments to test this hypothesis.

MATERIAL AND METHODS

Tumor. The hepatoma BW7756 lines utilized were maintained in our laboratory *in vivo* and *in vitro*. For tumor challenge, *in vitro*-line cells, trypsinized and washed, were used in a concentration containing the desired dose per mouse in 0.1 ml. These cells were inoculated subcutaneously in one flank, either alone or mixed with spleen cells.

Mice. C57L/J males (The Jackson Laboratories, Bar Harbor, Maine), 6 to 8 weeks old at the beginning of the experiments were used throughout. Mice comprised two groups: 1) Donors of spleen cells: Normal or bearing the *in vivo* hepatoma line; and 2) Recipients of spleen cells with or without AFP, and then challenged with a calibrated dose of the *in vitro* line.

Spleen cells. They were obtained from normal and tumor-bearing donors weekly after tumor transplantation by screen-teasing the spleens and suspending the cells in RPMI 1640 with antibiotics and fetal bovine serum. The cells were then washed three times in medium devoid of serum, counted (trypan blue exclusion method) and adjusted to a final concentration which contained the desired cell-dose per mouse in 0.1 ml, which was given intraperitoneally 48 hours prior to tumor challenge, or subcutaneously along with the tumor cells.

Amniotic fluid and AFP. Amniotic fluid (MAF) was obtained by puncturing the amniotic sacs of 15 to 18-day pregnant mice. To obtain pure AFP the MAF was chromatographed on DE52 anion exchange resin and fractionated on sephadex G-100. The appropriate fraction was then applied to an antibody affinity column prepared by coupling rabbit anti-mouse serum antibodies to cyanogen bromide-activated sepharose 4B (Pharmacia Fine Chemicals, Piscataway, New Jersey). The void fraction was concentrated and purity checked by immunodiffusion and microimmunoelectrophoresis using rabbit anti-mouse whole serum antibodies, and by polyacrilamide gel electrophoresis.

Pretreatment of spleen cells and injection of mice with AFP and control solutions. To investigate whether AFP modifies the effect of spleen cells upon tumor incidence, the spleen cells were incubated in AFP just before inoculation into the recipient mice. Control solutions included mouse amniotic fluid deprived of AFP, normal mouse serum and tissue culture medium. These solutions and AFP were also administered intraperitoneally to the cell-recipients daily, beginning at the time of spleen-cell innoculation and continuing over 7 to 10 days.

Evaluation of tumor-growth. The consequence of inoculating spleen cells -- with or without treatment of the cells or the recipients with AFP -- upon tumor growth was measured by determining the tumor incidence at various time-points after challenge with the hepatoma cells and the tumor weight at the end of the experiments.

I. Search for suppressor cells in the spleen of tumor-bearing mice

As mentioned in the Introduction our initial working hypothesis was that the enlarged spleen of tumor-bearing mice has an excess of suppressor cells due to the continuous stimulation by the high serum-AFP levels which occur during tumor growth. We tried to demonstrate such suppressor cells by attempting at conferring adoptive suppression to normal mice with spleen cells from syngeneic, tumor-bearing donors. Fifty million cells from normal (control) and tumor-bearing mice (obtained at various intervals after tumor implantation) were inoculated into the recipients 48 hours prior to tumor challenge. The latter was done with a dose of hepatoma cells giving 50% tumor takes on day 14 post-challenge. The data are shown in Table 1. Spleen cells from mice bearing tumors implanted 14 and 21 days before cell transfer diminish the rate of tumor growth, which is shown by the decrease in tumor incidence as compared with control mice preinoculated with normal spleen or liver cells (data not shown), and with mice receiving only the tumor-cell challenge. Seven, 28 and 35-day spleen cells do not significantly influence tumor growth, although a tendency to acceleration is caused by the latter. Continuation of this work with "older" spleens was not done because tumor-bearing mice die before day 42.

II. Search for a suppressive effect of AFP upon the anti-tumor capacity of spleen cells from hepatoma-bearing mice

In view of the results described above, the hypothesis was elaborated that spleen cells from tumor-bearing animals can display anti-tumor capacity only in an environment devoid of AFP. This would explain both the failure to reject the tumor by the original host (with elevated serum AFP levels) and the success of the spleen cells in the putative environment of the host, where no high AFP levels occur at the time of cell transfer, tumor challenge and for at least one week thereafter. To test this hypothesis we administered AFP to the spleen-cell recipients from the day of cell transfer on, for 7 or 10 days.

TABLE 1 Tumor Incidence in C57L/J Mice Pretransfused with Spleen Cells from Tumor-bearing and Normal Mice

Spleen day[a]	Tumor incidence Days after tumor challenge[b]		
	14	18	28
7	6/16(31)[c]	7/16(44)	10/16(63)
14	4/20(20)	4/20(20)	8/20(40)
21	0/12(0)	3/12(25)	7/12(58)
28	10/24(42)	12/24(50)	17/24(71)
35	6/8(75)	6/8(75)	7/8(88)
-[d]	7/13(54)	7/13(54)	8/13(62)
-[e]	35/63(56)	40/63(63)	46/63(73)

a: Day after tumor transplantation at which the spleen cells to be transfused were obtained from tumor-bearing mice. b: 10^5 hepatoma cells inoculated subcutaneously 48 hours after intraperitoneal transfusion of 50×10^6 spleen cells. c: Mice with palpable tumor/mice studied (percent of mice with palpable tumor). Tumor incidence was significantly lower at the three time-points shown, in mice receiving 14-day spleen cells, and on day 14 and 18, in mice receiving 21-day spleen cells. d: Mice in this group were transfused with spleen cells from normal mice. e: Mice in this group did not receive spleen cells. Mice inoculated intraperitoneally or subcutaneously with spleen cells from tumor-bearing mice did not develop tumors and survived as normal mice (data not shown).

Table 2 shows the data of two experiments in which two doses of hepatoma cells producing different tumor incidence were used for challenge. Although the number of animals in each group is too small for an appropriate statistical evaluation, it can be seen that the results shown in Table 1 are confirmed. The tumor incicence is lower in the mice which were pretreated with spleen cells from donors bearing a hepatoma transplanted 14 days earlier, than in the control group which did not receive spleen cells. This adoptive protection occurs despite AFP and normal mouse serum administration. However, in experiment II the mice treated with AFP show a tumor incidence on days 18 and 28 which is as high as in the control group (C). This suggests that AFP exerted some immunosuppressive effect capable of nullifying the protective potential of the spleen cells. However the interpretation of the results is obscured by the fact that normal mouse serum also seems to have displayed a similar immunosuppressive action since tumor incidence on day 28 is as high in mice treated with serum as in the mice treated with AFP. Additional experiments are being conducted in our laboratory to clarify this point.

TABLE 2 Tumor Incidence in C57L/J Mice Pretransfused with Protective Spleen Cells from Tumor-bearing Mice and Subsequently Treated with AFP or Normal Mouse Serum

Experiment	Set[a]	Treatment of spleen-cell recipients with:[b]	Tumor incidence Days after tumor challenge[c]		
			14	18	28
I	A	AFP	1/3(33)[d]	1/3(33)	1/3(33)
	B	NMS[e]	1/6(17)	2/6(33)	2/6(33)
	C	---	5/6(83)	5/6(83)	6/6(100)
II	A	AFP	0/7(0)	2/7(29)	4/7(57)
	B	NMS	0/6(0)	0/6(0)	4/6(67)
	C	---	3/12(25)	4/12(33)	6/12(50)

a: Mice in sets A and B were transfused intraperitoneally with 50×10^6 spleen cells from mice bearing an hepatoma transplanted 14 days before the excision of the spleen. Cell transfusion was done 48 hours prior to tumor challenge. Mice in set C did not receive spleen cells. b: One daily intraperitoneal injection of 0.1 ml of the indicated fluid for 7 and 10 days (experiments I and II, respectively), beginning the day of the spleen-cell transfusion. c: 1.5 and 0.5×10^5 hepatoma cells (experiments I and II, respectively) inoculated subcutaneously. d: Mice with palpable tumor/mice studied (percent of mice with palpable tumor). e: Normal mouse serum diluted 1:5 in phosphate buffered saline.

III. In situ tumor-enhancing effect mediated by spleen cells from tumor-bearing mice

The results described above show that if spleen suppressor cells and/or AFP play any role in anti-hepatoma immunity, this effect can not be easily demonstrated by inoculating the spleen cells and the challenging tumor-cell inoculum apart in time and space, regardless of whether AFP is also supplied. Thus, we undertook experiments to establish whether spleen cells administered simultaneously with the tumor cells and at the same site, would enhance tumor growth. The results are shown in Table 3. A tendency to acceleration of tumor growth is observed in the three groups that received spleen cells along with the hepatoma cells as compared with the control group which received only the latter cells. The tumor-enhancing effect is more pronounced ($p=0.05$) in the group which was inoculated with spleen cells preincubated with AFP. These results were confirmed in another experiment involving 8 mice in each group, and are, together with the observations described in the preceding section, the only indication we have thus far obtained of a suppressive effect due to AFP. Inoculation of hepatoma cells mixed with spleen cells from normal mice does not enhance tumor growth but tends to diminish it (Macario and coworkers, 1978).

TABLE 3 Tumor Incidence in C57L/J Mice Inoculated with Spleen Cells from Tumor-bearing Mice At the Site of Tumor Challenge

Set[a]	Pretreatment of spleen cells with:[b]	Tumor incidence Days after the cell-mixture inoculation		
		14	18	28
A	AFP	7/16(44)[c]	10/16(63)	13/16(81)
B	NMS[d]	3/16(19)	9/16(56)	14/16(88)
C	TCM[e]	4/16(25)	8/16(50)	14/16(88)
D	---	0/16(0)	5/16(31)	10/16(63)

a: Mice in sets A, B and C were inoculated subcutaneously with a mixture of 5 x 10^6 spleen cells from mice bearing a hepatoma transplanted 14 days before excision of the spleen and 10^5 hepatoma cells. Mice in set D did not receive spleen cells.
b: The spleen cells were incubated in the fluids indicated for 45 minutes at 23°C just before mixing them with the hepatoma cells for subcutaneous inoculation. c: Mice with palpable tumor/mice studied (percent of mice with palpable tumor). d: Normal mouse serum diluted 1:5 in phosphate buffered saline. e: Tissue culture medium.

DISCUSSION

Data in this paper show two important features: 1) Spleen cells from hepatoma-bearing mice can display anti-tumor as well as pro-tumor growth effects in putative hosts depending upon (a) the time post-tumor transplantation at which the spleen cells are obtained, and (b) whether the spleen cells are given intraperitoneally prior to tumor challenge, or subcutaneously, along with the challenging tumor cells, respectively. 2) AFP does not inhibit, or does so slightly, the adoptive protection conferred by spleen cells from tumor bearing mice given intraperitoneally to normal syngeneic hosts challenged two days later with tumor. However, AFP tends to amplify the tumor-enhancing effect of spleen cells from tumor-bearing mice administered subcutaneously along with the challenging hepatoma cell inoculum.

Identification of the spleen cell subpopulations responsible for the opposite effects observed, protection and enhancement, under the 2 experimental conditions tested, is currently being done in our laboratory.

REFERENCES

Alpert, E., J. L. Dienstag, S. Sepersky, B. Littman, and R. Rocklin (1978). Immunological Commun., 7, 163-186.
Macario, A. J. L., and co-workers (1978). (in preparation).
Mizejewski, G. J. and W. R. Dillon (1978). Arch. Immunol. and Exp. Therap. (in press).
Murgita, R. A., E. A. Goidl, S. Kontiainen and H. Wigzell (1977). Nature, 276, 257-259.
Murgita, R. A. and T. B. Tomasi, Jr., (1975a). J. Exp. Med., 141, 269-286.
Murgita, R. A. and T. B. Tomasi, Jr., (1975b). J. Exp. Med., 141, 440-452.
Sell, S., H. W. Sheppard, Jr., and M. Poler (1977). J. Immunol., 119, 98-103.
Sheppard, H. W., Jr., S. Sell, P. Trefts, and R. Bahu (1977). J. Immunol., 119, 91-97.

Studies of Natural Killer Cell-mediated Cytotoxicity in Normal Individuals and Cancer Patients

E. Lotzová, J. A. Maroun, K. B. McCredie, B. Drewinko and K. A. Dicke

Department of Developmental Therapeutics,
The University of Texas System Cancer Center,
M.D. Anderson Hospital and Tumor Institute, Houston, Texas, U.S.A.

ABSTRACT

Since NK cells may be important cell type in resistance and immunosurveillance to malignancies as well as in rejection of allogeneic bone marrow grafts, we have studied several parameters of human NK cell-mediated cytotoxicity. We have found that NK cell cytotoxicity was expressed not only in the blood of normal individuals but also in their bone marrow. However, the efficiency of bone marrow NK cells was approximately 50% of that observed in peripheral blood. AML and CML-diseased patients showed significantly lower NK cell activities in both peripheral blood and bone marrow. The lower NK cell activity in leukemic patients suggest involvement of NK cells in resistance to leukemias. In a few patients low NK cell activity could be caused by the presence of suppressor cells. However, in the majority of patients suppressor cells were not detected and thus another mechanism underlaying low NK cell activity must be involved. Bone marrow NK cell function was sensitive to cryopreservation since cryopreserved NK cells expressed only 50% of cytotoxicity when compared to fresh bone marrow cells. The bone marrow NK cell activity could be concentrated by discontinuous albumin gradient technique. The highest NK cell activities were found in fraction 1 and 2; NK cells activities in these fractions were 3 to 6 times higher than that of unfractionated bone marrow. Other fractions showed either comparable or lower NK cell activities than unfractionated bone marrow.

Key words: Natural killer cells, leukemia, bone marrow transplantation

INTRODUCTION

The relatively recently described phenomonen of natural killing of leukemias, lymphomas and other types of tumors *in vitro* by lymphoid-like cells from normal individuals has received increased attention, particularly because such cells may possibly play an important role in resistance and immunosurveillance to malignancies *in vivo* (for review see Pross and Bains, 1977; Lotzová and McCredie, 1978). The effector cells involved in this kind of cytotoxicity were designated natural killer (NK) cells to indicate their natural occurence (without any previous sensitization) and tumor-killing capability (Kiessling and co-workers, 1975). It has been shown that the majority of NK cells carry receptors for Fc portion of IgG molecule and that some have receptors for the activated third complement component (Pross and Bains, 1977). One group of investigators have found human NK cell activity to reside in the erythrocyte rosette-forming T cell fraction (West and co-workers, 1977) whereas, other investigators did not confirm this observation (Jondal and Pross, 1975; Peter and co-workers, 1976; Cooper and co-workers, 1977). Some controversy has arisen as to whether NK cells are members of the T cell, B cell or macrophage-monocyte lineage. Even though some suggested that human NK cells belong to a T cell population (West and co-workers, 1977; Kaplan and Callewaert, 1978), the factual data to support this suggestion are still missing. Currently available data indicate that NK cells do not belong to the mature T cell, B cell or macrophage populations since they do not posses easily detectable T and B lymphocyte markers and are nonadherent and nonphagocytic. Thus, the precise nature of NK cells remains to be determined. Beside the obscure nature of NK cells, little is known about their specificity, recognition, and the nature of antigens triggering their reactivity. The fact that the NK cell react to a variety of syngeneic, allogeneic and xenogeneic tumors of various histological types is indicative that NK cells reactivity is not directed to a single antigen, but to a rather broad spectrum of antigenic determinants.

In addition to their possible significance in tumor immunity, NK cells may be involved in rejection of allogeneic bone marrow grafts. This statement is based on the observation that bone marrow effector cells express characteristics of NK cells (Lotzová and Savary, 1977).

Possible involvement of NK cells in two important biological functions prompted us to study various parameters of human NK cell cytotoxicity, especially with regard to their resistance to leukemias.

MATERIALS AND METHODS

TARGET CELLS

Cultured human lymphoblastoid cell line, CEM, with T cell characteristics, was used in all our experiments as target. For cytotoxicity studies, ten million CEM cells in 0.5 ml of RPMI 1640 medium with 10% FCS were labelled with 100 μCi of radioactive sodium chromate (^{51}Cr) and incubated for 30 minutes at 37°C. The labelled cells were washed three times in 25 ml of RPMI 1640 and adjusted to required concentration.

EFFECTOR CELLS

Peripheral blood lymphocytes were separated from heparinized whole blood obtained from healthy volunteer donors or from patients on a Ficoll-Hypaque gradient. Bone marrow cells were obtained from three groups of donors: 1) healthy individuals serving as bone marrow donors in allogeneic bone marrow transplantation, 2) non-hematopoietic cancer patients in remission, 3) acute myeloid leukemia patients in relapse and 4) acute myeloid leukemia patients in remission. The first two groups of individuals were classified as normal donors. The bone marrow cells were collected from illiac crest. Bone marrow cells were separated from red blood cells either by Ficoll-Hypaque gradient or by centrifugation and subsequent buffy coat collection. Unless otherwise stated, the target-to-effector (T:E) cells ratio was 1:50. The ^{51}Cr cytotoxicity assay was performed as described previously (Lotzová and Savary, 1977).

Discontinuous albumin gradient technique of Dicke (1970) was used for fractionation of NK cells. Cryopreservation of bone marrow cells was performed as described previously (see for review Dicke and co-workers, 1978); 10% dimethyl sulfoxide was used as a cryoprotective agent and the cells were stored at -192°C.

STATISTICAL ANALYSIS

The differences between individual groups studied were evaluated statistically with a Student t-test, and the probability (p) was calculated.

RESULTS

NATURAL KILLER CELL ACTIVITIES IN PERIPHERAL BLOOD AND BONE MARROW OF NORMAL INDIVIDUALS AND AML AND CML PATIENTS

Since NK cells may represent an important factor in resistance and immunosurveillance to leukemias, it was of utmost interest to determine the NK cell activities in leukemia-diseased patients. Peripheral blood of 7 patients with chronic myeloid leukemia (CML), 16 patients with acute myeloid leukemia (AML) and 11 AML patients in remission was tested for natural killer cell-mediated cytotoxicity against T-lymphoblastoid cell line, CEM. Forty-five normal donors served as controls in these studies. T:E cell ratio was 1:50. The results of these studies are illustrated in Table 1. It can be seen that there was a significant decrease in peripheral blood NK cell activities in CML-diseased and AML-diseased patients. Even though the NK cell cytotoxicity of AML patients in remission was higher than that of AML-diseased patients, it was still significantly lower when compared to NK cell cytotoxicity of normal individuals. The low NK cell activities in AML patients cannot be explained by dilution effect caused by much larger number of blast cells present in the peripheral blood since the increase of T:E cell ratio to 1:100 or 1:200 was not accompanied by increase in NK cell activities. To determine whether low NK cell activities in AML patients could be caused by the presence of suppressor cells or blast cells that would interfere with NK cell functions, mixing experiments were performed. In these experiments, peripheral blood cells from AML-diseased patients were mixed with peripheral blood cells of normal individuals in 1:1 ratio and tested for NK cell activities. Peripheral blood cells of 12 AML patients did not exert any suppressive effect on NK cell cytotoxicity of normal individuals. However, peripheral blood cells of 4 AML patients significantly suppressed (50% reduction) the NK cells cytotoxicity of normal individuals, indicating that in the latter situation suppressor cells could be present. At the present time we have no knowledge which cell population(s) is involved in suppression. These studies are currently underway.

TABLE 1 Natural Killer Cell-Mediated Cytotoxicity in Peripheral Blood of Normal Donors and CML and AML Patients

Source of Effector Cells	% of Cytotoxicity Mean ± s.e. (n)*	p Value
Normal Donors	31.3 ± 1.2 (45)	
CML Patients	15.0 ± 2.8 (7)	<0.001
AML Patients	13.5 ± 2.9 (16)	<0.002
AML Patients in Remission	24.2 ± 1.9 (11)	<0.02

* n-number of individuals.

In the next series of experiments NK cell activity in the bone marrow of 13 normal individuals, 7 AML-diseased patients and 11 AML patients in remission was tested. As indicated in Table 2, NK cell activity in bone marrow of AML-diseased patients was also significantly lower than that of normal individuals. In contrast to peripheral blood, bone marrow NK cell activities of AML patients in remission did not show inferior activity in comparison to normal donors.

The significantly lower NK cell activities in CML and AML patients are in accordance with the involvement of NK cells in resistance to leukemias. Further studies, however, have to be done to determine the causes of this NK cell anomaly in AML patients and to understand its mechanism.

TABLE 2 Natural Killer Cell-Mediated Cytotoxicity in Bone Marrow of Normal Donors and AML Patients

Source of Effector Cells	% of Cytotoxicity Mean ± s.e. (n)*	p Value
Normal Donors	14.1 ± 1.2 (13)	
AML Patients	6.0 ± 0.9 (7)	<0.001
AML Patients in Remission	15.7 ± 1.9 (11)	<0.5

* n-number of individuals

EFFECT OF CRYOPRESERVATION AND DISCONTINUOUS ALBUMIN GRADIENT FRACTIONATION ON NK CELL ACTIVITIES.

If one assumes that NK cells are involved in resistance to leukemias and considers the autologous bone marrow transplantation as a therapy for leukemic patients, then it is important to determine the NK cell activities in cryopreserved bone marrow, since cryopreserved bone marrow cells, collected in the remissions of the disease, are transplanted back into the patients in the case of the relapse of the disease. For this reason, 11 samples of fresh bone marrow cells and 5 samples of cryopreserved bone marrow cells of AML patients in remission were tested for NK cell activities. The data shown in Table 3, clearly indicate that cryopreserved bone marrow NK cells expressed 2.5 times less activity than fresh bone marrow NK cells.

TABLE 3 Activities of Cryopreserved and Fresh Bone Marrow NK Cells of AML Patients in Remission

Type of Bone Marrow Cells	% of Cytotoxicity Mean ± s.e. (n)*	p Value
Fresh	15.7 ± 1.9 (11)	
Cryopreserved	6.1 ± 1.1 (5)	<0.01

* n-number of individuals.

This fact together with the observation of low NK cell activities in bone marrow in general, is rather disappointing in the view of possible NK cell importance in resistance to leukemia. The aim would be then to supply the leukemic patients not only with hemopoietic stem cells but also with maximum number of NK cells so as to strengthen their defence to leukemia. Thus, we have searched for the techniques which would allow enrichment of NK cells. The discontinuous albumin gradient technique was found useful in this regard. Unfractionated and fractionated bone marrow cells of 6 AML patients in remission were tested for NK cell cytotoxicity. As can be seen in Table 4, NK cell activities in fractions 1 & 2 were 3 to 6 times higher than that of unfractionated bone marrow. In fraction 3, the NK cell activities were comparable to those of unfractionated bone marrow with the exception of donor 4 and 6 where the NK cell activities were doubled in comparison to unfractionated bone marrow. Very low NK cell activities were found in fraction 4 and virtually no NK cell activities in Fraction 5. Thus, discontinuous albumin gradient fractionation technique represents a promising tool for NK cells enrichment.

TABLE 4 NK Cell Activities of Unfractionated and Discontinuous Albumin Gradient Fractionated Bone Marrow Cells

Donor*	Unfractionated	Fractions			
		1 & 2	3	4	5
1	14	65	10	4	1
2	13	58	17	0	3
3	14	44	21	12	2
4	10	61	29	11	4
5	15	46	12	11	0
6	16	54	36	8	2

* Bone marrow cells from AML patients in remission were used in these studies.

CONCLUSIONS

We have demonstrated that normal individuals expressed natural killer cell-mediated cytotoxicity not only in peripheral blood but also in bone marrow. However, the efficiency of bone marrow NK cells was 50% of that observed in peripheral blood. CML and AML-diseased patients as well as AML patients in remission, expressed some, but significantly lower NK cell activities in peripheral blood than normal individuals. NK cells activities were also significantly decreased in bone marrow of AML-diseased patients, but were normal in AML patients in remission. The cryopreserved bone marrow NK cells expressed 50% less activities than fresh NK cells. Low NK cell activities expressed by bone marrow could be significantly increased by discontinuous albumin gradient separation technique. The highest NK cell activities were found in fractions 1 & 2 (3-6 times higher than in unfractionated bone marrow). NK cell activities in fraction 3 were, in general, comparable to those of unfractionated bone marrow. Fraction 4 showed low and fraction 5 virtually no NK cell activities. The lower NK cell activity in leukemic patients suggests involvement of NK cells in resistance to leukemias. Even though the underlaying mechanism of low NK cell activity in leukemic patients is unknown, in 4 out of 16 patients the involvement of suppressor cells was indicated. However, since in a majority of AML patients suppressor cells were not detected, another mechanism must have been operating.

Although the etiology and biological significance of NK cells remains obscure, there is a suggestive evidence that these cells may be involved in resistance and immunosurveillance to leukemias and perhaps other malignancies. Beside their important function in tumor immunity, NK cells may also play a role in bone marrow transplantation. These two possibilities strongly justify further research on NK cells.

REFERENCES

Cooper, S.M., D.J. Hirsen, and G.J. Frion (1977). Spontaneous cell-mediated cytotoxicity against Chang cells by non-adherent, non-thymus derived, Fc receptor bearing lymphocytes. Cell Immunol., 32, 135-145.

Dicke, K.A. (1971). Bone marrow transplantation after separation by discontinuous albumin density gradient centrifugation: Thesis. Radiobiological Institute TNO, Rijswijk, The Netherlands, pp., 73-80.

Dicke, K.A., E. Lotzová, G. Spitzer, and K.B. McCredie (1978). Immunobiology of bone marrow transplantation. Semin. Hematol., 15, 263-282.

Jondal, M., and H. Pross (1975). Surface markers on human B and T lymphocytes. VI. Cytotoxicity against cell lines as a functional marker for lymphocyte subpopulations. Int. J. Cancer, 15, 596-605.

Kaplan, J., and D.M. Callewaert (1978). Expression of human T-lymphocyte antigens by natural killer cells. J. Natl. Cancer Inst., 60, 961-964.

Kiessling, R., E. Klein, and H. Wigzell (1975). "Natural" killer cells in the mouse. I. Cytotoxic cells with specificity for mouse Moloney leukemic cells. Specificity and distribution according to genotype. Europ. J. Immunol., 5, 112-117.

Lotzová, E., and C.A. Savary (1977). Possible involvement of natural killer cells in bone marrow graft rejection. Biomedicine, 27, 341-344.

Lotzová, E., and K.B. McCredie (1978). Natural killer cells in mice and man and their possible biological significance. Cancer Immunol. Immunothera., In Press.

Peter, H.H., R.F. Eife, and J.R. Kalden (1976). Spontaneous cytotoxicity (SCMC) of normal human lymphocytes against a melanoma cell line: A phenomenon due to a lymphotoxin - like mediator. J. Immunol., 116, 342-348.

Pross, H.F., and M. Baines (1977). Spontaneous human lymphocyte-mediated cytotoxicity against tumor target cells. VI. A brief review. Cancer Immunol. Immunthera., 3, 75-85.

West, W.H., G.B. Cannon, H.D. Kay, G.D. Bonnard, and R.B. Herberman (1977). Natural cytotoxic reactivity of human lymphocytes against a myeloid cell line. Characterization of effector cells. J. Immunol., 118, 355-361.

Progress in Radiobiology
and Cancer

Progress in Radiobiology Related to Cancer
Introduction to Symposium No. 8

G. W. Barendsen* and L. Révész

*Radiobiological Institute of the Organisation for Health Research TNO,
Rijswijk, The Netherlands
**Institute for Tumorbiology, Karolinska Institute, 104 01 Stockholm, Sweden

Radiobiology is an interdisciplinary field of research that is related in several aspects to the area of cancer research. Firstly ionizing radiations are recognized as agents that can induce cancer. The analysis of dose-effect relationships for the induction of various types of cancer in animals and men can provide insights in mechanisms of cancer induction and development. These insights and results of epidemiological studies are essential for the development of radiation protection standards. Secondly effects of ionizing radiations are of interest to cancer research, because they are employed for tumour therapy. In this field progress during the past ten years has led to a better understanding of responses of tumours and normal tissues and this has stimulated and guided new developments in the treatments of different types of cancer, e.g., fast neutron therapy, combined chemo-radiotherapy, etc. Thirdly, methods, models and techniques developed for radiobiological studies involving responses of cultured cells, transplantable tumours and animal tissues, have led to their application in other areas of fundamental cancer research. Of special interest in this respect are developments of in vitro models and assessments of their relevance to tumour cell proliferation kinetic characteristics.

For the present symposium Dr. Révész and I in our preliminary discussions decided that we would not include tumour induction by radiation as a topic, but confine discussions to the latter two areas of research and their relevance to tumour radiotherapy. After consultations with Dr. Gaitan-Yanguas, chairman of the symposium on "Progress in Clinical Radiotherapy in Cancer", a selection was made to avoid overlap of the two programs. We felt that as a common theme we would like to concentrate on discussions of effects of fractionated irradiations as contrasted to single doses. Furthermore, emphasis has been given to effects of combined treatments, with radiation as one of the two components. It is evident that due to limitation of the overall time, many topics of considerable interest could not be included.

These considerations have led to the composition of the program as it is now completed. I sincerely hope that our decisions will prove to provide a stimulus to interesting discussions and a fruitfull symposium and ultimately to progress in radiobiology and in the treatment of cancer.

The Influence of Radiation Induced Recruitment of Resting Cells in Tumours into the Compartment of Proliferating Cells on the Effectiveness of Chemotherapeutic Agents

G. W. Barendsen

Radiobiological Institute of the Organisation for Health Research TNO, Rijswijk, The Netherlands

ABSTRACT

Data on responses to irradiation with respect to cell survival and cell proliferation kinetics of experimental tumours have been used to analyse the influence on tumour growth of various chemotherapeutic agents administered at different time intervals after irradiation. It is shown that changes in proliferation kinetics induced by irradiation of some types of tumours can render these tumours more responsive to subsequent chemotherapy.

INTRODUCTION

Treatments of tumours with ionizing radiation or chemotherapeutic agents must in principle be aimed at a complete impairment of the proliferative capacity of all tumour cells, while the critical normal tissues must retain their integrity and function. This objective cannot in general be attained by the administration of a single large dose of radiation or of a drug, because of the limitations imposed by the tolerance of the various normal tissues. In order to increase the therapeutic margin, treatments are commonly spread over a considerable period of time by multiple administrations of doses with intervals of one or more days or by protracted continuous administration, e.g., by irradiation at low dose rate or slow infusions of drugs. In such treatments given over periods of one or more weeks, the overall effectiveness of the doses given must be expected to depend on the proliferative activity of cells in tumours as well as in the limiting normal tissues, because during the intervals, cell production may compensate partly for the lethal effect of the preceding doses. In addition, the sensitivity of cells to subsequent doses in a series, may be modified by the damage induced by the earlier doses of a protracted treatment.
Evidently, one of the aims of investigations of cell proliferation kinetics in tumours and normal tissues after treatments should be to predict the optimal timing for administration of doses of radiation or drugs in order to yield the best therapeutic response.

Responses of an Experimental Tumour and its Cell Kinetics to Irradiation

In extensive studies of a transplantable rat rhabdomyosarcoma, denoted R-1, it has been shown that volume changes induced by irradiation are not a direct measure of the damage induced with respect to the proliferative capacity of the tumour cells, but that a complex series of events occurs (Barendsen, 1970; Hermens, 1969). In Fig. 1 data are presented obtained from treatments of R-1 tumours with a single dose of 2000 rad of 300 kV X-rays. This dose is not large enough to cause local control and the surviving cells cause a recurrent growth through proliferation. In general, tumour volume changes depend mainly on cell production due to proliferation and on cell loss. The rate of cell production is a function of the fraction of proliferating cells and the cell cycle time. Both of these parameters may vary after a dose of radiation. The importance of these factors can be illustrated with Fig. 1. Curves 1 and 2 in part A represent the growth curves for untreated tumours and tumours treated with 2000 rad of 300 kV X-rays. Curve 2 shows that after irradiation the volume continues to increase during two days after irradiation, with a subsequent decrease. The minimum relative volume of 0.7 is attained at six to eight days after irradiation, but this decrease is relatively small and the pre-irradiation volume is attained again quite rapidly. This is in apparent contrast with the much larger effect at the cellular level as measured from experiments on the clonogenic capacity.

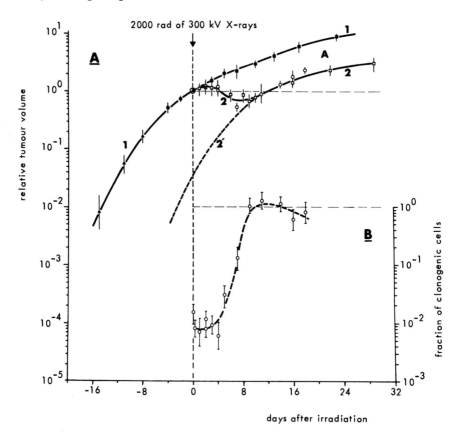

Fig. 1. Growth curves of R-1 rhabdomyosarcoma and the recurrence after a dose of 2000 rad of 300 kV X-rays. Part A, curve 1: growth curve of control tumours. The relative tumour volume of 1.0 corresponds to an average tumour volume of 1.5 cm^3; part A, curve 2: growth curve of tumours irradiated at t = 0; part A, curve 2': hypothetical curve extrapolating the growth of the recurrent tumour parallel to the growth of an untreated tumour; part B: variation of the fraction of clonogenic cells as a function of the time after irradiation.

Figure 1-B shows that immediately after a dose of 2000 rad of X-rays the fraction of clonogenic cells was reduced to 0.7 per cent relative to unirradiated tumours. This fraction remains approximately constant for three days, but starting at day four, a rapid increase is observed, and at day 10 a value is attained equal to that in unirradiated tumours of the same volume. This rapid rise between day 4 and day 10 is characterized by a mean doubling time of approximately 18 hours. This doubling time of the clonogenic fraction is much shorter than the doubling time of the tumour volume during the period after resumption of growth. As analyzed in detail by Hermens (1969), the cell cycle time in this tumour is shortened from 20 hours before irradiation to 12 hours at day four up to day 14 after irradiation with a dose of 2000 rad of X-rays; furthermore, the growth fraction first decreased and subsequently increased again during this period, while the cell loss factor first increased greatly and subsequently decreased again. Cell production during the first eight days after irradiation is due mainly to proliferation of cells which are not capable of producing an unlimited number of descendants and the clonogenic cells start to determine cell production only after growth of the tumour is resumed at day eight. It will be clear that the changes of tumour volume after irradiation depend in a complex way on cell proliferation parameters. In some other tumours analyzed with respect to cell kinetics after irradiation, shortening of the cell cycle has not been observed (Denekamp, 1970). From data on volume changes after irradiation of pulmonary metastases in patients with single relatively small doses, a short period of accelerated growth has also been observed (van Peperzeel, 1972). The main conclusion which can be drawn from these data is that volume changes of tumours after irradiation may be much less significant than changes in the basic cellular proliferation kinetics. Volume changes do not correlate directly with cell survival but depend on a complex pattern of cell production and cell loss variations.

During and after a course of fractionated radiotherapy it is likely that the cell kinetics may vary in a similarly complex way as observed for responses after a single treatment, but little experimental data are available to demonstrate these phenomena quantitatively.

Responses of Experimental Tumours to Combined Radio- and Chemotherapy

On the basis of these data on cell kinetics of irradiated tumours, showing that several days after treatment the growth fraction may increase and that the resting cells may be induced to start progressing through a new proliferative cycle, experiments were started to investigate the response of these tumours to a cell cycle phase specific drug, vinblastine. In Fig. 2 data are presented from an experiment in which volume changes of R-1 tumours were measured after treatments with different doses of radiation, after treatment with a dose of 1.5 mg of vinblastine per kilogram of rat body weight, and after combined treatments with 2000 rad of 300 kV of X-rays 48 hours before or 8 hours after the administration of vinblastine. For each combination, five animals per group, with one tumour per animal, were treated identically. The dose of 2000 rad of X-rays caused a growth delay of about 15 days. The combined treatment with vinblastine given 8 hours before irradiation produced within experimental variability the same effect as irradiation alone, but vinblastine given 48 hours after irradiation produced a delay of about 25 days. This value of 25 days is significantly larger than the value of 15 days obtained for a dose of 2000 rad of X-rays alone.

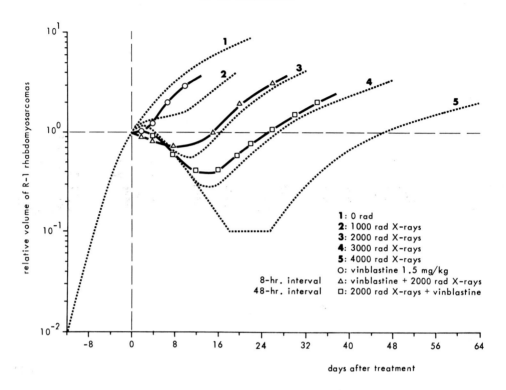

Fig. 2. Volume changes of transplantable rat rhabdomyosarcomas growing in flanks of WAG/Rij rats, after treatment with different doses of 300 kV X-rays and after combined treatments with 2000 rad of X-rays and 1.5 mg vinblastine per kilogram of rat body weight, administered 8 hours prior to or 48 hours after irradiation.

In a subsequent series of experiments the influence of the time interval between irradiation and administration of vinblastine on growth delay of the R-1 tumours was studied more extensively. The results of these experiments are summarized in Fig. 3. Values have been derived from the experimental data for <u>excess growth delay</u>, defined as the difference between the tumour growth delay obtained in the combined treatment and the sum of the delays of the individual treatments with 2000 rad of X-rays or 1.5 mg of vinblastine per kilogram of body weight. These values are presented as a function of the time interval in Fig. 3. The data yield a mean excess delay of 7.5 days for intervals between 48 and 288 hours.
Results of similar experiments with a dose of 1000 rad of X-rays followed by a dose of 1.5 mg of vinblastine per kilogram of rat body weight yielded an average excess growth delay of 8.0 days for intervals between 48 and 192 hours.
These results indicate that the effectiveness of administration of this dose of vinblastine is considerably enhanced for R-1 tumours which have been treated with a dose of at least 1000 rad of X-rays. Since this R-1 sarcoma is known to contain 60% of non-proliferating Q-cells, which are much more resistant to vinblastine than P-cells, only a small effect of vinblastine on R-1 sarcoma growth was expected and indeed observed experimentally. Furthermore, investigations of cell kinetics after irradiation have shown that Q-cells are recruited into the compartment of P-cells after a dose of 2000 rad of X-rays and cause a rapid repopulation of proliferating clonogenic cells. The enhanced effectiveness of vinblastine on growth delay of R-1 tumours after irradiation may therefore be ascribed to altered proliferation kinetics, involving recruitment of Q-cells. Similar studies with other types of transplanted tumours have

shown, however, that this is not a general phenomenon but that it might only occur in sarcomas and, at least not to the same extent, in carcinomas (Barendsen, 1978).

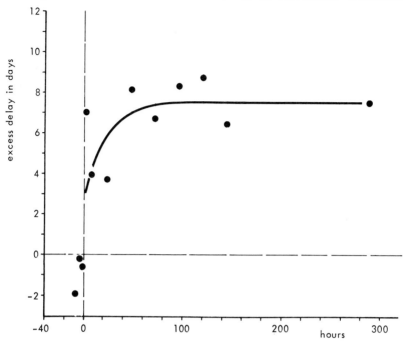

Fig. 3. Excess growth delay after combined treatments of R-1 tumours with 2000 rad of 300 kV X-rays and 1.5 mg of vinblastine per kg rat body weight as a function of the time interval between irradiation and vinblastine administration. Excess growth delay was derived as the difference between observed growth delay for combined treatments and the sum of growth delays of individual treatments with X-rays and vinblastine respectively.

REFERENCES

Barendsen, G. W. (1970). Experimental radiotherapy of a rat rhabdomyosarcoma with 15 MeV neutrons and 300 kV X-rays. I. Effects of single exposures. Europ. Jour. Cancer., 6, 89-109.
Barendsen, G. W., and H. C. Janse (1978). Differences in effectiveness of combined treatments with ionizing radiation and vinblastine, evaluated for experimental sarcomas and squamous cell carcinomas in rats. Int. J. Radiation Oncology Biol. Phys., 4, 95-102.
Denekamp, J. (1970). The cellular proliferation kinetics of animal tumours. Cancer Res., 30, 393-400.
Hermens, A. F., and G. W. Barendsen (1969). Changes of cell proliferation characteristics in a rat rhabdomyosarcoma before and after X-irradiation. Europ. Jour. Cancer, 5, 173-189.
Peperzeel, H.A. van (1972). Effects of single doses of radiation on lung metastases in man and experimental animals. Europ. Jour. Cancer, 8, 665-675.

The Relevance of Radiobiological Data on Mammalian Cells in Culture for the Prediction of Tumor Responses to Fractionated Radiation

Herman D. Suit and Leo E. Gerweck

Edwin L. Steele Laboratory of Radiation Biology,
Department of Radiation Medicine,
Massachusetts General Hospital and Harvard Medical School, Boston,
Massachusetts 02114, U.S.A.

ABSTRACT

The radiobiological characteristics of mammalian cells as determined from studies of cells in vitro have been of value in predicting response of tissues. The value has been in permitting general principles to be employed in devising strategies of treatment rather than direct use of values of D_0, n, D_q, etc. Namely, designations have been made of situations where low dose rate, reduced overall treatment time, large dose per fraction, etc. would likely be of greater efficacy than conventional schedules. There are many remaining difficulties with the use of in vitro data, especially the problem of low plating efficiency. Nonetheless, we wish to see more in vitro data and with emphasis on cells from resistant tumors, shoulder width, and PLDR.

INTRODUCTION

Dose fractionation has been a subject of serious interest to radiation therapy for virtually the entire history of the specialty. This has been due to the clear and great need for treatment strategies which would yield a greater differential radiation effect on tumor and normal tissues. Until recently the main radiobiologic fact which was employed in planning diverse fractionation schema was that cells in mitosis were more readily killed than interphase cells. The various schedules employed included very short or very long overall treatment time, low dose rate, small dose per fraction, increasing or decreasing dose size during the treatment course, split course, etc. Despite the sustained effort at unusual fractionation plans over several decades, virtually all radical radiation therapy has been based, at least in the United States, on 5 treatments of $\simeq$180-200 rad per week. For the past 10-15 years there has been a much expanded interest in clinical as well as laboratory testing of new fractionation schedules because of the important expansion of our understanding of radiobiologic and cell kinetic parameters of tumor and normal tissues.

In simple terms the new knowledge may be described as follows: 1) cells of diverse origin vary with respect to values for D_0 and n, e.g. n is $\simeq$1 for cells of germinal lymphoid, and bone marrow origin but may be quite high, e.g. >6-10 for cells of intestinal mucosa, melanoma, glioblastoma and that D_0 is $\simeq$150 rad for most cell lines but may be $\leq$100 rad for marrow or lymphoid cells or cells from patients with certain diseases, e.g. ataxia; 2) cell sensitivity varies with position of the cell in the cell replication cycle, last S being most resistant and G2-M and the G1-S interphase being the most sensitive; 3) cell proliferative activities for most normal tissues have been described in quantitative terms and they exhibit a great diversity with respect to cell cycle times, age density distributions, proliferative response to injury, etc. (for some tissues, the cells enter an intense proliferative phase after injury is registered (skin mucosa, intestinal mucosa) while there is essentially no proliferative response in other tissues (capillary endothelium, muscle, brain)); 4) cell proliferative activities in tumor tissues have also been investigated and they exhibit a marked variation, with respect to distribution of cell cycle times, growth fraction and cell loss factor; the values for these parameters change during the course of radiation. (Hall, 1973; Suit, 1973).

These factors may be utilized with success in modifying the treatment protocol in order to increase the response to tumor and/or diminish the response of normal tissue. That is, these considerations are different from the major efforts made by Ellis (1968), Strandquist (1944) and others to develop formulae which will allow the radiation therapist to achieve the same response in normal tissue, viz. a close approach to "tolerance", when a different fractionation schedule is employed. Specifically, the present interest is based upon maximizing the impact of differences in D_0, n, and proliferation in designing fractionation protocols in order to effect a therapeutic gain. That is, with reference to exponents for fractionation and total time, utilization will be made of the different values for these parameters for the various tissues concerned.

Radiation Sensitivity of Normal and Tumor Tissue Cells

The first quantitative radiobiological studies of human tissue cells by Puck and associates (1956) indicated that the radiation survival curves for fibroblastic cells from several human tissue systems were similar and that all were sensitive to radiation. Normal "epithelial like" cells from several tissues were also about equally sensitive to radiation. These investigators also showed that the radiation survival curve for HeLa cells (a human cervical carcinoma cell line) was not substantially different from the normal epithelial cells. In contrast, studies by Barranco (1971) and others (for example, Malaise et al, 1976) demonstrated that cells from some melanomas were relatively resistant to radiation, especially at low radiation doses, i.e. 50-400 rad. This resistance was due to the very broad shoulder of the radiation survival curve.

Weischelbaum (1976, 1977a) determined values for parameters of radiosensitivity for cell lines derived from a human osteosarcoma and a glioblastoma to be D_0 = 144 rad and n = 1.9 and D_0 = 143 rad and n = 1.4 respectively. Both of these tumors are regarded clinically as being quite radiation resistant. They demonstrated a very large repair of PLD for the osteosarcoma cell line. The same workers have reported similar radiosensitivity for cells derived from a medulloblastoma, a tumor often destroyed by radiation. The low D_0 and n values for all 3 lines are not consistent with clinical experience.

Further studies on the radiosensitivity of human glioblastoma cells were performed by Gerweck et al (1977). They studied the long term cultured glioblastoma cell line previously examined by Weischelbaum et al and two additional glioblastoma lines after short term in vitro passage. Two of the 3 glial cell lines, A3 and A7, (plating efficiencies of 6 and 30%) are considerably more radioresistant than most normal and malignant cells lines studied in vitro. These differences are most apparent in the dose range of 100 to 500 rad, i.e. the region determined by the shoulder of the survival curve. The broad shoulder and hence small response to doses of the order of 200 rad is consistent with the very poor response to conventional radiation therapy.

The radiosensitivity of a variety of explanted human carcinoma cells was examined by Wells et al (1977) after 1 to 19 in vitro passages. Cell lines from carcinomas of the bronchus, breast, endometrium, ovary, stomach and melanoma of the skin were all relatively radioresistant in that the extrapolation numbers ranged from 10 to 120 while D_0 values were standard, e.g. 120 to 160 rad. Cells derived from a seminoma were more sensitive with an extrapolation number of 2.3. These authors found no significant changes in radiosensitivity during the first 10-15 in vitro subcultures.

In summary, cells lines derived from various human tissues appear to differ markedly in their radiosensitivity, particularly the shoulder width of the survival curve. However, because of the variability in the experimental conditions employed in the culture procedures, plating efficiencies, time for colony formation, it is difficult to categorize cell sensitivity from different tissue systems. Interlaboratory comparisons would be facilitated by employment of a standard cell line. Despite all of these reservations, this type of study should in our opinion be refined and extended. Potentially lethal damage repair may also vary considerably for certain human cell lines. As in vitro studies are in vitro, results of such investigation can at best provide guides as to general principles and perhaps estimates of relative sensitivities, repair advantages, etc. They are not expected to provide quantitative estimates of the values for the various parameters which can be applied directly at the clinical level.

Genetic Disorders and Radiosensitivity

Enhanced cellular radiosensitivity has been shown to be associated with certain genetic disorders (Setlow, 1978). In 1975 Taylor and co-workers demonstrated that fibroblasts derived from patients with ataxia teleangectasia were abnormally sensitive to ionizing radiation. The laboratory and clinical observations on patients with this disease correlate well in that patients exhibit an abnormal sensitivity to radiation. Weischelbaum (1977b) has shown that fibroblasts from patients with retinoblastoma with the D type deletion are sensitive to X-irradiation. Studies of cells of these types should increase our knowledge of radiation molecular damage and repair and its relation to cancer induction. Also data from such investigation may influence the radiation treatment of patients who have certain pre-existing diseases.

Low Dose Rate Irradiation

Thomas et al (1977) employed 4-8 rad min^{-1} in whole body irradiation of patients with leukemia and other hematologic disorders as treatment and in preparation for marrow allograft. Previous estimates of the LD50 for man were in the range of 250 rad given as a single dose at a high dose rate with the cause of death being bone marrow suppression; even when employed to prepare patients for marrow transplant much higher doses (at high dose rate) could not be tolerated because of severe nausea and vomiting. Thomas used the low dose rate of 4 rad min^{-1} because of technical requirements and found that the dose level could be increased to 1000 rad due to a much lesser reaction of the GI system. There were three advantages to the low dose rate and markedly higher total dose: 1) as n values for marrow and lymphoid cell systems were ≃1 the reduced dose rate would have only a slight effect to reduce cell kill efficiency of the radiation; therefore the dose of 1000 rad should yield an increased likelihood of graft acceptance; 2) a much greater effect against the leukemic cells and; 3) because n values for cells of intestinal mucosa were high the patients would be able to tolerate the increased dose with acceptable levels of gastrointestinal disturbance. Successful marrow transplants have been made using the low dose rate whole body irradiation in combination with intensive chemotherapy. Radiobiological knowledge of the marrow, lymphoid, and intestinal tissues provides a rational explanation for the improved clinical effectiveness of the low dose rate regimen. Significantly, this clinical fact of differential response can be useful in devising other protocols. For example low dose rate treatment of abdominal lymphoma.

There is an additional factor which pertains to the use of low dose radiation: OER is probably less for low dose rate than for high, viz ≈2.0 vs 2.5-3.0 (Hall, 1973). Further, at the low dose rate there would be increased opportunity to irradiate cells in the G2-M phase. According to this consideration, low dose rate treatment would be predicted to be of greater effect than standard treatment when employed against a tumor cell population characterized by small or no shoulder, active proliferation and a hypoxic fraction where that tumor was growing in a normal tissue whose constituent cells have a large capacity for repair of radiation damage and are non-proliferating. Such conditions might obtain for certain lymphomas or seminoma growing in the lung, brain, etc. We do not know of trials for such situations. Low dose rate radiation therapy is now being investigated in several clinics following the work of Pierquin (1976). Most patients being treated in those studies are squamous cell carcinomas of the head and neck region and adenocarcinomas of the abdominal region (pancreas, stomach). Presumably there would be no differential repair or proliferative activity which would provide a basis for improved results and the reduced OER would be the principal factor which could contribute to a clinical gain. An additional factor would be that these treatments have been given over a reduced total time which might be advantageous, *vide infra*.

From radiobiological considerations, the efficacy of radiation administered continuously at low dose rate (40-100 rad hr) would be expected to be equivalent to that given in a series of small doses (e.g. 50-100 rad per fraction) at a high dose rate but over the same total time period. The latter would have major technical advantages, viz feasibility of precise patient immobilization, use of complex field arrangement, reduced strain on patient, and increased patient load per treatment unit.

The excellent therapeutic results associated with conventional interstitial therapy are attributed to be the consequence of the near idealized dose distribution. The low dose rate used in such treatment would probably be a small factor in the results achieved. This opinion is supported by the reported results obtained in the treatment of cancer of the uterine cervix by high dose rate - remote control afterloading applicator systems; the survival data are apparently comparable to those from conventional low dose rate intracavitary techniques (unpublished data of Joslin, 1978).

High Dose Per Fraction

The tumor-normal tissue situation where high dose per fraction would appear to be a superior treatment protocol would be essentially the converse of the one described for low dose rate treatment. Namely, tumor cell population would be characterized by a broad shoulder (high n value) growing in normal tissues whose cells had either a normal or small shoulder. Malignant melanoma is a human tumor whose cells have a very broad shoulder to their survival curve, at least for some of the tumors. As such, standard doses per fraction would be of much lesser effect that quite large ones. Indeed, Habermalz et al (1976) have reported that complete regression of melanoma was not observed at 200 rad per fraction but at 600-1000 rad per fraction a very high proportion of lesions did in fact regress. In our own hospital, some excellent responses have been achieved at 600-800 rad per fraction but not in every case. This appears to represent another clinical problem which is more effectively managed by use of a non-standard fractionation schedule.

Reduced Overall Time of Treatment

For rapidly growing tumors surrounded by normal tissues which are slowly or non-proliferative, there may be an important advantage to the tumor when the standard fractionation schedule of 25-35 daily treatments given over 5-7 weeks is employed. The conventional treatment pattern of ≃200 rads on a Monday-Friday basis with no treatments on Saturday and Sunday means intertreatment intervals (t_i) of 24 hours and 72 hours. During each such period the surviving cells repair radiation damage; further, cells which are in active proliferation would after a dose dependent delay continue to progress through the cell replication cycle and a proportion would undergo division before the next irradiation. For some normal tissues there would be no or negligible proliferation during the total time period of 5-7 weeks e.g. brain, muscle, vascular, endothelium, etc. In contrast, for high grade tumors (many mitotic figures seen) there would almost certainly be some proliferation during the intertreatment intervals. Accordingly for those situations where the tumor cells are actively dividing and the normal tissues are nonproliferative there should be a clear advantage to the tumor by the use of the long (24-72 hour) intertreatment intervals. Radiation damage (sublethal and potentially lethal) is essentially repaired within 3-4 hours, repair half time being of the order of 20-60 minutes for mammalian cells. These repair rates pertain to cells of both normal and malignant tissue (Hall, 1973). Split dose recovery curves for CHO cells irradiated in exponential growth phase show a rapid increase in survival fraction as intertreatment interval (t_i) is increased up to 2-4 hours; then survival fraction exhibits some variation due to changing age density distribution. At later times survival fraction increases regularly as intertreatment interval is prolonged beyond ≃8 hours because of cell proliferation during the intertreatment interval. For similar cells maintained at reduced temperature or reduced pO_2 during the intertreatment interval or for cells in the plateau phase, the split dose recovery curve shows the same initial rapid increase, but after reaching the maximum within 4 hours it remains flat. This also obtains for nonproliferative tissue systems in vivo. This can be illustrated by reference to the recovery curve for periosteal cell of the mouse femur (Hayashi et al, 1971). The endpoint employed was inhibition of callus formation by radiation given prior to fracture; this is described as the CID50 or the radiation dose which inhibits callus formation in the mouse femur by a specified amount in half of the treated subjects. In that experiment the CID50 was 1750 rad for single dose irradiation but for radiation given in two equal doses separated by 4 hours was 2180 rad. Relevant to the consideration here is that the CID50 increased with increasing t_i only slightly, viz the slope of the line connecting the CID50 values for t_i of 4 hours to 28 days was 0.02.

According to this line of reasoning, reduction of t_i from the standard 24-72 hours to some 4-8 hours would result in only a slight increase in the response of a non-proliferating system for a constant total dose given in the same number of equal dose fractions. In contrast for an actively dividing cell system, such shortening of the overall time would be expected to result in a marked increase in the level of cell kill because there would be no or much reduced increase in cell number during the intertreatment interval. A further attractive feature is that the sensitivity of the tumor cells would be expected to be greater at some 8 hours after irradiation because of shifting age distribution. Thus, appropriately selected intertreatment intervals might yield an advantage in addition to that obtained by overcoming the effect of proliferation. Accordingly, these simple radiobiological considerations permit one to predict for rapidly dividing tumor cell systems growing in non-proliferative or very slowly proliferating normal tissues that reduction only of the intertreatment interval (dose per fraction held constant) diminish or eliminate the advantage which accrues to the tumor with standard fractionation, viz a therapeutic gain. In actual practice there will be a need to reduce dose per fraction and total dose because there is some low level of cell proliferation and repair of radiation damage during the period of 4-24 hours. In a study of the response of normal human skin to radiation given in one or two fractions per day we found that in order to maintain the same gross response the total dose should be reduced by about 7% (Choi et al, 1977).

Clinical Experience with Two or Three Fractions per Day.

Evidence that such strategies for appropriately selected tumor tissue systems will achieve better results is evident by the report of Noran and Unyanga (1976). Their actual data showed only one complete regression of 9 lesions when treated with daily fraction of 220 rad for a total dose of 3100 rad in 20 days. In the group treated with mean doses of 2900 rad given in 27 fractions of 120 rad in 13 days 25-34 lesions regressed completely. We have employed an accelerated treatment for high grade sarcoma of soft tissue of the extremity in 7 patients in the past 9 years. Treatment was given as 180-200 rad twice daily with ≥ 4 hours between fractions. Of these 7 patients, two have had local failure. In 6 of the 7 patients the tumors were truly massive. There may be a positive clinical advantage for utilization of such accelerated treatment in a number of clinical situations. These might include the anaplastic sarcomas of soft tissue, rapidly growing lymphomas at diverse anatomic sites but specifically excluding abdominal area, Ewing's sarcoma, Oat Cell carcinoma and highly anaplastic tumors with high mitotic indices at a variety of sites. To assess the efficacy of this approach would need results from a clinical trial where patients were randomly assigned to treatment by conventional fractionation schedule or an accelerated schedule. Treatments have been given in multiple small dose fractions per day in several clinics. Reports indicate that higher local control rates are being achieved at acceptable normal tissue reactions. For example, two groups have reported experience with two fractions per day for head and neck cancer. Schukovsky et al (1976) and Jampolis et al (1976) have used 110-120 rads per fraction and two treatments per day; each group was impressed with their results. Littbrand et al (1975) have used 3 fractions of 100 rad per day to a total dose of 8400 rad given in 8.4 weeks (a two week rest period in the midportion of the treatment) in the treatment of bladder carcinoma. They are likewise impressed.* If these schedules are ultimately proven to be of greater clinical efficacy than conventional treatment, the advantage might be due to a lesser OER, larger number of fractions meaning more opportunities to irradiate G2-M cells, or perhaps to small differences in slope of the initial region of the cell survival curve which would favor the critical normal tissue cell.

Greatly Protracted Treatment

For tumors which are judged to be extremely slowly growing and histopathologically be characterized by few or no mitotic figures and growing in normal tissue situation where there would be some proliferative activity, there would be expected a clinical advantage by protracting the treatment. This would be true provided the tumor did not undergo a proliferative response to the radiation damage.

The idea in this treatment would be to give approximately the same dose per fraction but have the fractions spread over a longer time period. The expected increment in dose to achieve the same level of effect should be modest only to allow for some of the slow "repair" and a small amount of proliferation. It is well established in clinical radiation therapy that some normal tissues do make a proliferative response; for example, skin, mucous membrane, etc. and that there will be partial reconstitution of the irradiated normal tissues during the course of radiation at least for the tissues mentioned. This is commonly seen in the treatment of carcinoma of the breast by the "Baclesse" technique (Fletcher, 1972). It was also observed quite dramatically in the instance of the treatment of head and neck cancer over a 14 week period (Andrews et al, 1965). In the latter cases the total dose was increased to very high levels (10000r) which probably accounts for the morbidity encountered in that group of patients. Nonetheless, the acute mucous membrane reactions were extremely slight.

This work was supported by DHEW Grants #CA13311 and CA22860

REFERENCES

Andrews, J.R. (1965). Dose-time relationships in cancer radiotherapy. Amer. J. Roentg., Rad.Ther., and Nuc. Med. 93:56-74.
Barranco, S.C., M.M. Romsdahl, and R.M. Humphrey (1971). The radiation response of human malignant melanoma cells grown in vitro. Cancer Research 31: 830-833.
Choi, C.H. and H.D. Suit (1975). Evaluation of rapid radiation treatment schedules utilizing two treatment sessions per day. Radiology 116:703-707.
Ellis, F. (1968). Time, fractionation and dose rate in radiotherapy. In J.M. Vaeth (Ed.) Frontiers of Radiation Therapy and Oncology. S. Karger, Basel/New York. pp. 131-140.
Fletcher, G.H. (1973). Management of localized breast cancer. In G.H. Fletcher (Ed.) Textbook of Radiotherapy. Lea & Febiger, Philadelphia. pp.457-492.
Gerweck, L.E., P.L. Kornblith, P. Burlett, J. Wang, and S. Sweigert (1977). Radiation sensitivity of cultured human glioblastoma cells. Radiology 125:231-234.
Habermalz, H.J. and J.J. Fischer (1976). Radiation therapy of malignant melanoma. Cancer 38:2258-2262.
Hall, E.J. (1973). Radiobiology for the Radiologist. Harper and Row, Inc., Hagerstown, Maryland.
Hayashi, S. and H.D. Suit (1971). Effect of fractionation of radiation dose on callus formation at site fracture. Radiology 101:181-186.
Jampolis, S., G. Pipard, J.C. Horiot, M. Bolla and C. LeDorze (1977). Preliminary results using twice-a-day fractionation in the radiotherapeutic management of advanced cancers of the head and neck. Amer.J. Roentg, Rad. Ther. and Nucl. Med. 129:1091-1093.
Joslin, C.A. (1978). Unpublished data.
Littbrand, B. and F. Edsmyr (1976). Preliminary results of bladder carcinoma irradiated with low individual doses and a high total dose. Int. J. Rad. Onc., Biol.,Phys. 1:1059-1062.
Malaise, E.P., J. Weininger, A-M Joly and M. Guichard: Measurements in vitro with three cell lines derived from melanomas. In T. Alper (Ed.)Cell Survival after Low Doses of Radiation. J. Wiley & Sons, Great Britain, pp.223-225.
Norin, T. and J. Onyango (1977). Radiotherapy in Burkitt's lymphoma. Int. J. Rad. Oncol., Biol., Phys. 2:399-406.
Pierquin, B., F. Baillet, W. Mueller, and F. Wilson (1976). Irradiation semi-continue a baible deluit par telecobalt. J. Radiol. Electr. Med. Nucl. 57:841-843.
Puck, T.T. and P.I. Marcus (1956). Action of x-rays on mammalian cells. J. Exp. Med. 103:653.
Setlow, R.B. (1978). Repair deficient human disorders and cancer. Nature 271: 713.
Shukovsky, L.J., G.H. Fletcher, E.D. Montague and H.R. Withers (1976). Experience with twice-a-day fractionation in clinical radiotherapy. Amer. J. Roentg. 126:155-162.

Strandquist, M. (1944). Studien uber die kumulative Wirkung der rontgenstrahlen bie Fraktionierung. Acta Radiol. Suppl. 55:1.

Suit, H.D. (1973). Radiation Biology: A basis for radiotherapy. In G.H. Fletcher (Ed.) Textbook of Radiotherapy. Lea & Febiger, Philadelphia. pp.75-120.

Taylor, A.M.R., D.G. Harnden, C.F. Arlett, S.A. Harcourt, A.R. Lehmann, S. Stevens and B.A. Bridges (1975). Ataxia telangiectasia: a human mutation with abnormal sensitivity. Nature 258:427-429.

Thomas, E.D., C.D. Bruckner, Et al (1977). One hundred patients with acute leukemia treated by chemotherapy, total body irradiation and allogenic bone marrow transplantation. Blood 49:511-533.

Weichselbaum, R.R., J. Epstein, J.B. Little and P.L. Kornblith (1976). In vitro cellular radiosensitivity of human malignant tumors. Europ. J. Cancer 12:47-51.

Weichselbaum, R.R., J.B. Little and J. Nove (1977). Response of human osteosarcoma in vitro to irradiation: Evidence for unusual cellular repair activity. Int. J. Radiation. Biol. 31:295-299.

Weichselbaum, R.R., J. Nove and J.B. Little (1977). Skin fibroblasts from a D-deletion type retinoblastoma patient are abnormally X-ray sensitive. Nature 266:726-727.

Wells, J., R.J. Berry and A.H. Laing (1977). Reproductive survival of explanted human tumor cells after exposure to nitrogen mustard or X-irradiation; Differences in response with subsequent subculture in vitro. Radiation Research 69:90-98.

*As reported in this Congress by Edsmyr et al, the results of their study do not in fact demonstrate an advantage for the 100 rad per fraction group in terms of local control, survival, or complications.

Relevance of Radiobiological Data on Mammalian Cells in Culture for the Prediction of Tumor Responses to Fractionated Irradiation: Effects of Intercellular Contact

Ralph E. Durand

Radiobiology Section of The Johns Hopkins Oncology Center and Department of Environmental Health Sciences, The Johns Hopkins University, Baltimore, Maryland 21205, U.S.A.

ABSTRACT

Radiobiological data on mammalian cells grown in tissue culture often are used as a basis for the prediction of the radiation response of normal and neoplastic tissues *in vivo*, but such extrapolations are clearly subject to numerous uncertainties. In addition to the obvious differences of cell cycle kinetics and cell cycle distributions, more subtle factors including the influence of intercellular contact on repair processes may obtain *in vivo*. Intercellular contact influences the accumulation and repair of sublethal radiation damage *in vitro* for several strains of mammalian cells, and such effects seem likely to be of even greater importance *in vivo* with the multifraction irradiation schemes generally used in cancer treatment. Using Chinese hamster cells grown *in vitro* as single-cell monolayers or as small multicell "spheroids", we have attempted to evaluate the influence of intercellular contact on the accumulation and repair of sublethal damage in the multifraction situation. Our data suggest that the multifraction response of single cells in culture may provide a qualitative, but not quantitative model for the cellular response in organized tissues *in situ*.

Keywords

Intercellular contact, mammalian cells, multifraction irradiation, repair, spheroids, time-dose effects.

INTRODUCTION

As Dr. Suit has just stressed in his review (1978), mammalian cells in tissue culture have many differences from their counterparts growing as a tumor *in vivo*. Perhaps the most evident of these differences results from the usual method of assaying radiation response *in vitro*, the "survival curve" based on proliferative potential, where every cell is judged for its ability to replicate as a function of radiation dose. This endpoint itself demands that all cells of the population must initially be capable of unlimited proliferation, a situation which may not occur in the tumor. Further, due to many factors including convenience to the investigator, cell lines utilized *in vitro* often have a growth rate that is much more rapid than that of even the fastest-growing experimental tumors (Steel, 1977).

Consequently, it is clear that radiobiological data obtained with mammalian cells in culture should not necessarily be expected to provide insight into the clinical response of a tumor to many fractions of radiation delivered over a relatively long treatment time. Nonetheless, mammalian cells in culture are certainly one of the most convenient systems available, and in selected situations, may adequately model the responses of more complex systems.

In addition to their growth in conventional monolayer cultures, certain mammalian cells will also spontaneously grow as multicell clusters, or "spheroids", in liquid suspension cultures (Sutherland and Durand, 1976). With the Chinese hamster V79 lung cell system, the cells initially aggregate into small clusters, then grow (by cell division) to sizes exceeding 1 mm in diameter, containing $> 10^7$ cells (Fig. 1). Largely as a result of the geometry of this growth situation, the nutrient supply to the internal cells of the spheroid is dependent upon diffusion processes, resulting in gradual development of central necrosis.

Fig. 1. Central sections from V79 Chinese hamster cell spheroids grown 5, 7, 11, 15 or 19 days respectively. Insert is 0.5 mm full scale.

Additionally, our previous work has demonstrated a gradual accumulation of a population of G_1-like non-cycling cells, of which some even become hypoxic as the spheroid increases in size (Sutherland and Durand, 1976). The radiation response of the spheroid is predictably altered by this induced hypoxia, but, in addition, alterations in radiation response are seen even in very small spheroids containing no hypoxic cells. These alterations have been characterized in some detail (Durand and Sutherland, 1972; Durand and Sutherland, 1973; Durand and Sutherland, 1975; Durand and Olive, 1978), and have been attributed to an as yet unidentified process which is, however, closely linked to intercellular contact.

As might be expected, multifraction irradiation of large spheroids results in reoxygenation and redistribution of cells (Durand and Sutherland, 1976) from the hypoxic and non-cycling compartments--a situation similar to that observed in tumors. These effects are dramatic, particularly at large doses per fraction, but in the more clinically relevant situation of numerous small dose fractions, these processes are dominated by repair and repopulation. As with most systems $in\ vitro$, the spheroid is characterized by a very rapid potential growth rate, since the generation time of the cycling cells is of the order of 16 hours (Durand, 1976). Consequently, in the experiments reported here, we have attempted to determine the contribution of intercellular contact alone to the repair and recovery of spheroids exposed to multifraction irradiation, within an experimental framework kept relatively short in respect to the potential growth rate. To specifically address the question of the role of intercellular contact, we used small spheroids (one-day old clusters of 6-20 cells per spheroid), irradiated at a net dose-rate of approximately 100 rads/minute. This dose-rate was dictated by previous observations by ourselves and others that cell proliferation continues for dose-rates of less than ~ 50 rads/minute, and by exceeding this rate, we hoped to simplify interpretation of our results.

MATERIALS AND METHODS

Spheroids were grown one day, using the Chinese hamster V79-171 cell line. The growth medium used was Eagles minimal essential medium (MEM) supplemented with 5% fetal calf serum and 2 mM glutamine; all medium components were purchased from Grand Island Biological Company (GIBCO). Additional details regarding spheroid growth and assay techniques have been reported previously (Sutherland and Durand, 1976). Spheroids were irradiated in a water-jacketed growth flask at a distance of 80 cm from the target of a 4 MEV Varian linear accelerator, at a dose-rate of ~ 300 rads/minute. Single cells were irradiated as monolayers in Falcon plastic petri dishes, using an AECL dual source "Gammacell 40" cesium irradiator, with a dose-rate of ~ 150 rads/minute.

As samples were required for analysis, the spheroids or single cells were trypsinized to produce a single cell suspension, and an appropriate number of cells plated for colony formation.

RESULTS

Chinese hamster V79 cells grown as spheroids, but reduced to single cells for survival assay, show a markedly enhanced survival after irradiation (Fig. 2) due to an increased shoulder on the single-dose radiation survival curve. The different symbols shown in Fig. 2a represent different experiments, including the control populations of monolayer cells and cells from spheroids from the experiments included in subsequent figures. Note that the survival modification produced by the intercellular contact is fairly substantial; at high radiation doses, a survival increase by factor of ~ 10 can be seen. This increased ability of cells grown as spheroids to accumulate sublethal damage is also reflected by an increased repair of sublethal damage in the usual type of split-dose recovery analysis (Fig. 2b). Here, the two doses of radiation were delivered with the indicated time interval between them; the cells were held at room temperature (to inhibit cell progression) between the dose fractions. Radiation doses were chosen to give equal surviving fractions of cells for monolayers or spheroids, and the survival increase with time between fractions is thus a direct reflection of the cellular recovery under the two growth conditions.

As might be predicted on the basis of the response to the single fraction radiation, a greater recovery factor was observed for both single cells and spheroids at the larger doses. Concomitant with this observation, the recovery appeared essentially identical for single cells and spheroids at the lowest fractional

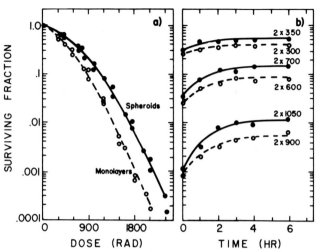

Fig. 2. Survival of V79 Chinese hamster cells from monolayers (-O-) or 1-day-old spheroids (-●-) exposed to single-dose (panel a) or two-fraction (panel b) irradiation schemes. In panel b, survival is plotted as a function of time between the two fractions; the indicated doses (in rads) were chosen to provide essentially identical survival for cells from the two growth conditions.

doses chosen, although the scatter of data within an experiment (or even groups of experiments) tends to increase uncertainties for such high survival levels. This again is a reflection of the colony formation technique as an indicator of radiation damage, insofar as small differences in data can be demonstrated only with large numbers of repeat observations.

Based on the single-dose and two-dose responses of monolayer and spheroid cells (Fig. 2), one can easily predict the multifraction response, at least in the situation where "complete" repair is permitted (ie, the entire shoulder of the survival curve reappears), and no repopulation occurs (Fig. 3). In this case, using the curves of Fig. 2a, the differential response of spheroids and single cells to a "large" and "small" dose per fraction is projected. Note that with the "large" dose per fraction, the predicted survival curves clearly diverge, and thus the protective effect of intercellular contact would be augmented in the multifraction regimen. However, the prediction of the response to "small" doses per fraction is entirely dependent upon the assumptions (or models) one chooses to define the low-dose (shoulder) regions of each survival curve, and this is, in essence, the subject addressed by Dr. Suit (1978) and myself. By assuming a non-zero initial slope that is essentially equal for both single cells and spheroids, one predicts only small differences in the multifraction response (Fig. 3). However, any difference in this "initial" slope of the survival curves resulting from intercellular contact would be amplified in the multifraction experiment with small spheroids and the corresponding monolayer controls.

As previously mentioned, any multifraction experiment with a tissue culture system requires a number of compromises. To simplify interpretation, a sufficiently high net dose-rate can be used so that cellular proliferation is inhibited. Also pertaining to this point, the total dose of radiation should ideally be delivered sufficiently rapidly that no changes occur in the cell population. Conversely, multifraction experiments, by definition, require protracted irradiation schemes.

Based on preliminary experiments of our own as well as considerable data of others, dose-rates in the range of 65-125 rads/hour were chosen, with fractionation intervals of 1, 2, 4, or 6 hours. As predicted, cell proliferation during these irradiation schemes was inhibited, as all populations showed constant cell numbers at all assay intervals.

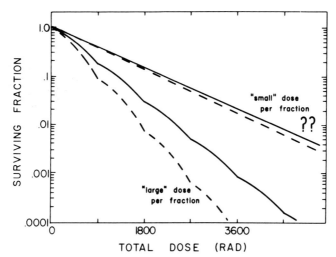

Fig. 3. Predicted multifraction survival of monolayer (- - -) or spheroid (———) cells, based on single-dose survival curves of Fig. 2a, and assuming complete recovery with no progression between doses. The "small" dose per fraction response is predicted on the basis of the apparent initial slope (0-100 rad) of the survival curves of Fig. 2a.

In all dose regimens studied, we found that spheroid cell survival, as a function of total accumulated dose, was always greater than that of monolayer cells exposed to the multifraction treatments (Fig. 4). In Fig. 4, the differential response of the two populations to radiation schemes of 100 rads/hour (panel a), 200 rads every 2 hours (panel b), 400 rads every 4 hours (panel c), or 600 rads every 6 hours (panel d) is compared to the corresponding single fraction response (light lines in all panels). Several features are evident. First, data collected for all fractionation schemes were fit best by survival curves with an initial shoulder, suggesting that sublethal damage was accumulated even in the multifraction regimens. This observation, of course, is highly dependent upon the shortness of some of the fractionation intervals, which clearly would not allow "complete" repair of sublethal damage with the kinetics suggested in Fig. 2b. Second, as predicted, larger doses per fraction produced more cell kill, despite the fact that the "effective" dose-rate was 100 rads/hour in all cases. Even though the spheroids clearly survived better in all cases, the data were still somewhat equivocal as to whether the relative survival increase (repair capacity?) was reduced when large numbers of small fractions were used (Fig. 4a).

This latter question, whether the relative amount of repair exhibited by the cells grown in contact was decreased in the multifraction situation, was more directly addressed by plotting the data as "isoeffect" curves, where the total dose of radiation required to produce a given endpoint was plotted as a function of the number of fractions (or extrapolated partial fractions) required to achieve each endpoint. In Fig. 5, endpoints of 10% survival, 1% survival, and .01% survival were chosen, and the number of fractions required to reduce the best fit survival curve of each radiation scheme to these levels was plotted.

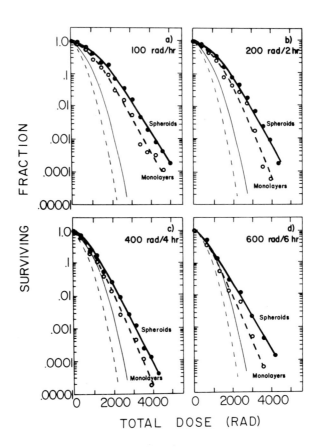

Fig. 4. Experimentally-determined survival of monolayer (-O-) or spheroid (-●-) cells after multifraction irradiation, compared to acute exposure (corresponding light curves). Dose-regimens shown include hourly 100 rad fractions (panel a), 200 rads every second hour (panel b), 400 rads every fourth hour (panel c), and 600 rads every 6 hours (panel d). Since all data for monolayers and spheroids suggested curves with an initial shoulder region, the multifraction data for survival levels < 10% were subjected to weighted least squares regression analysis, and the best-fit curves thus obtained are shown.

Again two points are particularly evident. In all cases the slopes of the curves are not significantly different than the value of ∿ .22 usually obtained with such Strandquist-type plots for tumors, even for clinical data over much longer time frames (Hall, 1973). Secondly, in all cases, the slope found for the cells irradiated as spheroids (influenced by intercellular contact and other tissue-like factors) showed a somewhat reduced dependency on fraction number at all levels of effect. This decreasing effectiveness with increasing fraction number is

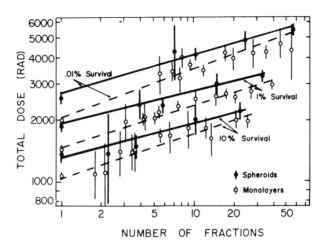

Fig. 5. Isoeffect curves for monolayer (-O-) or spheroid (-●-) cells, with total dose required to achieve .01%, 1% or 10% survival respectively plotted as a function of the number of (complete or "partial") fractions required to achieve this dose. For monolayer cells, dose regimens of 75, 100, 125 rad/hr; 150, 200, or 250 rad every second hour; 300, 400, or 500 rad every 4 hours, and 400, 500, or 600 rads every 6 hours were used. Spheroids were exposed to 100 rad/hr, 200 rad/2 hr, 400 rad/ 4 hr, or 600 rad/6 hr. For each dose regimen, the best-fit regression line for the weighted data between 10% and .01% survival was determined, and from this, the required dose and its associated 95% confidence interval for each isoeffect level was determined and plotted. The lines shown represent the best-fit regression lines for the weighted data. For the single cells, slopes of .215 ± .043; .204 ± .025; and .227 ± .044 were found for 10%, 1% and .01% survival; the corresponding values for the spheroids were .174 ± .108; .174 ± .040; and .183 ± .075. All uncertainties are 95% limits.

consistent with the hypothesis that the "initial slope" of the single dose survival curve for single cells and spheroids is identical. Further, this result is in good agreement with other data we have previously presented (Wathen, Durand, and Kademian, 1978) where we found that with low dose-rate continuous irradiation (88 rads/hour), there was no difference in the response of single cells and small spheroids.

DISCUSSION

The data presented here suggest that intercellular contact can produce a quantitative change in the repair capacity of mammalian cells, and consequently a quantitative change in the response of cells grown under such conditions to multifraction irradiation. However, this enhancement of survival is apparently maximal for the single dose situation, and is decreased in the multifraction situation.

These conclusions, however, may be highly dependent on the experimental protocols used in obtaining these results. Both spheroids and single cell populations were asynchronous at the start of treatment; since little cell growth was observed during the treatment intervals, cell cycle alterations were obviously produced, and it seems likely that such alterations in cell cycle kinetics and possibly distributions may have affected the overall response. The choice of effective dose-rate utilized may also be questioned, although each of the single cell isoeffect curves in Fig. 5 includes effective dose-rates differing by as much as a factor of 2.

In conclusion, our purpose was to examine the influence of intercellular contact on the radiation response of mammalian cells in tissue culture, and to attempt to evaluate whether this effect has any relevance to the question of predicting tumor responses to fractionated radiation on the basis of the response of cultured single cells. From the data presented here, it seems safe to conclude that the recovery exhibited by monolayer cells seems qualitatively similar to that which would be predicted for cells enjoying any benefit of cell-cell contact. Thus, the response of tumors to fractionated irradiation may well be more dependent on other factors such as reoxygenation and cell redistribution, and these can not easily be simulated by conventional tissue culture methods.

ACKNOWLEDGEMENTS

The technical assistance of Ms. S. Brown is gratefully acknowledged. This research was supported by grants CA-23489 and CA-23511 awarded by the National Cancer Institute, DHEW.

REFERENCES

Durand, R.E. (1976). Cell Cycle Kinetics in an In Vitro Tumor Model. Cell Tiss. Kinet., 9, 403-412.

Durand, R.E. and Olive, P.L. (1978). Radiation-Induced DNA Damage in V79 Spheroids and Monolayers. Radiat. Res., submitted.

Durand, R.E. and Sutherland, R.M. (1972). Effects of Intercellular Contact on Repair of Radiation Damage. Exp. Cell. Res., 71, 75-80.

Durand, R.E. and Sutherland, R.M. (1973). Growth and Radiation Survival Characteristics of V79-171b Chinese Hamster Cells: A Possible Influence of Intercellular Contact. Radiat. Res., 56, 513-527.

Durand, R.E. and Sutherland, R.M. (1975). Intercellular Contact: Its Influence on the Dq of Mammalian Cell Survival Curves. In Tikvah Alper (Ed.), Cell Survival After Low Doses of Radiation: Theoretical and Clinical Implications. The Institute of Physics and John Wiley & Sons, Ltd. 237-247.

Durand, R.E. and Sutherland, R.M. (1976). Repair and Reoxygenation Following Irradiation of an In Vitro Tumor Model. Int. J. Radiat. Onc., Biol., Phys., 1, 1119-1124.

Hall, E.J. (1973). Time Dose and Fractionation in Radiotherapy. In Radiobiology for the Radiologist, Harper & Row Publishers, Hagerstown, Maryland. pp. 251-258.

Steel, G.G. (1977). Growth Kinetics of Tumors, Clarendon Press, Oxford.

Suit, H.D. (1978). Relevance of Radiobiological Data on Mammalian Cells in Culture for the Prediction of Tumor Responses to Fractionated Irradiation. These Proceedings.

Sutherland, R.M. and Durand, R.E. (1976). Radiation Response of Multicell Spheroids--an in Vitro Tumor Model. Curr. Topics Radiat. Res. Q., 11, 87-139.

Wathen, P., Durand, R.E. and Kademian, M.T. (1978). Low Dose-Rate Irradiation of Multicell Spheroids. Int. J. Radiat. Onc., Biol., Phys., in press.

Rationale for the Selection of Combined Treatment Schedules Using Fractionated Radiation and Chemotherapy

Theodore L. Phillips

*Department of Radiation Oncology, University of California,
San Francisco, California 94143, U.S.A.*

ABSTRACT

Experimental data so far available from in vitro, in vivo tumor, and in vivo normal tissue systems suggest that the major interaction is additive cell killing. There is no clear-cut clinical evidence for an optimum time or mechanism for combining chemicals and radiation to enhance tumor response without the enhancement of normal tissue response. Indeed, normal tissue response seems more uniformly affected than is tumor response, since many tumors are resistant to specific chemotherapeutic agents, but most normal tissues are damaged to some extent by them.

Clinical studies have shown improved survival in a few sites, such as pediatric tumors, oat cell carcinoma of the lung, ovarian carcinoma, colorectal and pancreas lesions, selected head and neck lesions, and brain tumors. Combined treatment at all sites, however, is associated with enhanced normal tissue damage. In essentially all trials beneficial effects appear due to multiple courses of chemotherapy, not necessarily administered during radiotherapy. Much further investigation is required before it can be concluded that simultaneous radiotherapy and chemotherapy are of benefit.

INTRODUCTION

Soon after the early use of chemotherapy in the treatment of Wilms' tumor, it was noted that increased radiation reactions occurred in normal tissues, particularly skin and lung. In these early trials increased tumor response was also noted in both the local site and by means of control of distant metastases. The combined use of radiotherapy and chemotherapy has been widespread in pediatric tumors, with enhanced cure rates noted in Wilms' tumor, Ewing's sarcoma, rhabdomyosarcoma, and acute lymphocytic leukemia (Muggia, Cortes-Funes, and Wasserman, 1978).

Three major reasons for combining radiotherapy and chemotherapy have been put forward. In the first situation, the combination is administered in order to enhance local control which may not be optimum with radiotherapy alone. In the second situation, chemotherapy is administered to sterilize distant metastases outside the radiation volume and usually consisting of microscopic foci. In the third situation, radiation is used to sterilize a sanctuary from adequate chemotherapeutic dosage or effect (Carter and Wasserman, 1975).

More recently, controlled clinical trials and extensive laboratory investigations with in vitro cells, in vivo tumor systems, and in vivo normal tissue systems have been used to determine a more rational combination of radiotherapy and chemotherapy.

The initial favorable results in certain specific sites have led to much optimism as to the potential for combining radiotherapy and chemotherapy, but on the other hand marked enhancement of local normal tissue damage has been observed, tempering such optimism.

EXPERIMENTAL OBSERVATIONS WITH SINGLE DOSES IN VIVO AND IN VITRO

Studies have been performed using chemotherapeutic agents alone with various cell lines cultured in vitro. Many drugs show exponential survival curves, including alkylating agents and the nitrosoureas, while others show a typical biphasic curve dependent on resistant populations due to cell age distribution or other partitioning of resistant cells (Phillips, 1979b). In most cases radiation survival curves, when corrected for cell kill by the chemotherapeutic agent and for changes in cell age distribution, have been unaltered in terms of both shoulder and slope. Actinomycin D appears to alter the width of the shoulder and to inhibit repair of sublethal damage as measured in vitro. Adriamycin reduces the width of the shoulder, but does not change sublethal damage repair. BCNU and cis-platinum have been reported to reduce the width of the shoulder and perhaps reduce repair capacity, although work is still ongoing with these and other agents. Combined radiotherapy and bleomycin appear to enhance cell kill by modification of the slope of the dose response curve.

We have evaluated a number of cancer chemotherapeutic agents for their effect on the killing of intestinal crypt cells, survival after whole lung irradiation, and survival after esophageal irradiation (Table 1).

TABLE 1 Normal Tissue DEF^{o} Values†

Drug	Dose (MTD) mg/kg	DEF Values		
		Crypts	Lung	Esophagus
Actinomycin	0.3	1.2	$1.0^{\ddagger}$ - 1.6^{*}	1
Adriamycin	8	1.2		1
BCNU	25	1.2		1.2
Bleomycin	100	1.9	$0.9^{\ddagger}$	$1.1^{\ddagger}$
Cyclophosphamide	250	0.9	$1.3^{\ddagger}$	1.05
5-Fluorouracil	140	1.2		--
Methotrexate	700	1.5		--
Cis-platinum	13	1.3		1.5
Vincristine	3	0.9	$1.2^{\ddagger}$	1.0

$^{o}DEF = \dfrac{\text{Radiation dose for effect without drug}}{\text{Radiation dose for effect with drug}}$

†Measured 2 hrs after drug dose.
‡At less than the MTD.
*At more than the MTD.

In these studies the intestinal crypt cell system of Withers and the pulmonary and esophageal lethality systems of Phillips were employed (Phillips, Wharam, and Margolis, 1975). Drug was administered at the maximum tolerated dose (less than 1% kill) 2 hrs intraperitoneally prior to irradiation. It can be seen in Table 1 that all 9 agents tested enhance radiation effect in at least one organ as measured by the dose effect factor. Some tissue specificity is seen, with actinomycin particularly effective in lung and intestinal crypts and cyclophosphamide effective in lung but not in intestinal crypts or esophagus. Cis-platinum was effective in enhancing damage in both tissues tested, and methotrexate and 5-FU were particularly effective in the intestine. To a certain extent these enhancements follow the known cytotoxicity of the specific drugs in normal tissues, although the cytotoxicity of these drugs appears more generalized in normal tissues than in experimental tumors.

The intestinal crypt cell system can be used to generate survival curves for radiation delivered two hours after an intraperitoneal injection of the maximum tolerated drug dose. The cell kill by drug alone can be estimated by survival curves created by a fixed dose of radiation and increasing graded doses of drug.

TABLE 2 Effect of Chemical Agents on D_q (Intestinal Crypt Cells)

Drug	MTD mg/kg	Estimated Drug Kill S/S_o	Estimated D_q*
Actinomycin	.3	0.4	300
Adriamycin	8	0.3	310
BCNU	25	0.35	270
Bleomycin	100	0.4	0
Cytoxan	250	0.6	400
5-FU	140	0.1	300
Methotrexate	700	0.1	300
Platinum	13	0.2	280
Vincristine	3	0.7	320

*D_q derived from extrapolation minus cell kill.

It can be seen in Table 2 that drugs at the maximum tolerated dose cause cell kill ranging from 30-90%. These drug doses cause only slight changes in the D_q, with the exception of bleomycin where the D_q is reduced to 0. Drugs might impair the repair of sublethal damage to a slight extent, but might eliminate it completely with bleomycin.

It is difficult to determine whether these changes in D_q have been adequately corrected for drug cell kill. One method for determining the actual influence on sublethal damage repair would include a comparison of drug administration before to drug administration after irradiation, since administration before might influence the width of the shoulder, but administration after would not. Results of these experiments are shown in Table 3.

TABLE 3 Comparison of Crypt Cell Survival - MTD of Drug Given Before or After Radiation (1100 rad)

Drug	No. Crypt Cells Drug -2 hrs	No. Crypt Cells Drug +3 hrs	Ratio
Actinomycin	30	65	2.2
Adriamycin	50	50	1
BCNU	38	55	1.4
Bleomycin	1	65	65
Cyclophosphamide	90	90	1
5-Fluorouracil	26	15	0.6
Methotrexate	2	15	7.5
Cis-platinum	24	32	1.3
Vincristine	65	100	1.5

The number of crypt cells surviving after a radiation dose of 1100 rad are shown when drug was given 2 hrs before the irradiation or 3 hrs after, when repair is essentially complete. It can be seen that with certain drugs the number of cells surviving when drug was given after irradiation is far higher than when drug was given before. This is particularly evident for actinomycin D, bleomycin, methotrexate, and vincristine. Only Adriamycin and cyclophosphamide caused no change, while 5-FU drug administration after irradiation was more effective.

Although one might infer from this that decreased survival when drug is given before, rather than after, irradiation is due to inhibition of sublethal damage repair, this may not be the case. Recent experiments using high LET radiations reveal that most agents, particularly BCNU, bleomycin, and cis-platinum, also have the same result with high LET (Phillips, 1979a). The increased response is abolished for actinomycin D and methotrexate, and it would suggest that sublethal damage repair inhibition by actinomycin D and synchronization by methotrexate do not occur with high LET, but that the effects seen with bleomycin, BCNU, and other drugs are due to other modes of action, possibly radiation modification of the drug injury.

Information is more limited with drug-radiation interactions in experimental tumors. We have studied extensively the interaction of 9 chemotherapeutic agents and radiation using the EMT6 mouse mammary carcinoma (Begg and co-workers, 1979).

TABLE 4 Drug-Radiation Interactions in EMT6 Tumor

Drug	Percent Cell Kill at MTD	Days Growth Delay @ MTD	Presence of > Additive Effect	Time of > Additive Effect
Actinomycin	10	2.6	--	--
Adriamycin	40-90†	2	No	--
BCNU	99.9	5	Yes?	+1, +8 hr
Bleomycin	75	3	No	Protection -8 hr
Cyclophosphamide	99.9	10	Yes?	+1, +8 hr
5-Fluorouracil	60	4.7	No	--
Methotrexate	24	2.1	No	--
Cis-platinum	84	3	--	--
Vincristine	53	1.7	--	--

†2 mm lung nodules.

The results shown in Table 4 show the percent cell kill caused by a maximum tolerated dose of drug as measured by the _in vivo_ treatment, _in vitro_ assay method. Drug was given intraperitoneally 24 hrs before assay, conducted by removal of the tumor and preparation of cell suspensions for plating. It was noted that repair of potentially lethal damage and elimination of artifacts was maximal at the 24 hr point. Also shown in Table 4 are the days growth delay caused in the tumor at a size of approximately 8 mm diameter by the same drug dose. There was good agreement between cell kill and growth delay, although for BCNU the delay seemed rather short for the amount of kill and for actinomycin D rather long for the amount of kill. Experiments using combined radiation and drug were conducted for some of the agents in which drug and radiation were separated by intervals from 2 - 64 hrs before or after irradiation. Cell kill was no more than additive, with the exception of BCNU and cyclophosphamide, the most effective agents on their own. With these two agents slight degrees of response greater than additive were seen when drug was given 1 and 8 hrs following irradiation. Although a number of the agents have not yet been studied for these combined effects, it can be generalized that there is little suggestion of a marked enhancement of radiation effects on this tumor by any of the agents studied at any of the times between ±64 hrs relative to radiation.

OBSERVATIONS WITH MULTIPLE DOSES IN NORMAL TISSUE AND TUMORS

The amount of information available on fractionated irradiation with chemotherapeutic agents is quite limited. We have conducted experiments with the intestinal crypt cell system using two radiation doses divided by 3 hrs for 9 chemotherapeutic agents and for radiation alone. The results are shown in Table 5.

TABLE 5 Split Dose Survival Ratios - Intestinal Crypt Cells (B_6AF_1 Mice)

Drug	Dose mg/kg	Dose (rad)	No. Cells 1 Dose	No. Cells 2 Dose	Survival Ratio
Radiation Only		900	149	--	--
"	"	1000	34	104	3.1
"	"	1170	23	49	2.1
"	"	1200	9	53	6
Actinomycin D	0.2	1170	4.2	67	16
Adriamycin	6	1260	2	17	8.5
BCNU	19	1170	4	44	11
Bleomycin	25	1000	1	4	4
Cytoxan	180	1200	5	74	15
Cis-platinum	9	1000	15	44	3
5-Fluorouracil	100	1000	27	103	3.8
Methotrexate	1300	900	1.4	3	2.1
Vincristine	2.2	1200	9	59	6.5

It can be seen that with radiation alone there is a marked increase in the number of surviving crypt cells with 2 doses as compared to 1 with a 3 hr interval, with ratios ranging from 2.1 to 6. When drug doses were given at the maximum tolerated dose, it can be seen that there was no significant suppression of sublethal damage repair as evidenced by increased survival with split doses. The dose in rad, shown in the third column, was given as either one or two doses, and the results shown in the next columns. The survival ratios are not markedly reduced, although one might infer that with the very large dose of methotrexate used, the survival ratio is somewhat low. It can be generalized that with none of these agents was sublethal damage repair markedly inhibited. It should be pointed out that the actinomycin D dose used was rather low, as compared to that used in a number of in vitro studies in which impaired repair has been shown. It should also be pointed out that the intestinal crypt cell has a tremendous ability to repair sublethal damage and that much larger drug doses may be required or longer intervals of exposure to suppress sublethal damage repair. In these experiments drug was given 2 hrs prior to the first radiation dose and the time may be insufficient for maximum damage to this cell or to the DNA.

We have conducted multifraction studies with actinomycin D in the lung and have found evidence of suppression of sublethal damage repair (Phillips, Wharam, and Margolis, 1975). A limited number of split dose experiments have been conducted in vitro and suggest that actinomycin D and cis-platinum inhibit repair of sublethal damage. There is no convincing evidence that any of the active anti-tumor chemotherapeutic agents significantly inhibit repair of single strand breaks in DNA. There is a paucity of work available on their effect on the rejoining of chromosome breaks or double stranded DNA breaks. Much more work is required in this area to determine which, if any, chemotherapeutic agents inhibit repair of sublethal damage generally and under what conditions they may do it in specific situations.

CLINICAL OBSERVATIONS

Thousands of clinical trials have been conducted combining radiotherapy and chemotherapy for the reasons outlined in the beginning of this report. Only a very small fraction of these studies have been conducted in a controlled, randomized fashion and in even smaller fraction have shown positive results. The clinical sites and histologies in which generally accepted positive results have occurred are summarized in Table 6 (Muggia, Cortes-Funes, and Wasserman, 1978).

TABLE 6 Improved Clinical Results with Combined Radiation and Chemotherapy in Controlled Trials

Tumor Site	Drug(s)*	Timing†	No. of Courses	Site of Effect°
Wilms' tumor	Act D, VCR	D, A	6-12	M
Ewing's sarcoma	VAC	D, A	4-12	L, M
Rhabdomyosarcoma	VAC	D, A	4-12	L, M
Oat cell-lung	POCC, CMC, CAM	B, D, A	4-8	L, M
Ovary	Chlorambucil	A	6-12	L
Colorectal	5-FU	D, A	2-12	L, M
Pancreas	5-FU	D, A	6-12	L, M
Head and Neck	5-FU	D	1	L
Head and Neck	Bleomycin	D	1	L
Brain	BCNU	D, A	6-8	L

*Act D=actinomycin D; VCR=vincristine; VAC=vincristine, actinomycin D, & cyclophosphamide; POCC=procarbazine, oncovin, CCNU, & cyclophosphamide; CMC=cyclophosphamide, methotrexate, & CCNU; CAM=cyclophosphamide, Adriamycin, & methotrexate; 5-FU=5-fluorouracil; BCNU=carmustine.
†B=before; D=during; A=after.
°L=local; M=metastases.

Not tabulated here but also important is the improved response seen in acute lymphocytic leukemia when patients receive whole brain radiation prophylactically after induction of remission with multidrug chemotherapy. In this situation the radiation sterilizes cells in a sanctuary, the central nervous system, not reached by the systemic chemotherapy. Some trials have suggested that a similar sterilization could be achieved by intrathecal methotrexate, but this has not been generally accepted.

Some generalizations may be made from Table 6. It is evident that the primary result has been an improvement in both local control and in survival due to control of distant metastases in the specific situations in which improved survival has been seen. In almost all cases drug has been administered during, but particularly after, radiation in multiple courses. Only in two limited studies of the head and neck region have single courses of drug given during radiotherapy been apparently effective and only in a very limited number of sites, not yet confirmed by larger clinical trials.

In all of the situations in which accepted improved results have occurred, the chemotherapeutic agents were quite active on their own, with 30-60% complete plus partial responses observed in advanced disease. The drug was given in multiple courses and it appears that the major effect was an additive killing of chemotherapy to that of the radiation in the local site and chemotherapeutic sterilization of microscopic distant metastases. Concomitant administration of drug was carried out in many of the trials, but does not appear to have been a major factor and may have been in some a complicating factor.

In these positive trials and in many, many negative trials, combinations of radiation and chemotherapy have led to enhanced normal tissue damage by radiation in the irradiated volume (Phillips and Fu, 1976, 1977, 1978). A number of agents are summarized in Table 7.

TABLE 7 Enhanced Effects in Normal Tissues

Drug	Organs Showing Increased Toxicity	No. of Reports	Severity*
Actinomycin D	Skin, esophagus, lung, GI, GU, liver, CNS, bone and soft tissue.	24	++, +++
Adriamycin	Skin, esophagus, lung, heart, GI, GU, bone and soft tissue.	20	++, +++
Bleomycin	Skin, esophagus, GI, lung.	6	+, ++, +++
Cyclophosphamide	Lung, bladder.	5	+, ++
Hydroxyurea	Skin, esophagus, lung.	4	+
5-Fluorouracil	Skin, GI, liver, eye.	7	+, ++
Methotrexate	Skin, CNS.	9	+, ++, +++
Vincristine/ vinblastine	Esophagus, lung.	2	+
BCNU	GI.	1	+
Cytosine arabinoside	Optic nerve	1	++

* + = mild, noticeable; ++ = marked; +++ = severe, fatal.

The greatest number of reports has centered around actinomycin D and Adriamycin with many of the reactions severe or fatal. Bleomycin has caused fatal reactions in the lung and methotrexate fatal reactions in the central nervous system. Most of these increased responses appear with drug administered immediately before or during radiotherapy.

GENERAL PRINCIPLES

The _in vitro_ and _in vivo_ experimental information and the clinical trial data now available lead to some general conclusions and hypotheses. Unfortunately the amount of experimental data available is quite limited, particularly the range of cell types studied, the range of drug doses and drug types, and the intervals between drug and radiation. Based on this limited information, a number of statements can be made:

1) Chemotherapeutic agents are generally cytotoxic in most _in vitro_ cell systems with variations depending primarily on suspension methods and cell age distributions.
2) Chemotherapeutic agents are cytotoxic in many normal tissue systems _in vivo_, particularly rapid renewal systems, although certain notable exceptions exist, i.e., intestinal crypt cell resistance to cyclophosphamide and bone marrow resistance to a number of agents including bleomycin.
3) There is limited data to suggest that cancer chemotherapeutic agents modify radiation damage accumulation or repair. Most agents cause only additive cell kill.
4) Radiation may fix damage or inhibit the repair of chemotherapeutic damage, as seen in intestinal crypt cells, particularly with BCNU and bleomycin.
5) Chemotherapeutic agents appear much more uniform in their injury to proliferating normal tissues than in their efficacy in killing a wide range of tumor cells.
6) A number of single agents and combinations have proven effective in enhancing local control and survival in selected human tumors. In almost all cases this

has occurred with multiple drug administration regimens with no clear-cut advantage of simultaneous radiation-drug administration.
7) At this point it would appear that the best use of fractionated radiotherapy and cancer chemotherapeutic agents would be high dose local radiotherapy to known disease, followed by multiple courses of chemotherapy utilizing agents with high response rates in advanced disease.

REFERENCES

Begg, A. C., Fu, K. K., Kane, L. J., and Phillips, T. L. (in press). Single agent chemotherapy of a solid murine tumor: Comparison of growth delay and cell survival assays. Cancer Research.

Carter, S. K., and Wasserman, T. H. (1975). Interaction of experimental and clinical studies in combined modality treatment. Cancer Chemother. Rep., 5, 235-241.

Muggia, F. M., Cortes-Funes, H., and Wasserman, T. H. (1978). Radiotherapy and chemotherapy in combined clinical trials: Problems and promise. Int. J. Radiation Oncology Biol. Phys., 4, 161-171.

Phillips, T. L. (1979a). Combined Chemo/Radiotherapy of Cancer: Present state and prospects for use with high LET radiotherapy. Europ. J. Cancer (in press).

Phillips, T. L. (1979b). Editor - Combined Modalities: Chemotherapy/Radiotherapy, Report of the Hilton Head Conference, November 1978. Int. J. Radiation Oncology, Biol. Phys. (in press).

Phillips, T. L., and Fu, K. K. (1976). Quantification of combined radiation therapy and chemotherapy effects on critical normal tissues. Cancer, 37, 1186-1200.

Phillips, T. L., and Fu, K. K. (1977). Acute and late effects of multimodal therapy on normal tissues. Cancer, 40, 489-494.

Phillips, T. L., and Fu, K. K. (1978). The interaction of drug and radiation effects on normal tissues. Int. J. Radiation Oncology Biol. Phys., 4, 59-64.

Phillips, T. L., Wharam, M. D., and Margolis, L. W. (1975). Modification of radiation injury to normal tissues by chemotherapeutic agents. Cancer, 35, 1678-1684.

ACKNOWLEDGEMENTS

Supported by National Cancer Institute Grants CA17227 and CA20529.

Possibilities and Limitations in the Use of Hypoxic Sensitizers and Hyperbaric Oxygen for Improvement of Fractionated Radiotherapy of Cancer

J. F. Fowler

Gray Laboratory of the Cancer Research Campaign, Mount Vernon Hospital, Northwood, Middlesex, U.K.

ABSTRACT

Results of hyperbaric oxygen radiotherapy (HBO) have shown significant improvements over conventional radiotherapy in several clinical trials but not in others. The results indicate that hypoxic cells are indeed a problem in some tumours, and HBO can help even when 25 - 30 small daily doses are used. Electron-affinic radiosensitizers reach hypoxic cells more readily than HBO. Their radiosensitizing potential is compared with that of oxygen. The question of fraction size is discussed. One prospect is that shorter overall treatment times might give better results when hypoxic cells are eliminated using these radiosensitizers.

Key words: Radiotherapy, radiosensitizers, hyperbaric oxygen, fractionation.

INTRODUCTION

Hypoxic cells are present in many tumours which exceed 1 or 2 mm in diameter. They require nearly three times more X-ray dose to kill them than well-oxygenated cells do. They are present as annular "sleeves" about 150 micrometers radial distance from capillary vessels (Fig. 1).

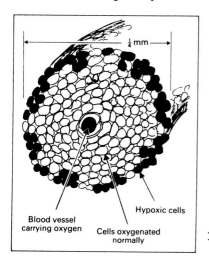

Fig. 1.

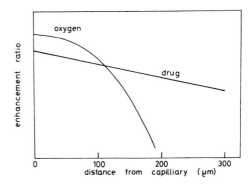

Fig. 2.

Oxygen has normally been used up within this radius, in the metabolism of the cells through which it passes. Immersing the patient in a hyperbaric oxygen tank would increase the oxygen dissolved in the blood and therefore the range of radial diffusion. The snag is that high oxygen concentrations in the blood cause vasoconstriction so that the method is to some extent self-limiting. Nevertheless, evidence has been presented that the average oxygen tension inside human tumours rises after 15 or more minutes whole-body immersion in hyperbaric oxygen, HBO, (Cater and others, 1957). The patient would then be treated with radiation through the walls of the HBO tank (Churchill-Davidson and others, 1957).

Electron-affinic radiosensitizers provide an alternative method of dealing with hypoxic cells. They were developed by the pioneering work of Adams & Dewey (1963); see Asquith and others (1974); Adams (1973, 1976); Adams and others (1976); and Fowler and others (1976). They mimic the radiosensitizing action of oxygen, but are not used up in the metabolism of cells. Therefore they will diffuse further, radially outwards from the blood capillaries. Figure 2 illustrates this. Even if the concentration in the blood of radiosensitizer is effectively less than that of oxygen, the radiosensitization will be greater at the radius of the hypoxic cells. This is the principle of the method. Accessibility of cells in solid tumours to these substances has in practice been found to be good.

In this paper I shall first review the clinical results of hyperbaric oxygen. From these results certain conclusions can be drawn that are relevant to the application of hypoxic cell radiosensitizers.

HYPERBARIC OXYGEN

Two recent reviews of the clinical results are available: Dische and others (1978) and Suit (1978). Of the 3,500 or so patients treated in clinical trials reported up to now, half of them in HBO and half controls in air, the MRC Clinical Trial in the UK and South Africa includes about 1400 and will be summarised first. Table 1 shows results for Carcinoma of the Cervix, Stage III.

TABLE 1   MRC Ca Cervix Stage III

	HBO	Air	Comments
5-YEAR SURVIVAL			
6F) Portsmouth (37 pts)	42%	17%	In-air results
6F) Oxford (23 pts)	46%	8%	were poor
25F) Glasgow (127 pts)	50%	37%	Combined,
30F) Mt.Vernon (56 pts)	39%	28%	P < .05
LOCAL CONTROL, 5 YRS			
Glasgow	87%	60%	Highly sig.
Mt.Vernon	76%	50%	
SEVERE MORBIDITY			
Bowel	12%	4%	

The first interesting conclusion from Table 1 is that 6 fractions in air in 3 weeks appears to be a very poor regime which is however improved greatly by the use of HBO, up to the average level for all four centres. This is an important clinical illustration of the "bringing up to a uniform good level" of normally poor schedules when hypoxic cells are dealt with. In the centres using 25 to 30 small daily fractions, there is a just significant improvement of 11-13% in survival, but the local control of tumours showed a highly significant increase of 26 or 27%. These results in HBO are the best ever achieved by any method of treating Ca Cervix Stage III anywhere. However, an increase in severe morbidity was also found.

Table 2 shows the MRC Trial results for advanced cancer of the head and neck (Henk and others, 1977). In both of these trials an increase of about 20% in

TABLE 2   MRC Head and Neck Ca

	HBO	Air	Comments
1st Trial (276 pts)			
Local control (5y)	53%	30%	P < .001
Survival (5y)	40%	40%	Saved by Surgery
2nd Trial (103 pts)			+ Leeds
Local control (2y)	65%	47%	P < .05
Survival (2y)	71%	50%	P < .01

No difference in complications.
10% lower dose to larynx fields only.

local control was found. In the first trial however there was no difference in 5-year survival because the locally recurrent tumours were removed surgically. In the second trial, as yet reported only to 2 years, there was a significant increase in both survival and local control, with no increase in morbidity. These patients had clearly benefited from the use of hyperbaric oxygen.

In the MRC trial no advantage was obtained with HBO in the treatment of <u>bladder carcinoma</u>. In the treatment of <u>Ca bronchus</u> at Portsmouth, however, 2-year survival was increased from 12% to 25% (123 cases - just sig.diff.) although with small daily fractions no advantage had been seen in about 80 cases (Cade & McEwen, 1978).

Tables 3 and 4 summarise results of previously published HBO trials compared with the MRC trials.

TABLE 3   Other Cervix Survival

Author	(Fracts)	Patients	HBO	Air	Sig?
Johnson 1969	(30F)	50	44%	25%	No
Bates 1969	(6F)	36	56%	25%	Yes, but not random
Ward 1973	(10F)	49	58%	71%	No
Brady 1973	(25F)	40	41%	39%	No
Fletcher 1973	(20-30F)	233	33%	41%	No
MRC Trial		320	33%	27%	.08

Only the Fletcher trial has significant numbers and in this one the serious morbidity doubled from 12 to 24%.

TABLE 4 Other Head and Neck – Local Control

Author	(Fracts)	Patients	HBO	Air	Sig?
Van den Brenk 1968	(4F)	29	71%	15%	No
Chang 1973	(30F)	42	65%	55%	No
Shigematsu 1973	(8-10F)	38	76%	71%	No
Plenk 1974	(4-30F)	60	66%	–	No
RTOG 1978		48	58%	38%	Just
MRC 1st		273	53%	30%	< .001
" 2nd		103	65%	47%	Just

Whilst not all the trials show a significant difference, the trend is in favour of HBO in 11 out of the 13 trials. There seems no longer any doubt that hypoxic cells are a problem in radiotherapy and that hyperbaric oxygen can, to some extent, reduce the problem. What is in doubt however is whether this is the best way of dealing with hypoxic cells. A further doubt is whether HBO improves the very best, as it obviously can improve poor schedules.

The Dose Enhancement Ratio corresponding to an increase in local control of say 20%, that is from 45% to 65% as in Henk's head and neck HBO trials lies between 1.1 and 1.3, depending upon which clinical dose response curve for tumour control one prefers. This can be compared with the Dose Enhancement Ratio for normal tissue damage of about 1.1. These ranges overlap but there is a tendency for the tumour responses to be increased more than the normal tissue responses. Thus other methods of dealing with hypoxic cells should continue to be explored. It is noteworthy that treatments employing multiple small fractions were improved by the use of HBO.

It can be concluded that "if Dr. Henk's results for head and neck cancer are confirmed (a significant improvement in both local control and survival for no extra normal tissue complications), then a real advance in radiotherapy will have been made" (Suit, 1978).

RADIOSENSITIZERS OF HYPOXIC CELLS

Electron-affinic sensitizers sensitize hypoxic but not well-oxygenated cells. This is their crucial advantage over other anti-cancer drugs, because most tumours have hypoxic cells and most normal tissues do not. There are two such drugs in clinical use at present and the next generation of such sensitizers is being developed but is several years away. Misonidazole[1] is a 2-nitroimidazole and metronidazole[2] a 5-nitroimidazole. They give maximum enhancement ratios, at

---

[1] Ro-07-0582, Roche Products Ltd., Welwyn Garden City, England
[2] Flagyl,   May & Baker Ltd., Dagenham, Essex, England

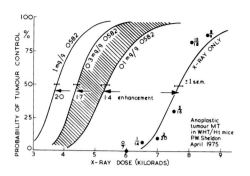
Fig. 3.

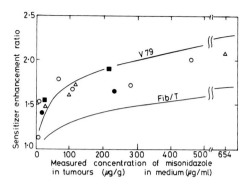
Fig. 4.

very high concentrations in vitro, of 2.5 and 1.9 respectively as compared with the oxygen enhancement ratio of 2.8 in the mammalian cells used for the tests.

The effect in animal tumours of using various doses of misonidazole is shown in Fig. 3 (Sheldon & Hill, 1977). Local control of a mouse tumour (i.e. cure of the mice) is plotted against X-ray dose, given as single doses. The response curve is displaced to the left with increasing doses of the drug. This is radiosensitization. Since there is no enhancement of effect in normal tissues (unless they are hypoxic), it also represents the therapeutic gain. The shaded area reprssents the expected range of enhancement corresponding to the amounts of drug that can be administered clinically. The Dose Modifying Factors are here 1.4 to 1.7. These would be expected to correspond to the use of 20 fractions of 1g each, or 6 fractions of 3.3g each, respectively, of misonidazole.

The same enhancement ratios are also indicated in Fig. 4, which shows single-dose Sensitizer Enhancement Ratios[3] for four types of solid mouse tumour in vivo and two types of cell line in vitro. The SER values for the tumours in vivo are at least as high as those for chinese hamster cells in vitro at the low concentration used clinically. For example, serum concentrations of about 20-25 μg/ml are obtained after daily doses of 1g misonidazole. Concentrations of 70 μg/ml have been obtained after each of six large (3.5g) doses spaced over 3 weeks. These correspond on Fig. 4 to SER's of 1.4 and 1.7 respectively. Tumour concentrations have ranged between 40% and 107% of the plasma levels, mostly 70 - 90%, in human tumours. There is therefore an excellent laboratory basis for expecting a significant effect clinically, even though the amount of misonidazole must be limited to about $12g/m^2$ (20 - 24g total) over a few weeks in order to avoid neurotoxicity (Dische and others, 1977).

Clinical results have also shown positive enhancement. Figure 5 shows that the growth delay in skin nodules treated with 800 rads after misonidazole was the same as in nodules treated with 960 rads of X-rays alone (Thomlinson and others, 1976). This means an SER of 1.2 for a single dose. From this ratio Denekamp and others (1978) estimate that the proportion of hypoxic cells in these human nodules was probably between 5 and 35%.

---

[3]Sensitizer Enhancement Ratio = SER

SER = $\frac{\text{Dose without sensitizer}}{\text{Dose with sensitizer}}$ to produce a given effect.

Figure 6 shows Urtasun's (1976) clinical trial of metronidazole in glioblastomas. The radiotherapy schedule was a modest one : 9 x 330 rads in 3 weeks. A very significant increase in mean survival time was obtained with the metronidazole. However, it was only improving a rather poor radiotherapy schedule up to the same level as the good ones. This is therefore the same type of performance as was shown for HBO at the top of Table 1.

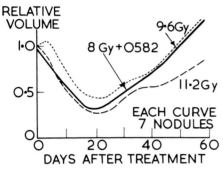

Fig. 5.

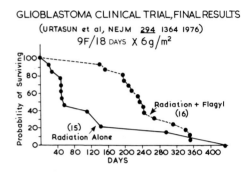

Fig. 6.

Since metronidazole can only be acting on the hypoxic cells, however - either by radiosensitizing them or by killing them directly, as a specific cytotoxic drug for hypoxic cells - this result indicates clearly that hypoxic cells are a problem in glioblastomata.

No enhancement of the normal-tissue response has been seen in patients with metronidazole or misonidazole, but it is expected that for tissues which are hypoxic, such as laryngeal cartilage, radiation doses will have to be reduced by 5 or 10% (Hendry & Sutton, 1978), i.e. probably less than with HBO.

FRACTIONATION WITH MISONIDAZOLE

We have carried out comprehensive sets of fractionation experiments on two contrasting types of mouse tumour. The results illustrate the capabilities of radiosensitizers.

First, a transplanted mammary tumour in C3H mice was used because it was known to reoxygenate rapidly and extensively and was therefore a challenging test for fractionated treatments with misonidazole. Figure 7 shows the results of a number of fractionation schedules, one for each point. Tumour control is plotted against overall treatment time, always for the same degree of normal tissue skin reaction. The volume doubling times of the tumours ranged from 4 to 12 days, so for "days" in Fig. 7 perhaps "weeks" might be considered instead on a clinical scale, if volume doubling is the relevant comparison.

Figure 7 shows that for X-rays only there is an optimum overall time of 9 or 10 days. Beyond that, the treatment was too slow to succeed. At shorter overall times than 9, the results were very variable. The poor results are explained by failure of the tumour to re-oxygenate. Between 5 and 9 days the tumour probably becomes fully reoxygenated. The effect of giving misonidazole before each fraction, in several of the schedules, is shown in Fig. 8 by the vertical arrows. In each case the tumour control <u>has been brought up to an acceptable level</u>. At the optimum overall time, the increase was not significant. At shorter overall times, enhancement was greater because reoxygenation was less complete. Even the single dose was improved to the same level as the other results, in this tumour.

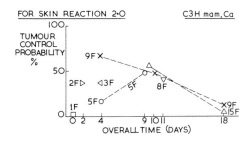
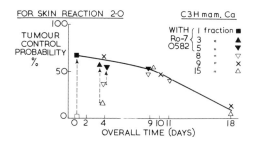

Fig. 7. Fig. 8.

Although it was possible to find an optimum overall time before which hypoxic cells had been eliminated by reoxygenation, this cannot be guaranteed in human tumours. The use of misonidazole brought the poor schedules up to the same level as the good ones. Just such an improvement was demonstrated in Urtasun's clinical use of metronidazole (Fig. 6) as well as in the use of HBO with 6 fractions (Table 1). All this is strong evidence to suggest that if hypoxic cells are eliminated, shorter fractionation schedules will become more reliable. The economic consequences are obvious. (An interesting alternative with this C3H tumour was the twice-a-day schedule, 9 fractions in 4 days) (Fowler & colleagues, 1976).

The second mouse tumour we investigated was a slowly-shrinking sarcoma that reoxygenated much less than the previously described tumour. The results are shown in Fig. 9, again all for estimated equal normal tissue damage. With X-rays only, 20 fractions were better than 5 at all overall times (full lines), the shorter times being better than the longer because this tumour grew fast, with a volume doubling time of just under 2 days (Sheldon & Fowler, 1978).

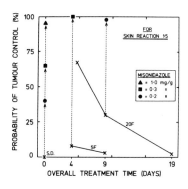

Fig. 9.

When misonidazole was given before each fraction, the 20-fraction and also the 5-fraction schedule was increased (dotted lines). The single doses were not in this tumour enhanced to as high a level as the multiple fractions. The poor 5-fraction schedule was improved to a higher level than the best 20-fraction X-ray only schedule. So was the mediocre 20F schedule. This is an illustration of something we have yet to see demonstrated clinically: a mediocre X-ray schedule being improved beyond the best that the optimum X-ray-alone schedule could achieve.

If human tumours are more like the second than the first type of mouse tumour, then significant clinical improvements will be observed readily.

SIZE OF FRACTION

When single doses are given to a mixed population of oxic and hypoxic cells, the SER increases with size of X-ray dose. This is obvious because at very low doses the oxic cells will mask the effect on the hypoxic ones. If tumours reoxygenate in such an "artificial model" way that the proportion of hypoxic cells always returns to its original value before each of the next doses, then again the SER will increase with size of X-ray dose.

However, if reoxygenation in the tumour is better than this, a modest amount of extra cell killing by the radiosensitizer will eliminate them all. The observed SER may not be very big, but the result on the tumour (for a given normal-tissue injury as in Figs. 7, 8, 9) cannot be improved, because all the hypoxic cells have been removed. In this situation - when total success is readily achieved - the observed SER depends only on reoxygenation; fraction size with the sensitizer is not important. It only matters in a secondary way: if reoxygenation is speeded by using one fractionation schedule rather than another, then of course it matters. But we cannot predict the balance of normal tissue injury versus more reoxygenation in human tumours when fractionation is varied.

If total success is not being achieved, and a substantial proportion of hypoxic cells is present late in the fractionation schedule, then a few large fractions will be the best way to use radiosensitizers. A smaller effect would be observed from a large number of small fractions, as in the lung cases from Portsmouth treated with HBO for 30 or 40 fractions. In the same centre they have found a significant effect of HBO with 6 fractions. By contrast, the improved results in HBO for Ca Cervix treated with 25-30 fractions show that in Ca Cervix, success is easier to achieve and multiple small fractions, plus HBO as a way of dealing with hypoxic cells, can achieve it.

In summary, a few large doses each with radiosensitizer is likely to give a larger enhancement than many small doses; but it might enhance from a poorer control schedule so that the result may be no better than with multiple small fractions.

A separate reason for wishing to use multiple small fractions is the direct cytotoxicity of these sensitizers to hypoxic cells specifically (Stratford & Adams, 1977). The effect depends more on time of exposure than on concentration, so a series of small drug doses would amplify this potentially useful aspect. However, it would be cancelled by cyclic reoxygenation and is not yet so well established as radiosensitization, as a significant mechanism in vivo.

GIVE SENSITIZERS LATE OR EARLY?

If radiosensitizers cannot be used throughout a course of fractionated therapy, should they be given early or late in the schedule? A superficial consideration might suggest that they should be kept to the end, to be sure of helping to eliminate the last remaining hypoxic cells. This is wrong. They should be given at the beginning, except in the unlikely event that the number (not the proportion) of hypoxic cells increases towards the end of treatment. In that case the treatment is grossly failing anyway.

The reason for giving sensitizers early is that the number of hypoxic cells should be reduced as quickly as possible, so that there are fewer left to become reoxygenated. Figures 10 and 11 give results of calculations which show the relative cell killing in the hypoxic and oxic populations of a mixed tumour. If we consider only the hypoxic cells, it makes no difference whether the sensitizer is given late or early. But if the sensitizer is given late there are of course more hypoxic cells present early, which produce extra oxic cells if they become reoxygenated. The differences are not large, but are in the direction of better cell kill when the sensitizer is used early. The one exception was mentioned above and corresponds to a tumour which continues to grow significantly throughout the fractionated schedule.

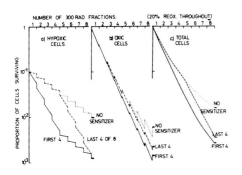
Fig. 10.

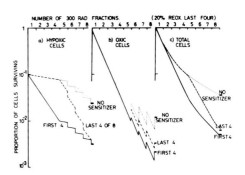
Fig. 11.

CONCLUSIONS

Hyperbaric oxygen results have demonstrated that there is a problem of hypoxic cells in some types of tumour, and that even small daily fractions with HBO can give improved results. Reasons are given for expecting radiosensitizing drugs to penetrate better than HBO to the hypoxic cells, which exist in tumours exceeding 1 or 2 mm diameter. Preliminary clinical studies have demonstrated significant effects on human tumours of misonidazole and metronidazole. They also indicated that peripheral neurotoxicity sets a limit of $12g/m^2$ of misonidazole administered within a period of 3 to 6 weeks.

A number of clinical trials are being initiated using misonidazole. Future clinical results of hypoxic-cell radiosensitizers will clarify further the role of hypoxic cells in causing resistance to radiotherapy.

ACKNOWLEDGEMENTS

I have pleasure in thanking my colleagues for stimulating discussions, especially Dr. S. Dische, Prof. G.E. Adams, Drs. J. Denekamp, R.C. Urtasun, O.C.A. Scott, J.M. Hendry and H. Suit. Thanks are also due to the Editors of Spectrum for Fig. 1; Cancer Treatment Review for Fig. 2; British Journal of Cancer for Figs. 3, 9, 10 and 11; British Journal of Radiology for Fig. 4; Clinical Radiology for Fig. 5; International Journal of Radiation Oncology, Biology, Physics for the basis of Figs. 7 and 8 and New England Journal of Medicine for Fig. 6.

REFERENCES

Adams, G.E. (1973). Chemical radiosensitizing of hypoxic cells. Brit. Med. Bull., 29, 48-53.
Adams, G.E. (1976). Hypoxic cell radiosensitizers for radiotherapy. In F.F. Becker (Ed.), Cancer: A Comprehensive Treatise, Vol. 6, Plenum Press, New York. Chap. 6, pp. 181-219.
Adams, G.E. and D.L. Dewey (1963). Hydrated electrons and radiobiological sensitization. Biochem. Biophys. Res. Comm., 12, 473-477.
Adams, G.E., J. Denekamp, J.F. Fowler (1976). Biological basis of radiosensitization by hypoxic-cell radiosensitizers. In K. Hellmann and T.A. Connors (Eds.) Chemotherapy, Vol. 7, Plenum Press, New York. pp. 187-206.
Asquith, J.C., M.E. Watts, K. Patel, C.E. Smithen and G.E. Adams (1974). Electron affinic radiosensitization : V. Radiosensitization of hypoxic bacteria and mammalian cells in vitro by some nitro imidazoles and nitro pyrazoles. Radiat. Res., 60, 108-118.

Cade, I.S. and J.B. McEwen (1978). Clinical trials of radiotherapy in hyperbaric oxygen at Portsmouth 1964-1976. Clin. Radiol., 29, 333-338.

Cater, D.B., A.F. Phillips and I.A. Silver (1957). Proc. Roy.Soc.B146, 289.

Churchill-Davidson, I., C. Sanger and R.H. Thomlinson (1957). Oxygenation in radiotherapy. II Clinical application. Brit. J. Radiol., 30, 406-422.

Dische, S., M.I. Saunders, M. Lee, G.E. Adams and I.R. Flockhart (1977). Clinical testing of the radiosensitizer Ro-07-0582. Experience with multiple doses. Brit. J. Cancer, 35, 567-579.

Dische, S. and others (1978). Summary of results of the MRC trials of hyperbaric oxygen. Brit. J. Radiol. (in press). Also to be published in The Lancet, after July 1978.

Denekamp, J., J.F. Fowler and S. Dische (1977). The proportion of hypoxic cells in a human tumor. Int. J. Rad. Oncol. Biol. Phys., 2, 1227-1228.

Fowler, J.F., G.E. Adams and J. Denekamp (1976). Radiosensitizers of hypoxic cells in solid tumours. Cancer Treatment Reviews, 3, 227-256.

Fowler, J.F., P.W. Sheldon, J. Denekamp and S.B. Field (1976). Optimum fractionation of the C3H mouse mammary carcinoma using X-rays, the hypoxic cell radiosensitizer Ro-07-0582, or fast neutrons. Int. J. Rad. Oncol. Biol. Phys., 1, 579-592.

Gray, L.H., A.G. Conger, M. Ebert, S. Hornsey and O.C.A. Scott (1953). The concentration of oxygen dissolved in tissues at the time of irradiation as a factor in radiotherapy. Brit. J. Radiol., 26, 638-648.

Hendry, J. . and M.L. Sutton (1978). Care with radiosensitizers. Brit. J. Radiol. (in press).

Henk, J.M., P.B. Kunkler and C.W. Smith (1977). Radiotherapy and hyperbaric oxygen in head and neck cancer. The Lancet, July 16, 101-105.

McNally, N.J., J. Denekamp, P.W. Sheldon and I.R. Flockhart (1978). Hypoxic cell sensitization by misonidazole in vivo and in vitro. Brit. J. Radiol., 51, 317-318.

Sheldon, P.W. and J.F. Fowler (1978). Radiosensitization by misonidazole (Ro-07-0582) of fractionated X-rays in a murine tumour. Brit. J. Cancer, 37, Supple. III, 242-245.

Sheldon, P.W. and S.A. Hill (1977). Hypoxic cell radiosensitizers and tumour control by X-ray of a transplanted tumour in mice. Brit. J. Cancer, 35, 795-808.

Stratford, I.J. and G.E. Adams (1977). The effect of hyperthermia on the differential cytotoxicity of a hypoxic cell radiosensitizer, the 2 nitroimidazole Ro-07-0582, on mammalian cells in vitro. Brit. J. Cancer, 35, 307-313.

Suit, H.D. (1978). Hyperbaric oxygen and irradiation. Review of laboratory experimental and clinical data. In Proc. International Meeting for Radio-Oncology at Baden, Austria, May 1978. To be published by Georg Thieme Publishers, Stuttgart.

Thomlinson, R.H., S. Dische, A.J. Gray and L.M. Errington (1976). Clinical testing of the radiosensitizer Ro-07-0582. III. Response of tumours. Clin. Radiol. 27, 167-174.

Urtasun, R.C., P. Band, J.D. Chapman, M.L. Feldstein, B. Mielke and C. Fryer (1976). Radiation and high dose metronidazole (Flagyl) in supratentorial glioblastomas. New England J. Med., 294, 1364-1367.

The Skin as a Model for the Analysis of Radiation Injury

R. L. Cabrini

*Gerencia de Investigaciones, Comisión Nacional de Energía Atómica,
Buenos Aires, Argentina*

ABSTRACT

The skin of the rat is presented as a model of experimental irradiation. Biological material was irradiated with a wide range of dosis and analysed after several periods within the acute response. Several sources of irradiation were employed: X, γ, and β rays and particles. Epithelial cell response is caracterized by a fast volume increase wich is not clearly due to edema, as it was long been considered. Stereological studies at electron microscopy level revealed an increased content of fibrilar proteins (tonofilaments) and an increased number of mitochondrias.
Some biological aspects of irradiated epithelium were also interesting particularly those related to a substantial increase of pentose-shunt enzyme activities. These enzyme variations at cell level were evaluated by means of microspectrophotometric techniques.
There is a close relationship between stereological and biochemical data showing a metabolic response of epithelial cells subjected to a letal injury.

INTRODUCTION

Several factors account for the practical importance of the knowledge of the skin as a biological efector of irradiation. Generally, it is not feasible to perform an external irradiation without skin involvement. On the other hand, there are also numerous tumours of cutaneous origin or having a clear malpighian differentiation.
In the Pathology Laboratory at the National Atomic Energy Comission we have analysed a wide range of different skin reactions to irradiation. Up to the present, we have studied the responses to the radiobiologically so-called "high doses" which are comparable to the tumour doses used in therapeutics.
Our principal interest was focused on skin epithelium, since it is an easy structure to anlyse at cellular level.
Several irradiation sources were applied to our model of tail, sole or dorsal rat skin. They are detailed in Table 1.
Concurrent methods were employed to analyse structural alterations, proliferation systems and biochemical responses.

RESULTS

The first structural modification detected as initial effect of radiation is an epithelial volume increase.(Rey and Klein-Szanto,1972). Within certain limits,the volume increase is proportional to the dose and the post-irradiation time considered. (Fig.1)

TYPES OF IRRADIATION

X RAYS	250	Kw	10 mA	0.6-	16 Kr
X RAYS	30	Kw	20 mA	3 -	16 Kr
β (^{90}Sr)	2.27	MeV	20 r/sec	16 -	32 Kr
γ (^{60}Co)	1.30	MeV	300 r/min	1 -	10,000 Kr
DEUTERONS	27	MeV	150,000 r/min	50 -	10,000 Kr

Table 1. Irradiation sources, energy and dose range applied to the skin model.

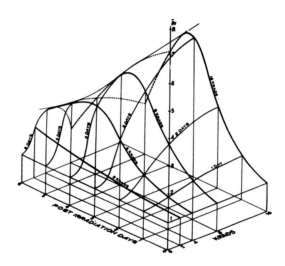

Fig.1. Post-irradiation time plotted against dose and relative epithelial thickness.

The increase in the epithelial thickness is much more evident in those epithelia which have hair follicles than in hairless areas, such as sole skin or mucous membranes.(Itoiz and Frasch,1977). A chasing after tritiated thimidyne injection showed that the observed acanthosis is largely due to a displacement of the hair

follicles which become included within the epithelium (unpublished data).
A detailed study showed an increased number of cells per surface unit but the
magnitude of the increase found was not enough to explain the total volume increase.
An augmented size of each individual cell or an increase of intercellular spaces
had to be admitted.(Fig.2).

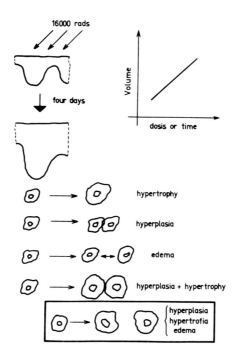

Fig.2. Schematic representation of different
possible mechanisms of epithelial thickness
increase.

The question then arose about which component of the epithelial tissue is modified
to account for the total volume increase. Answers to this question have been found
by applying stereological methods.(Rey, Klein-Szanto and Cabrini,1979).
The cellular volume increase was analysed by electron microscopy. The stereological
methods applied to a very rigorously controlled fixed material made it possible to
establish a clear differentiation between the intracellular and the extracellular
increase.(Klein-Szanto, Rey and Cabrini,1977). It was found that cells submitted
to high doses of radiation increase their volume and modify the number and distribu=
tion of the organelles.(Fig. 3 and 4). The generally admitted production of intra=
cellular edema was not always present. On the contrary, modifications and redistri=
bution of intracellular structures were observed.
The nucleous increases its volume and nuclear membrane invaginations appear. There
is a clear increase in the number and volume of mitochondrias and a very striking
increase of tonofibrils.(Fig.5).
These data support the idea of an accelerated cellular energy system (mitochondrial
factor) and of a steady increase in protein synthesis (tonofibrills increase).
The energy system was analysed by us by means of quantitative histochemical
techniques which were technologically developed in our laboratory and then applied

to our models (Cabrini, Viñuales and Itoiz,1969; Cabrini, Klein-Szanto and Itoiz, 1970; Frasch,Itoiz and Cabrini,1978). These techniques were mainly based, in principle, on measurements of biochemical and histochemical models of known activity using a microspectrophotometer as sensor to gauge a stoiquiometric reaction (Fig. 6)

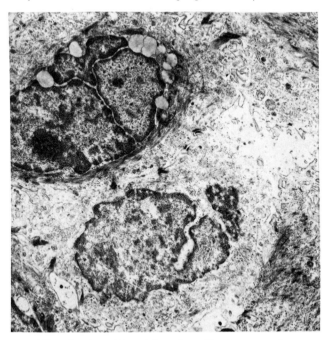

Fig. 3. Irradiated basal cell. Deep nuclear invaginations and formation of lipid vacuoles.

Microspectrophotometric evaluation showed a remarkable variation of oxidative enzymes. Fig 7 show results obtained subsequent to deuteron irradiation,(Cabrini and co-workers, 1970). The same general behaviour was observed after X and β radiations. Krebs cycle enzymes decreased their activities per volume of tissue and a shift towards pentose-shunt enzyme activities was observed (Klein-Szanto and Cabrini,1970, 1972; Itoiz and Frasch, 1977).

Total protein synthesis in irradiated cells was studied by microphotometry of the naphtol yellow S reaction. (Fig. 8) . An increment of total protein was detected which was coincident with the tonofilaments increase observed by electron microscopy. (Conti, Gimenez and Cabrini, 1976). Incorporation of tritiated aminoacids detected by liquid scintillation counting after microdisection of epidermal samples also pointed to an increase in protein synthesis. A remarkable increment in cystine up= take was evident.(unpublished data).

All these experiments have shown that a number of concurrent methods is to be applied to obtain some fairly complete information about radiation effects on tissues. Results obtained through different methodologies must be correlated to reach a basic global knowledge of the process.

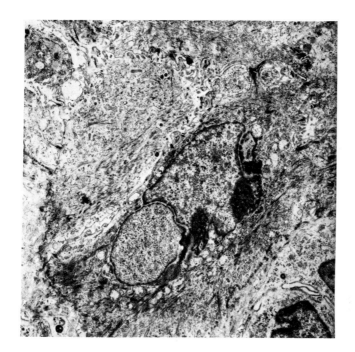

Fig.4. Irradiated spinous cell. Nuclear fragmentation, aggregations of tonofibrils bundles and enlarged nucleolous.

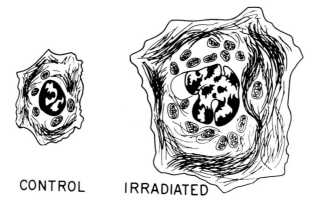

CONTROL   IRRADIATED

Fig. 5. Schematic representation of volume and intracellular changes of irradiated keratinocytes based on morphological and stereological observations.

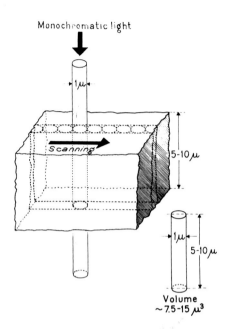

Fig. 6. Shematic representation of the microphotometric scanning method. Tissue is authomatically moved over a small light spot and optical densities are graphically registered and/or integrated.

Fig. 7. Mean enzymatic concentration in marginal lesion epithelium irradiated with 500 Krads. The microphotometric values of four enzyme activities studied are plotted against time post irradiation. Time axis on logarithmic scale.

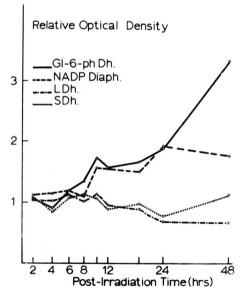

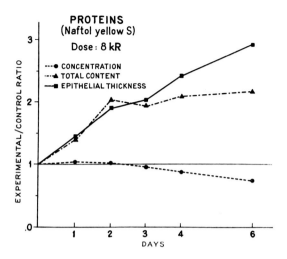

Fig. 8. Microspectrophotometric determination of proteins in irradiated epithelium.

REFERENCES

Cabrini,R.L., E.J. Viñuales, and M.E.Itoiz (1969). A microspectrophotometric method for histochemical quantitation of succinic dehydrogenase.Acta Histochem. 34, 287-291.
Cabrini, R.L., A.J.P. Klein-Szanto, and M.E. Itoiz (1970). Unidirectional scanning for the microspectrophotometric investigation of enzyme reactions in squamous epithelium. Acta Histochem., 36, 397-401.
Cabrini, R. L. , M. E. Itoiz, J. Mayo, E.E. Smolko, and A. J.P. Klein-Szanto (1970). Microspectrophotometric study of histoenzymatic reactions in rat epithelium subjected to 0.5 Mrad of deuteron irradiation. Int. J. Radiat. Biol.,18, 415-421.
Conti, C.J., I.B. Gimenez, and R. L. Cabrini (1976). Microspectrophotometric quantitation of nucleic acids and protein in irradiated epidermis. Strahlentherapie, 151, 236-239.
Frasch, A.C.C., M.E.Itoiz, and R.L.Cabrini (1978). Microspectrophotometric quantita= tion of diaminobenzidine reaction for histochemical demonstration of cytochrome oxidase. J. Histocem. Cytochem., 26, 157-162.
Itoiz,.E., and A.C.C.Frasch (1977). Variations in the enzymatic behaviour of irradiated epidermis as function of age and lesion localization. Int. J.Radiat. Biol., 31, 499-502.
Klein-Szanto, A.J.P., and R.L.Cabrini (1970). Microspectrophotometric study of oxidative enzymes in irradiated epidermis. Int. J.Radiat. Biol.,18, 235-241.
Klein-Szanto, A.J.P., and R. L. Cabrini (1972). Acute response of oxidative enzyme systems in epidermis subjected to beta radiation. Arch. Derm. Forsh.,243, 226-231.
Klein-Szanto, A.J.P., B.M.de Rey, and R.L.Cabrini (1977). Volumetric modifications of X-irradiated keratinocytes. J. Cutaneous Pathol.,4, 23-31.
Rey, B.M de, and Klein-Szanto, A.J.P. (1972). A histometric study of acute radiation effects on rat tail epidermis. Strahlentherapie, 6, 699-704.
Rey, B.M. de, A.J.P. Klein-Szanto, and R.L.Cabrini (1979). Ultrastructural stereology of X-irradiated epidermis. Rad. Res. 77, (in press).

Summary of Papers Presented at Symposium No. 8

L. Révész* and O. C. A. Scott**

*Department of Tumor Biology II, Karolinska Institutet, Stockholm, Sweden
**Division of Biophysics, Institute of Cancer Research, Sutton, Surrey, U.K.

In his introduction the Chairman pointed out that the title "Progress in Radiobiology in Cancer" covered such a wide area that some restriction was necessary, and the theme "fractionation" was used to tie the symposium together.

Dr. Suit opened the session with a survey of possible variables in time, dose, and dose-rate in radiotherapy, and went on to relate these ideas to knowledge gained from irradiation of cells in culture. If cells in culture reveal a high "extrapolation number" suggesting a great capacity for repair, two approaches can be considered: If we wish to kill such cells, it is logical to use a small number of large doses. If, on the other hand, we wish to spare them, small doses per fraction or low dose rate should be used. As an example he quoted the work of Thomas on whole body irradiation. The acute lethal dose for man is around 250 rad, but if the dose rate is reduced to 4 - 8 rad per minute, 1000 rad can be tolerated due to the great repair capacity of gut; on the other hand the number of lymphoid cells killed is increased. This is an example of sparing a normal tissue. If it is the tumour which has the great repair capacity, e.g. melanoma, large doses per fraction may be used with success.

If excessive cell-proliferation in the tumour is a cause of failure, it is logical to reduce the overall time of treatment as in the work of Norin et al. on Burkitt's lymphoma.

Cell survival curve considerations have been of the greatest value in understanding the connection between RBE and dose per fraction using neutrons and in avoiding the excessive normal tissue damage observed by Stone in the first neutron trial.

Dr. Durand described a work in which Chinese hamster cells grow in contact with one another in "spheroids". Since in tumors (or normal tissues) cells are in contact, the spheroid model will be closer to the situation in vivo than classical Puck technique. Cells in spheroids display a greater capacity to accumulate sublethal injury than the same cells separated by trypsinization. New work reported here suggested that the difference in cell survival between cells in spheroids and separated cells for a given dose of radiation is greatest for single doses, and becomes less marked as the number of fractions is increased.

Dr. Phillips discussed the combination of X-rays and chemotherapy. The combination might be of advantage in three situations: 1. the improvement of local cure; 2. the presence of metastases only reached by chemotherapy; and 3. the use of radiation to kill cells in "sanctuary sites" where chemotherapeutic agents cannot reach them.

The interaction of the two agents at one site may be subadditive, or additive, or potentiation may occur. In addition "spatial co-operation" (as in example 3 above) may be observed.

He reported experimental work in which the test system was the Withers intestinal crypt cell survival assay. He defined the "dose effect factor" (DEF) as the dose of radiation alone, divided by the dose of radiation in the presence of a drug, for a given effect.

The combination of Bleomycin with X-rays did not alter the slope of the survival curve, but may have reduced the repair capacity to zero ($D_q = 0$). The DEF was 1.9 for surviving crypts.

He was able to distinguish effects on repair capacity from other effects by using a low LET radiation and a high LET radiation. This led to interesting results. Actinomycin D was more effective given before irradiation, but this difference disappeared using high LET irradiation. In contrast, the difference observed with BCNU persisted when high LET was used. His general conclusion was a little pessimistic. Some tumors are resistant to some drugs but cells from normal tissues appear to be generally sensitive. Phillips also reported work of Barendsen on a rat rhabdomyosarcoma. If X-rays are given first, the surviving tumor cells may proliferate rapidly for a time, and during this period should be more sensitive to chemotherapy.

Dr. Fowler compared hypoxic cell sensititizers with hyperbaric oxygen (HPO) in radiotherapy.

Oxygen is potentially the best sensitizer but, since it is metabolized, it may fail to penetrate to all the hypoxic cells in tumor. Hypoxic sensitizers are less effective (as judged by the enhancement ratio) but may penetrate to all the hypoxic cells.

He summarized the results of the trials of HPO and concluded with Dr. Suit´s words: "If Henk´s results for head- and neck-cancer are confirmed (a significant improvement in both local control and survival with no extra normal tissue complications) then a real advance in radiotherapy will have been made". He pointed out that HPO is more effective if schedules are used which give bad results in air, but nevertheless treatments employing multiple small dose-fraction were also improved by HPO.

Electron-affinic sensitizers are of more potential value than other chemotherapeutic agents because most tumors contain hypoxic cells, and most normal tissues do not. The sensitizer misonidazole has proved to be an excellent sensitizer in a wide varity of mouse tumors. Clinical work has also revealed sensitization of skin nodules with an enhancement ratio of around 1.2. Urtasun´s clinical trial with Flagyl is encouraging.

Fractionation experiments with mouse tumors suggest that the least effective fractionation schemes are improved the most by misonidazole. Considering dose per fraction he suggested that the enhancement ratio is likely to be large if large doses per fraction are used; on the other hand, using prolonged treatment, the direct cytotoxic effect of misonidasole on hypoxic cells may be a bonus. If sensitizers are given during a part of a fractionation course, they should be given at the beginning. He concluded that HPO trials had demonstrated that hypoxic cells are a problem in radiotherapy, and the use of sensitizers will clarify the situation. He recommends that in future clinical trials two different levels of dose be used in the experimental arm. He hopes this will enable us to obtain answers more quickly.

Morphologic and some functional changes of the irradiated skin were analyzed by Cabrini. The study comprised detailed description of the structural changes of cellular micro-organelles, activity of the respiratory enzymes and protein synthesis.

The characteristics of radiofrequency and microwave interactions with biological materials were described by Dr. Portela. The factors influencing the propagation and absorption characteristics of the waves were explained, and the dosimetric problems of various parameters discussed in regard to therapeutic applicability. Current experimental and clinical work on thermotherapy of tumors, alone or in combination with radio- and chemo-therapy, was reviewed. Diagnostic possibilities were also pointed out, especially in detecting breast-cancer by microwave radiometry.

Progress in Clinical Radiotherapy in Cancer

Clinical Experience with Negative Pi Mesons and Other High-LET Radiations[1]

J. M. Sala and M. M. Kligerman*

*Cancer Research and Treatment Center, University of New Mexico,
Albuquerque, New Mexico 87131, U.S.A.
*CRTC/UNM and the Los Alamos Scientific Laboratory, Los Alamos,
New Mexico 87545, U.S.A.*

ABSTRACT

This paper summarizes results to date with high linear-energy-transfer (high-LET) particles (neutrons, pions, and heavy ions), being tested at centers throughout the world, with emphasis on preliminary information concerning negative pi mesons (pions) at Los Alamos. Results thus far in randomized Phase III clinical trials with neutrons indicate the injury rate is higher than with conventional radiation, but the possibility still exists that the net benefit in local tumor control and subsequent increased survival will outweigh the risk of injury. Pilot studies with pions designed to establish normal tissue tolerances prior to starting randomized Phase III clinical trials are nearing completion at Los Alamos. Results have shown relatively modest reactions of all pion-irradiated normal tissues, particularly those in the pelvis. Complete regressions with pions alone have been observed only at doses of 2700 peak pion rads or more. Subsequent surgery and conventional radiation have been well tolerated after pion therapy by those patients requiring these procedures. Pilot studies are also underway with heavy ions of helium, carbon, and neon to establish tolerance dose levels for Phase III clinical trials. Normal tissue reactions have been modest. A preliminary value for relative biological effectiveness (RBE) of 1.2 has been established for helium ions, although it appears a higher value will be experienced for carbon ions.

Keywords: particle radiotherapy, high linear energy transfer (high LET), low LET, neutrons, pions, carbon ions, helium ions, clinical trials, therapeutic gain.

INTRODUCTION

The clinical investigation of irradiation of tumors with heavy particles is of major interest throughout the world today. Because of their size and in some instances electrical charge, they create a biological response different from that

---

[1]These investigations were supported in part by U.S. Public Health Service Grants No. CA-16127 and CA-14052 from the National Cancer Institute, Division of Research Resources and Centers, and by the U.S. Department of Energy.

seen with conventional radiation, such as the gamma rays of cobalt, high energy x-rays, electron beams, and interstitial irradiation. Heavy particles currently under investigation include neutrons, negative pi mesons (pions), heavy ions, and protons.

Radiations can be defined by the density of the ionization they incur within a submicroscopic volume of tissue. Those which have a relatively low density of ionization (x-rays, gamma rays of cobalt, electrons, protons, and pions in their flight through tissue) are called "low linear energy transfer" (low-LET) radiations. Those which are characterized by a high density of ionization (neutrons, heavy ions and pions in their stopping region) are termed "high-LET" radiations. A beam may be considered high LET if it deposits more than 50 kilovolts per micron of tissue.

The greater density of ionization of high-LET radiations has a greater probability of creating double-strand breaks in DNA molecules, which are difficult to repair, as compared to the more limited injury caused by low-LET radiations. Because of this inhibition of repair, high-LET particles have the potential of overcoming the protective effects of hypoxia and variations in cell cycle sensitivity which inhibit the effectiveness of low-LET radiation.

Charged particles have an additional characteristic which makes them attractive for clinical investigation. While x-rays, gamma rays of cobalt, and neutrons (which have no electrical charge) deposit dose exponentially as they pass through tissue, charged particles have a finite range in tissue. As these particles are slowed down by the tissue they traverse, a greater density of ionizing events occurs, thus depositing a greater dose at the end of their range (the "Bragg peak"). By superimposing a number of contiguous Bragg peaks within a defined tumor volume, the peak region can be spread to enhance effective dose deposition within the tumor volume, while largely sparing surrounding normal tissue. For negative pions, a unique secondary event occurs which increases their biological effectiveness. As they stop, the pions are absorbed by the positively charged nuclei of atoms. The presence of this additional energy causes the nuclei to become unstable and disintegrate, releasing large, densely ionizing subnuclear fragments of short range. This augments the dose and enhances the biological effectiveness in the stopping region.

The testing of these various particles at radiotherapy centers throughout the world is largely directed toward determining whether their clinical use will produce a therapeutic gain--greater effect on tumor than on normal tissues--thereby improving management of tumors. This paper is concerned with those heavy particles which are also characterized by high-LET radiation (i.e., neutrons, pions and heavy ions). Protons, while low-LET in character, are under investigation in the Soviet Union and the United States (at the Harvard cyclotron) because their dose localizing properties are the most advantageous of all particles. However, since they do not have the potential for increased biological effect within the tumor volume, they are not addressed here.

NEUTRONS

Clinical experimentation with neutrons was first reported by Stone and Larkin (1942), but the method lost favor because of the large number of injuries to normal tissues. The concepts of relative biological effectiveness (RBE) and inhibition of normal tissue repair by heavy particle irradiation were not appreciated at that time. Greater understanding of these biological phenomena has led to renewed interest in neutrons, and Phase III trials are underway at a number of institutions throughout the world.

Recently, Dutreix and Tubiana (1978) compared data from several of these facilities and concluded that either better or worse results are obtained with neutrons as compared to photons, depending on tumor types and sites. They noted that neutrons appeared most encouraging for carcinomas of the upper respiratory and digestive tracts (primary and lymph nodes), salivary gland tumors, advanced carcinoma of the cervix and soft tissue sarcoma. However, they stated that discouraging results have been obtained with brain tumors and that therapeutic benefit as compared to low-LET radiation is doubtful for adenocarcinoma of the breast and carcinoma of the gastrointestinal tract. They did note that two centers reported superior results for a mixed schedule of neutrons and photons (given on alternating days) as opposed to neutrons only.

These are tentative results, as less than 3000 patients throughout the world have been treated with neutrons, and follow-up is short for many of them. Dutreix and Tubiana caution that dose distribution and energy absorption in the tissues at stake may be contributing to the marginal success with these particles; thus, the conclusions should not be extended to high-LET radiations which can be localized, i.e., pions and heavy ions.

Halnan (1978) evaluated normal tissue responses to neutron irradiation at the same group of centers and concluded that adjustment of dosage seems essential, in view of the undesirable incidence of apparent high dose normal tissue damage. He stated that the reasons for these complications may include difficulties in treatment planning and dose distribution, need for reduced doses for large treatment volumes, physical errors in dose calculation, and inadequate allowance for the effects of fractionation and overall elapsed time.

Greater numbers of patients need to be treated in randomized clinical trials to determine the place of neutrons in clinical radiotherapy. While the current injury rate appears higher than desirable, the possibility still exists that the net benefit in local tumor control and subsequent increased survival will outweigh the risk of injury. The neutron centers of the European countries, the United States, and Japan have agreed to cooperate in common protocols so that a definitive assessment of the value of neutrons can be obtained as soon as possible.

NEGATIVE PI MESONS

Pions represent the binding energy which holds neutrons and protons together in the nuclei of atoms. They have a mass one-seventh that of a neutron. As noted previously, pions in their flight behave as low-LET radiation, while in their stopping region, subnuclear fragments are released, a portion of which are high-LET in character. Thus, pions in their stopping region, deposit a simultaneous mixture of high and low LET. Approximately 10 to 20 percent of the ionizing radiation in the stopping region of the large spread beams used for patient treatment is high LET (>50 kilovolts per micron) in character (Kligerman 1978).

Phase I-II trials designed to determine normal tissue tolerances prior to initiating randomized Phase III trials of pion radiotherapy are nearing completion at Los Alamos, New Mexico, in the United States. To date, a total of 136 tumors in 67 patients have been irradiated in these pilot studies (see Table 1). Radiobiological studies with pions are being conducted in Canada and Switzerland, preparatory to the start of Phase I-II studies. Since the authors' experience is with pions at Los Alamos, emphasis will be given to results thus far with those particles.

Preliminary studies with nine patients having multiple tumor nodules in skin and subcutaneous tissues established the RBE for acute injury of skin at 1.42 for pions versus 100 kVp x-rays (Kligerman and co-workers, 1977). A subsequent evaluation of the time to regrowth of 16 nodules in one patient participating in that experiment who could be followed for 346 days suggested the possibility of therapeutic gain of 37 percent for pions versus x-rays at 300 days (Kligerman and co-workers, 1978a).

TABLE 1   Types/Numbers of Tumors Treated with Pions at Los Alamos

Sites	Cases	Primaries	Nodes	Mestastases (Primary)
Brain	3	2		3 (Breast)
Skin/Subcutaneous Tissues	11	1	2	20 (Breast)
				21 (Melanoma)
				2 (Stomach)
Melanoma	1		1	
Breast	2	1*	1	2
Lung	5	3		5 (Melanoma)
				2 (Endometrium)
Oral Cavity	3	2	5	
Oropharynx	9	9	9	
Nasopharynx	4	4	1	
Larynx	4	4	4	
Stomach	2	1	1	
Pancreas	9	9	2	
Liver	3	1		5 (Rectum)
Bladder	2**	2		
Rectum	5	5	1	
Prostate	4**	4		
Sarcoma	1	1		
TOTALS	68	49	27	60
			136	

*Male patient
**One patient with two pelvic primaries

Subsequent to the skin studies, patients with locally advanced, recurrent or metastatic solid tumors in or near sites contemplated for Phase III clinical trials were selected for Phase I-II pilot studies to assess tumor response and normal tissue reaction. The primary objective was to obtain tolerance levels for various types of anatomic structures, notably oral and gastrointestinal mucosa, lung, and brain. Since the Phase III clinical trials will be addressed toward patients with tumors characterized by little chance of survival by any modern cancer treatment modality or combined modalities in any stage, or those controllable in early stages but not in advanced stages, the outlook for this group of patients in the Phase I-II studies was generally quite poor at the outset. Many had metastatic disease where the goal of treatment was palliative, but where information on tolerance could be obtained. Most had massive primaries with or without apparent regional adenopathy (all Stage III or IV). No patient had received previous conventional radiotherapy to pion-treated volumes. Of those who had received chemotherapy prior to pion radiotherapy, the tumor was either actively growing in spite of chemotherapy, or drugs were stopped at least two weeks prior to the start of pion radiotherapy.

The pion treatment was delivered generally in five fractions per week, with daily fractions ranging between 110 and 140 peak pion rads maximum and total doses ranging between 1000 and 4600 peak pion rads. Doses for each new tissue type were purposely low at the outset, as a safety measure, and were escalated with each successive group of patients as sufficient information was obtained to indicate that higher doses could be safely tolerated. Patients who appeared able to tolerate additional radiation at the end of pion treatment, particularly those in whom tumor regression was incomplete, were given additional conventional radiation. When appropriate, surgical excision of residual mass or regional adenopathy was performed after pion treatment. Chemotherapy was instituted as necessary for disseminated disease.

Patients in this series have now been followed for periods up to 15 months. Given the short follow-up time and the variety of tumor types and doses applied to this series of patients, it is not yet appropriate to report long-term survival. In patients followed for 6 to 15 months (Kligerman and co-workers, 1978b), no patient treated with pions alone who received less than 2700 peak pion rads exhibited complete tumor regression. Conventional radiation was well tolerated by those patients requiring that treatment after pion therapy, although tumor regression was less consistent and normal tissue reactions were greater in these patients. This tolerance, however, is an additional measure of the recuperative ability of normal tissues following pion radiation. Furthermore, when subsequent surgery was carried out, no difficulty was experienced in these operative procedures. Normal tissues within the peak pion radiation field tolerated operative procedures almost as well as if they had not been irradiated.

It is clear, however, that the maximum tolerated doses have not yet been reached in all tissues, as acute reactions to pion radiotherapy have been modest at the doses applied. It is estimated that doses are presently within 5 to 15 percent of tolerance. No patient in this series treated with pelvic portals experienced diarrhea, mucus, bleeding, or spasm. No abnormality of the rectal mucosa was apparent upon proctoscopy. Two patients had an inconstant minimal increased frequency of urination without burning and without blood, which rapidly cleared after pion therapy. The only intermediate reaction reported to date in a patient receiving pelvic radiation has been in a patient with a 13 cm prostatic carcinoma, who received 3300 peak pion rads, 23 fractions, 31 days, who developed two stools a day in the second month after treatment, which cleared within a month. Only one of 13 patients receiving abdominal irradiation experienced diarrhea, and only one displayed mild anorexia and weight loss which the radiotherapists felt could be attributed at least partly to pion radiotherapy.

In the head and neck, reactions have also been mild. Two of the earliest of the 40 cases exhibited a severe reaction of the gingival border, which was successfully avoided in subsequent patients by the oral application of a radioprotectant (acetylcysteine) to the gingival area free of tumor involvement (Kligerman 1977). Of 14 head and neck patients, one had no reaction, four displayed only injection, and five had patches of pseudodiphtheritic membrane over less than half the field. Four patients (the two treated prior to the application of radioprotectants, one with gingival involvement and one with a nasopharyngeal tumor) displayed a third-degree reaction (confluent pseudodiphtheritic membrane), over less than half the field. Five of the 14 patients displayed salivary reactions to pion radiotherapy with suppression of saliva and dryness of the mouth. This suppression has persisted in all five patients. Two patients treated for laryngeal tumors developed edema of the larynx and of the subcutaneous tissues of the anterior neck (dewlap) in the second month after the end of pion therapy. Although it mostly cleared at the end of one month, a small amount of dewlap and +1 edema of the larynx persists.

There have been no serious untoward effects in any patient treated thus far with pions. One patient in the skin series is now at 27 months follow-up. To date, there is no quantitative or qualitative difference in the tissues treated with pions as compared to those treated with 100 kVp x-rays, and there is no evidence of subcutaneous fibrosis.

Observations on the most recent patients treated with escalation of the dose to a minimum of 3300 and a maximum of 4100 peak pion rads to the target volume indicate this dose is close to normal tissue tolerance. This dose is delivered at the rate of 100 minimum peak pion rads per day in 33 treatments over six to seven weeks.

HEAVY IONS

At the Lawrence Berkeley Laboratory, pilot clinical studies are underway to evaluate the effectiveness of charged particle therapy with helium, carbon or neon ions. The ions are produced by the 184" synchrocyclotron and the Bevelac, both of which are high energy physics research accelerators. Similar to the studies with pions, the heavy ion studies are exploring two potentially useful attributes of charged particle radiotherapy: (1) increased biological effect on tumor, and (2) reduced radiation damage to adjacent normal tissues through precise localization of the heavy ion Bragg peak within the tumor volume.

More than 60 patients have been treated with heavy ions, including patients with tumors of the brain, head and neck, esophagus, pancreas, stomach, uterine cervix, and rectum. As with pion radiotherapy, no serious untoward effects have been encountered. Progress toward development of accurate treatment planning and acquisition of data concerning relative biological effectiveness and normal tissue effects has been encouraging, preparatory to initiation of randomized Phase III clinical trials. Although a preliminary RBE has been established for helium at 1.2 (Castro, 1978), data suggest a higher RBE may be experienced with carbon ions. As with pions, a five- to 10-year trial will be needed to determine the clinical applicability of these beams.

REFERENCES

Castro, J. (1978). Personal communication.

Dutreix, J., and Tubiana, M. (1978). Evaluation of clinical experience concerning tumor response to high LET radiation. In <u>Proceedings of the Third Meeting on Fundamental and Practical Aspects of the Application of Fast Neutrons and Other High LET Particles in Clinical Radiotherapy</u>, Pergamon Press, Oxford, in press.

Halnan, K. (1978). Evaluation of normal tissue responses to high LET radiations. In <u>Proceedings of the Third Meeting on Fundamental and Practical Aspects of the Application Fast Neutrons and Other High LET Particles in Clinical Radiotherapy</u>, Pergamon Press, Oxford, in press.

Kligerman, M.M. (1977). Personal communication.

Kligerman, M.M. (1978). Potential for therapeutic gain similar to pions by daily combinations of neutrons and low-LET radiations. Submitted to <u>Medical Hypotheses</u>.

Kligerman, M.M., Smith, A., Yuhas, J.M., Wilson, S., Sternhagen, C.J., Helland, J.A., and Sala, J.M. (1977). The relative biological effectiveness of pions in the acute response of human skin. Int. J. Radiat. Oncol. Biol. Phys., 3, 335-339.

Kligerman, M.M., Sala, J.M., Wilson, S., and Yuhas, J.M. (1978a). Investigation of pion treated human skin nodules for therapeutic gain. Int. J. Radiat. Oncol. Biol. Phys., 4, 263-265.

Kligerman, M.M., Wilson, S., Sala, J., von Essen, C., Tsujii, H., Dembo, J., and Khan, M. (1978b). Results of clinical applications of negative pions at Los Alamos. In Proceedings of the Third Meeting on Fundamental and Practical Aspects of the Application of Fast Neutrons and Other High LET Particles in Clinical Radiotherapy, Pergamon Press, Oxford, in press.

Stone, R.S., and Larkin, S.C., Jr. (1942). The treatment of cancer with fast neutrons. Radiol., 39, 608.

Hyperbaric Oxygen in Radiotherapy

R. Sealy

*Radiotherapy Department,
Groote Schuur Hospital and University of Cape Town, South Africa*

ABSTRACT

This paper is a review and personal experience. Some of the signal events leading to the concept of significant tumour hypoxia are noted and the perivascular tumour cord, venous infarction and stasis is recalled as being causes of this. The reasons for possible failure to reoxygenate cells under treatment conditions may include failure of initial tumour regression and superoxygenation, insufficient oxygenation or vascular spasm. The treatment of head and neck cancer shows improved local control both with primary and nodal disease and evidence of improved survival. This may also be the case in the lung. In cancer of the cervix uteri, improved results in oxygen tend to be associated with trials involving few fractions and not using radium. Prospects for the treatment of bladder cancer seem to be poor but palliation of malignant melanoma is good. Possible new avenues involve combinations with neutrons, vasodilators and hypoxic cell sensitisers. Promising results are put forward in the use of Misonidazole with hyperbaric oxygen.

Key words: Hyperbaric oxygen, sensitisers, Misonidazole (Ro-07-0582) head and neck cancer, cervix uteri cancer, malignant melanoma

Two reports published in 1942 by Lacassange, and by Evans, Goodriff and Slaughter (1942), which indicated that the radiosensitivity of normal structures was increased in the absence of oxygen, were followed by that of Thoday and Read (1947) who indicated that the frequency of chromosome damage was oxygen-dependent. Later, L.H. Gray and his colleagues in 1953, suggested that the concentration of oxygen dissolved in tissues at the time of x-irradiation, was a significant factor in radiotherapy and that cells deprived of oxygen at the time of irradiation were more resistant by a factor of 2.5. This effect was apparently absent in irradiation by neutrons. Two years later in 1955, Churchill-Davidson, Sanger and Thomlinson published the first study of the treatment of patients in high pressure oxygen.

The quarter of a century since these signal reports appeared has given rise to a body of knowledge which enables us to begin to evaluate some of the reports on the clinical significance of oxygenation in radiotherapy. Since some of the successes and failures may be concerned with the basic pathology of tumours and physiology of hyperbaric systems, some of the salient features should be considered.

Thomlinson and Gray in 1955, indicated that cords of human lung cancer greater than 150-200 microns in diameter, developed necrotic areas as a result of hypoxia. It is considered (Hewitt, 1959; Thomlinson, 1967) that cells on the periphery of such areas - perhaps 1% of the malignant cells present, are sufficiently hypoxic to have increased radiation resistance to low LET radiation. Such tumour cells will probably be exposed to oxygen tensions of less than about 3mm of mercury, a level at which radiation sensitivity of about 50% of that of fully oxygenated cells exists (Alper 1956; Deschner, 1959; Elkind 1965). Moreover, in radio-curable tumours, progressive fractions of radiation destroy the well oxygenated cells most proximal to the capillaries, which then leads to a decrease in the intercapillary distance, or a state of super-vascularization (Rubin and Casarette, 1968). It is possible however, that there is more than one reason for the existence of hypoxic cells in tumours. Thomlinson (1960) has described areas of old and new necrosis, the latter being nearer to capillaries. Another possibility is that areas of venous infarction or stasis (Churchill-Davidson, 1957) present in large tumours lead to necrosis and the presence of surviving hypoxic cells. It can be speculated that any cells managing to survive under such conditions, may not be subject to re-oxygenation by decrease in intercapillary distance. It may well be too, that in such areas or in large, hard tumours, the utilisation of oxygen by partially hypoxic cells would not allow complete oxygenation of all cells at the time of treatment (Scott, 1957) and so result in clinical resistance. Powers and Tolmach, in 1964, brought evidence that suggests all cells may not be oxygenated, even high pressure oxygen, and increased oxygenation may not take place during treatment (Hewitt, 1964; Jamieson, 1965).

In the arterial blood, the haemoglobin is about 95% oxygen saturated with a pO_2 value of about 100mm of mercury, whilst at the venous end of a capillary, the pO_2 is about 40mm; about ten times the level needed to increase the radiation sensitivity of hypoxic cells by about 50%. A steep gradient in oxygen tension therefore probably exists from the venous capillary to the significant hypoxic cell, perhaps with a factor of twenty or more. Increased transport of oxygen to overcome this can only take place by solution in the plasma, with about a twenty-fold increase at 3 atmospheres absolute. However, at this pressure, the total oxygen carried in the blood increases only by about one third, the rate of increase being about 2% per atmosphere (Churchill-Davidson, 1957). Therefore a possible advantage exists in treating patients with higher oxygen pressures, as has been done by van den Brenk (1968). The usually used pressure of 3 atmospheres absolute does not require anaesthesia to prevent convulsions, but might produce a significantly less efficient oxygenation of the critical cells.

However, perusal of the substantial published series of results by van den Brenk (1968) and the two series of Chang (1973) and Henk (1977a) and their colleagues, treating patients at 3 atmospheres absolute, show no obvious difference in the local clearance rates of the two methods of treatment. All, however, suggest that there is improved clearance in patients treated in hyperbaric oxygen. Moreover, Henk and Smith (1977b), have reported improved survival at all tumour sites and significantly higher recurrence-free rates in carcinoma of the larynx, when patients were irradiated with 10 fractions of 400 rads in oxygen, compared to 30 fractions of approximately 210 rads in air. Chang (1973), in a small carefully conducted trial, also reports an improved local clearance rate and survival of patients treated in oxygen when compared with two control groups of patients treated with large or small fractions in air, but perhaps owing to small sample size, this trial did not reach statistical significance. Our findings in cancer of the nasopharynx also suggest improved local control, with more of the patients treated in oxygen dying from distant metastases (Sealy,1977). Of especial interest are the improved results for early carcinoma of the mouth, where oxygen appears to be particularly beneficial (Henk, 1977a). Our own experience in treating 28 patients in Cape Town Stage t2, that is 2-4cm in diameter, squamous carcinoma of the mouth, with 6 fractions of 600 rads in 17 days in hyperbaric oxygen, mirrors this. At the present time, the actuarial 1-year local clearance rate is 89%. These findings are of special interest when compared with a series of patients with Cape Town t3 mouth cancer, that is 4-6cm in diameter, published some years ago (Sealy, 1972; Sealy,1974a). These patients were initially treated by intra-arterial infusion with either Methotrexate or Vinblastine and then irradiated in air or in hyperbaric oxygen. In these series of patients, the initial response rate after chemotherapy, as judged by reduction in tumour size, was greater in the Methotrexate groups than in the Vinblastine groups, but the 1-year local clearance rate was in each case better in the Vinblastine group. There is no apparent difference in the 1-year clearance rate in those patients treated in hyperbaric oxygen as compared with those treated in air.

TABLE 1 Head and neck cancer. Intra-arterial infusion and radiation in air or hyperbaric oxygen

	Thiotepa MTX	VLB	Thiotepa MTX	VLB
No. patients	84	23	10	23
Response rate after C.T. %	83	40	60	30
	250kV 200 rads x 23		Co^{60} 500 rads x 9 HPO 3 A.T.A.	
1-Year Clearance rate %	13	30	20	43

Since the reduction in size yielded residual volumes similar to those of t2 tumours, the local control rates at 1 year of 40% or less and similar to those patients treated with radiation alone, compare unfavourably with our recent 89% local control rates of t2 mouth cancers. Pre-radiation shrinkage with cytotoxic agents therefore seems unlikely to greatly improve the results of treatment of larger tumours in high pressure oxygen.

The radiation of lymph node metastases from head and neck cancer also indicates improved results. Van den Brenk's findings (1968), suggest improved local clearance but does not quite reach statistical significance. A similar trend has been noted by Chang (1973). Our finding on the operability rate, after irradiation, of fixed node metastases from carcinoma of the nasopharynx, favours hyperbaric oxygen at the 1% level of significance (Sealy, 1977). Henk and his colleagues (1977a) found the 5-year recurrence-free rate to be higher in nodal metastases treated in oxygen (56%), compared with those treated in air (35%).

Present conclusions therefore, on the treatment of carcinoma of the head and neck, are that there is firm evidence for improved local control rates, that early lesions especially, tend to benefit and there are improved results in the irradiation of lymph node metastases. It is clear that with a suitable choice of fractionation schemes, undue tissue effects are not seen and that there is an improvement in the therapeutic ratio.

Cade and McEwen (1978) have for the past 14 years conducted clinical trials in the treatment of carcinoma of the bronchus. They report two series of patients - those treated in 6000 rads in 40 fractions and those treated with 3600 rads maximum tissue dose in 6 fractions. Utilising this scheme, they report an improved survival for squamous cell carcinoma at this time, and the trial continues. Only a small proportion of patients seen with carcinoma of the bronchus at their centre have been included in this trial, but it is important to identify an improved method of irradiation for locally inoperable squamous carcinoma of the bronchus, where until now, evidence for prolongation of life has been scanty.

A considerable number of studies have been reported on the later stages of carcinoma of the cervix, including two non-randomised studies by Bates and Churchill-Davidson (1970) and Johnson and Walton (1974). Both involved comparatively small numbers of patients, did not use radium and suggested oxygen benefit with both large and small fractions. Randomised studies include three smaller studies by Ward (1974), and Cade and McEwen (1978). One of these, comparing small and large fractions without radium, finds benefit for oxygen. A study of intermediate size by Dische (1974), using small fractions, and a large one by Fletcher and colleagues (1977), both using radium, showed no statistical evidence of the value of oxygen for small fractions. However, a larger multicentre trial recently reported by Watson (1978), gives evidence of improved local control rate and survival.

Our own studies with Stage III cancer over 7 years (Bennett, 1978), now comprises 256 patients of whom 206 have been followed for more than 1 year. Patients are randomised to receive either treatment in air or oxygen and either treatment in air or oxygen and either with large or small (213 rads x 27) fractions, originally each of 1775 rets. The initial large fraction size (450 rads x 10) was associated with a high rate of bowel morbidity and therefore the dose was reduced to 400 rads per fraction, or 1550 rets. All patients received a single radium insertion of 2000 mgm hours at the end of treatment.

At the present time, there is no difference in the overall survival of all patients treated in oxygen compared with those in air, nor between those treated with 27 fractions in air or oxygen. There is however, a possible favouring of oxygen at 1 and 2 years when 10 fractions of 400 rads are considered (at present at the 7% level). There is also, at 2 years, a favouring of 27 fractions in air as compared with 10 fractions of 400 rads in air. There is as yet, no difference between 27 fractions in air and 10 in oxygen. This would suggest, at this time, that there is no advantage for oxygen multiple small fractions, but that there may be for few large fractions. It would appear too, that there is a significantly higher bowel morbidity, both in frequency and degree, for 450 rad fractions compared with 213 rad fractions; for 450 rad fractions compared with 400 rad fractions both in air and oxygen, and that the overall morbidity may be higher for the 400 rad fractions in oxygen than air. Perrins and Wiernik (1973), who studied serial biopsies of patients undergoing treatment in air and oxygen, found increased effects in oxygen. This trial is being completed and will be reported in detail at a later stage.

There has been extensive work on fraction size in relation to oxygenation. Suit and his colleagues (1967), Withers and Scott (1964), Revesz and Littbrand (1974), Nias (1974) and Rubin (1969), have all brought evidence that the O.E.R. increases with the size of individual fractions. This is of benefit provided efficient tumour oxygenation takes place. Should this not be so, but normal hypoxic cells be oxygenated, then increased morbidity may be expected and the therapeutic gain would fall. It is possible that we shall demonstrate this.

When the various series of cervical cancer are reviewed as a whole, two trends emerge. Firstly, there is a tendency for a positive result when large fractions in air and oxygen are compared, and secondly, when no radium is used.

TABLE 2   Cervix Uteri : All Studies
Fraction sizes compared

Air	Oxygen	No. studies positive for oxygen
Large	Large	3/5
Small	Large	1/2
Small	Small	1/4
Radium Compared		
Radium		1/5
No radium		3/4

These results could also support the contention that vasoconstriction in the presence of hyperbaric oxygen (Johnson, 1972; Milne, 1973) may militate against a favourable result, especially when small fractions are used (Johnson, 1974).

Several series of bladder tumours have been reported. The first, that of van den Brenk (1968), was small and suggested advantage. A similar conclusion was reached by Plenk (1972) but not Dische (1973). Cade and McEwen (1978) reported 3 series with no suggestion of benefit, as did Wiernik, reporting two multicentre trials in 1978. It is therefore concluded that hyperbaric oxygen is unlikely to be of value at this site.

Malignant melanoma has a reputation of radioresistance which is not entirely justified. We have previously reported that palliative irradiation in hyperbaric oxygen may be beneficial (Sealy, 1974b). Our further experience supports this. Twice-weekly fractions of 400 rads minimum tumour dose were given for 5 weeks. Twenty-three lesions were randomised to be treated in air or oxygen. Eight were treated in oxygen and 15 in air. Seven out of 8 oxygen-treated lesions responded, compared with 8 out of 15 treated in air. This suggests improved results in oxygen, but is not statistically significant. If however, the 12 patients previously reported are added, the result is in favour of oxygen (8/15 treated in air compared with 17/20 in oxygen).

There seem to be several new avenues to explore. Firstly, the OER of neutrons is about 1.6. Should the trials presently being conducted continue to show benefit for neutrons, a combination of neutrons and oxygen may be considered. This may be relatively cheap compared with the more advanced forms of particle therapy. Secondly, if there is indeed a significant oxygen-induced tumour vascular shut-down, then there is merit in considering possible vasodilators which could be administered after pressurisation. There may be a place for further studies on the relative haemodynamics of normal tissues and tumours under the influence of hyperbaric oxygen, radiation and possible vasodilators. Thirdly, hypoxic sensitisers offer attractive possibilities. The most effective of these is Misonidazole, which has an enhancement ratio about half that of oxygen but has the known major advantage of excellent tumour penetration, which oxygen may not have. Moreover, if given 4 hours before treatment, that $3\frac{1}{2}$ hours before pressurisation, the sensitiser will have reached the tumour before any possible vascular shut-down occurs. We are investigating this approach and find it very promising (Sealy, 1978).

We have to date completed treatment in 14 patients with Misonidazole and hyperbaric oxygen. There have been 72 compressions. Two patients declined to complete the treatment in oxygen. The treatment schedule is 6 fractions of 600 rads twice-weekly in 17 days at 3 A.T.A. with Misonidazole 2.0gm/m2 body surface 4 hours before irradiation. We have firstly, established that this is a safe method of treatment from the point of view of toxicity. One patient experienced three convulsions and although denied at the time, a past history of epilepsy was subsequently established. Another patient became agitated on one compression, did not convulse and successfully completed the course of treatment. Eight experienced mild neuropathy and one, transient ataxia. All patients had very gross disease and were considered to be unlikely to benefit from conventional treatment. One of two melanomas and one of two muco-epidermoid carcinomas regressed completely. Ten patients with squamous carcinoma of the head and neck all had gross recurrent disease after chemotherapy, gross destructive tumours or fixed nodes. Seven of the 8 patients who completed treatment according to plan, were cleared of malignant disease in the treated area 1-12 months after treatment. Two patients have unhealed sloughs at the tumour site, and 3 have developed disease elsewhere. Two have died; one of metastases and one of overwhelming haemorrhage from the original tumour site. One of the two patients with persisting disease refused to complete treatment in oxygen. This approach clearly requires further investigation.

In summary, we can say that there are improved results in the head and neck, and also in the cervix, especially where large fractions and when no radium is used, and in the palliative treatment of malignant melanoma. There is hope of improvement in the lot of some lung cancer patients, but not for bladder cancer. We have reached a time when we should not abandon the best sensitiser - molecular oxygen - but seek to find, probably in combination with other agents, the best way of using it.

The Trial on cancer of the cervix conducted at Groote Schuur Hospital was supported by a grant from the Medical Research Council of the United Kingdom of Great Britain and Northern Ireland. The Misonidazole was kindly donated by Messrs. Roche Products (Pty) Ltd.

REFERENCES

Alper, T., Howard-Flanders, P. (1956). The role of oxygen in modifying the radiosensitivity of E.Coli B. Nature, 178, 978.

Bates, T., Churchill-Davidson, I. (1970). Quoted by Bewley, D.K. in Brit. J. Radiol., 43, 498-499.

Bennett, M.B. (1978). Treatment of Stage III squamous carcinoma of the cervix in air and hyperbaric oxygen. Brit. J. Radiol., 51, 68.

Cade, I., McEwen, J.B. (1978). Clinical trials of radiotherapy in hyperbaric oxygen at Portsmouth. Clin. Radiol., 29, 333-338.

Chang, C.H., Conley, J.J., Herbert, C. (1973). Radiotherapy of advanced carcinoma of the oropharyngeal region under hyperbaric oxygenation. Am. J. Roentgenol., 117, 509-516.

Churchill-Davidson, I., Sanger, C., Thomlinson, R.H. (1955). High pressure oxygen and radiotherapy. Lancet, I, 1091-1095.

Churchill-Davidson, I., Sanger, C., Thomlinson, R.H. (1957). Oxygenation in radiotherapy. Brit. J. Radiol., 30, 406-422.

Deschner, E.E., Gray, L.H. (1959). Influence of oxygen tension on x-ray induced chromosomal damage in Ehrlich Ascites Tumour cells irradiated in vitro and vivo. Radiation Research, 11, 115-146.

Dische, S. (1973). The hyperbaric oxygen chamber in radiotherapy of carcinoma of the bladder. Brit. J. Radiol., 46, 13-17.

Dische, S. (1974). The hyperbaric oxygen chamber in the radiotherapy of carcinoma of the uterine cervix. Brit. J. Radiol., 47, 99-107.

Elkind, M.M., Swain, R.W., Alexcio, T., Sutton, H., Moses, W.B. (1965). Oxygen, Nitrogen recovery and Radiation Therapy. In Cellular Radiation Biology, Williams and Wilkins, Baltimore, 442-466.

Evans, T.C., Goodriff, J.P., Slaughter, J.C. (1942). Temperature and radiosensitivity of skin of newborn rats. IV. Effects of decreased circulation and breathing during radiation. Radiology, 38, 201-206.

Fletcher, G.H., Lindberg, R.D., Caderoa, J.B., Taylor-Wharton, J. (1977). Hyperbaric oxygen as a radiotherapeutic adjuvant in advanced cancer of the uterine cervix. Cancer, 39, 617-623.

Gray, L.H., Conger, A.D., Ebert, M., Hornsey, S., Scott, O.C.A. (1953). The concentration of oxygen dissolved in tissues at the time of irradiation as a factor in radiotherapy. Brit. J. Radiol., 26, 638-648.

Henk, J.M., Smith, C.W. (1977a). Radiotherapy and hyperbaric oxygen in head and neck cancer. Final report of first controlled clinical trial. Lancet, 2, 101-103.

Henk, J.M., Smith, C.W. (1977b). Radiotherapy and hyperbaric oxygen in head and neck cancer. Interim report of second clinical trial. Lancet, 2, 104-105.

Hewitt, H.B., Wilson, C.W. (1959). The effect of tissue oxygen tension in the radiosensitivity of leukaemia cells irradiated in situ in the livers of leukaemic mice. Brit. J. Cancer, 13, 675-684.

Hewitt, H.B. (1964). The oxygen status of tissue cells as revealed by radiation survival curves. Brit. J. Radiol., 37, 719-720.

Jamieson, Dana, van den Brenk, H.A.S. (1965). Oxygen tension in human malignant disease under hyperbaric conditions. Brit. J. Cancer, 19, 139-150.

Johnson, R.J.R., Wiseman, N., Wilson, L. (1972). Vascular response of tumour and normal tissue to hyperbaric oxygen. Lahey Clinic. Bull., 21, 128-135.

Johnson, R.J.R., Walton, R.J. (1974). Sequential study on the effect of the addition of hyperbaric oxygen in the 5-year survival rates of carcinoma of the cervix treated with conventional fractional irradiations. Am. J. Roentgenol., 120, 111-117.

Lacassagne, A. (1942). Chute de la Sensibilite aux Rayons X chez la Souris Nouveau-née en etat D'Asphyxié. Compt. Rend. Acad. Sci. (Paris), 215, 231-232.

Milne, N., Hill, R.P., Bush, R.S. (1973). Factors affecting hypoxic KHT tumour cells in mice breathing O_2, $O_2 + CO_2$ or hyperbaric oxygen with or without anaesthesia. Radiology, 106, 663-671.

Nias, A.H.W. (1974). The oxygen enhancement ratio of mammalian cells under different radiation conditions. Fifth International Hyperbaric Congress Proceedings, Vancouver, Canada, Vol.2, 650-659.

Perrins, D.J.D., Wiernik, G. (1973). The effect of irradiation on cell kinetics in the human rectal mucosa. Brit. J. Radiol., 46, 323.

Plenk, H.P. (1972). Hyperbaric Radiation Therapy. Preliminary results of a randomised study of cancer of the urinary bladder and review of the oxygen experience. Am. J. Roentgenol., 114, 152-157.

Powers, W.E., Tolmach, L.J. (1964). Demonstration of an anoxic component in a mouse tumour-cell population by in vivo assay of survival following irradiation. Radiology, 83, 328-335.

Revesz., L., Littbrand, B. (1974). Variation of the oxygen enhancement ratio at different x-ray dose levels and its possible significance. In Advances in Radiation Research Biology and Medicine, vol. 3, Eds. J.F. Duplan and A. Chapiro, Gordon & Breach, London, 1215-1224.

Rubin, P., Casarette, G.W. (1968). The radiopathologic basis for oxygen breathing in radiotherapy. In Clinical Radiation Pathology, vol.2, Saunders, Philadelphia, 934-972.

Rubin, P., Poulter, C.A., Quick, R.S. (1969). Changing perspectives in oxygen breathing and radiation therapy. Am. J. Roentgenol., 105, 655-681.

Scott, O.C.A. (1957). A model system for examining the radiosensitivity of metabolising layers of cells. Brit. J. Cancer, 11, 130-136.

Sealy, R., Helman, P. (1972). Treatment of head and neck cancer with intra-arterial cytotoxic drugs and radiotherapy. Cancer, 30, 187-189.

Sealy, R., Helman, P., Greenstein, A., Shepstone, B. (1974a). Treatment of locally advanced cancer of the head and neck with inter-arterial cytotoxics, cobalt and hyperbaric oxygen therapy. Cancer, 34, 497-500.

Sealy, R., Hockly, J., Shepstone, B. (1974b) The treatment of malignant melanoma with cobalt and hyperbaric oxygen. Clin. Radiol., 25, 211-215.

Sealy, R., Berry, R.J., Ryall, R.D.H., Mills, E.E.D., Sellars, S.L. (1977). Treatment of carcinoma of the nasopharynx in hyperbaric oxygen: An outside assessment. Internat. J. Rad. Oncol., Biol., Phys., 2, 711-714.

Sealy, R. (1978). A preliminary study in the use of Misonidazole in cancer of the head and neck. Brit. J. Cancer, 37, Suppl.3, 314-317.

Suit, H., Lindberg, R., Suchato, C., Ozenne, A. (1967). Radiation dose fractionation and high pressure oxygen in radiotherapy of the DBA mouse mammary carcinoma. Am. J. Roentgenol., 99, 895-899.

Thoday, J.M., Read, J. (1947). Effect of oxygen on the frequency of chromosome aberrations produced by x-rays. Nature, 160, 608.

Thomlinson, R.H., Gray, L.H. (1955). The histological structure of some human lung cancers and the possible implications for radiotherapy. Brit. J. Cancer, 9, 539-549.

Thomlinson, R.H. (1960). An experimental method for comparing treatments of intact malignant tumours in animals and its applications to the use of oxygen in radiotherapy. Brit. J. Cancer, 14, 555-576.

Thomlinson, R.H. (1967). Oxygen therapy - Biological considerations. In Modern Trends in Radiotherapy, I, Eds. Deeley and Wood, Butterworths, London, 52.

Van den Brenk, H.A.S. (1968) Hyperbaric oxygen in radiation therapy. Am. J. Roentgenol., 102, 8-26.

Ward, A.J., Stubbs, B., Dixon, B. (1974). Carcinoma of the cervix: Establishment of a hyperbaric oxygen trial associated with the use of the Cathetron. Brit. J. Radiol., 47, 319-325.

Watson, E.R. (1978). The First Cervix Trial, 1966-1973. Brit. J. Radiol., 51, 67.

Wiernik, G. (1978). Controlled trials in carcinoma of the bladder. Brit. J. Radiol., 51, 68.

Withers, R., Scott, O.C.A. (1964). The oxygen effect on normal tissues. Brit. J. Radiol., 37, 720.

Hyperthermia and Radiotherapy

Jens Overgaard

*The Institute of Cancer Research and the Radium Centre,
Radiumstationen, DK-8000 Aarhus C, Denmark*

ABSTRACT

Hyperthermia interacts with radiation in several ways. Heat has a radiosensitizing ability and furthermore appears to be selectively cytotoxic to radioresistant (hypoxic) tumour cells. Therefore, there has been a growing interest in the potential of using heat as an adjuvant in radiotherapy.

The technical problems related to a uniform application of sufficient heat to human tumours is far from solved, and also the temperature measurement technique is in a developing state. However, it appears that varying electromagnetic heating techniques and possibly also ultrasonic hyperthermia may prove to be the most suitable ways of heating tumours — especially due to the valuable possibility that several such beams can be focused on the tumour.

The early clinical experience with combined hyperthermia and radiation should be seen in the light of the technical insufficiencies. However, recent results support our experimental observations of an enhanced tumour response on combined treatment. Enhancement ratios of about 1.2 to 1.4 appear to be a realistic expectation when using temperatures below 44° C.

Hyperthermia may be a useful modality in combination with radiation. If selective tumour heating can be obtained the highest enhancement ratio may be due to the strongly radiosensitizing effect of heat administered simultaneously with radiation. The radiosensitizing effect is dependent not only on the time interval but also on the temperature, and increasing temperatures markedly enhance sensitization. However, as the heating technique is still insufficiently developed, a sequential therapy utilizing the hyperthermic cytotoxicity on radioresistant tumour cells appears to be the most attractive therapy for improvement of the therapeutic ratio.

KEYWORDS

Hyperthermia, radiation, clinical studies, effect on tumour and normal tissue, therapeutic ratio.

INTRODUCTION

The use of heat as an adjuvant to radiation therapy is far from a new idea. Already early in this century, hyperthermia was proposed as a potential modality in radiation therapy (Schmidt, 1909; Müller, 1913). During the following years there was a number of reports of more or less successful use of combined hyperthermia and radiation as described in the comprehensive review by Selawry, Carlson and Moore (1958). After the Second World War followed a period of weak interest which probably was related to the introduction of Mega voltage treatment. However, in the late sixties a renewed interest grew up (Cavaliere and others, 1967; Overgaard and Overgaard, 1972). This interest has been continuously growing and the potential usefulness of hyperthermia in combination with radiation has been substantiated by accumulating experimental data which appear to be confirmed in principle when applied to human cancer treatment.

The present review is an attempt to briefly update the biological rationale as well as the early clinical experience with the use of local hyperthermia in combination with irradiation in the treatment of solid malignant tumours.

BIOLOGICAL BACKGROUND

The rationale for using hyperthermia as a modality combined with ionizing radiation is complex due to the complexity of interaction between heat and radiation.

Radiosensitizing Effect of Heat

Moderately elevated temperatures (41-43° C) have a direct influence on the radiosensitivity of a cell population (Bronk, 1976; Dewey and others, 1977; Ben-Hur, Elkind and Riklis, 1978). This is caused by several mechanisms of which the most important are: (1) A direct hyperthermic radiosensitization as evidenced by reduction of the D_0. (2) Decreased repair of sublethal damage by reducing the shoulder of the survival curve (reduced D_q). (3) The variations in sensitivity throughout the cell cycle is different for radiation and hyperthermia in such a way that radio-resistant stages of the cell cycle are normally sensitive to hyperthermia. Thus, the combined treatment reduces the cell cycle-dependent sensitivity variations. (4) Hyperthermia may be more efficient in hypoxic than in oxic cells and thereby reduce the oxygen-enhancement ratio. (5) Experimental observations indicate that hyperthermia reduces the repair of potential lethal damage.

These interactions between heat and radiation appear to occur in both malignant and normal cells, and although the two latter (4 and 5) interactions should be expected to occur mainly in relatively central and poorly vascularized areas of solid tumours, the therapeutic advantage of a simultaneous heat-radiation treatment seems to be doubtful or even non-existent. Fig. 1 demonstrates the thermal enhancement ratio (TER) found in an experimental mammary carcinoma and the surrounding skin on simultaneous treatment with heat and radiation. Both tissues exhibit an increased TER with increasing temperature, but the thermal enhancement ratio is of the same degree and so suggests no gain in therapeutic ratio. Although a few studies point to a differential effect in favour of the tumour response (Robinson, Wizenberg and McCready, 1974; Gillette and Ensley, 1978), most investigations showed no constant significant improvement in tumour effect versus that seen in normal tissue (Thrall, Gillette and Dewey, 1975; Stewart and Denekamp, 1978; Overgaard, 1978b). Consequently, if surrounding normal tissue is heated together with the tumour, simultaneous treatment may not be preferable to radiation alone. Only if the heat dose (combination of temperature and time) is high, a difference may become observable. This is mainly due to direct hyperthermic cell destruction, but such high doses of heat introduce the risk of direct heat damage to surrounding normal tissues.

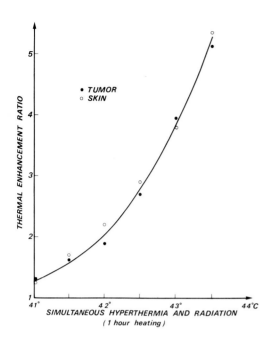

Fig. 1. Thermal enhancement ratio in an experimental tumour and overlaying skin as a function of temperature. No difference was observed between the TER for skin and tumour response when hyperthermia and radiation were given simultaneously.

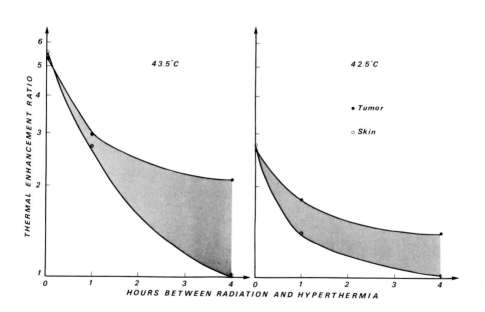

Fig. 2. Thermal enhancement ratio in an experimental tumour and its surrounding normal tissue as a function of interval between radiation and hyperthermia. With increasing split time, normal tissue TER shows a steeper fall than tumour TER. This causes an improved therapeutic effect (grey area).

Hyperthermic Cytotoxicity

Besides the ability to enhance directly the effect of irradiation, heat has also a direct cytotoxic effect (Overgaard 1977, 1978a). This effect is strongly enhanced by certain environmental factors among which an increased tissue acidity appears to be one of the most prominent. Therefore is it possible to give a hyperthermic treatment which nearly selectively destroys cells in a more acidic environment. Such conditions are typical for large areas of solid tumours. In such tumours, poor vascularization induces hypoxia and anaerobic glycolysis resulting in lactic acid production. The excess lactic acid cannot be removed due to the insufficient blood supply and therefore accumulates in the tissue and acidificates the environment. Cells which are so situated may be destroyed by moderate heat treatment, and the remaining surviving proportion of tumour cells which are more oxic and less acidic can be eradicated by smaller total radiation doses. In turn, this will result in an improved therapeutic ratio as the normal tissue is unaffected by the applied heat doses if direct radiosensitization can be avoided. Fig. 2 shows the thermal enhancement ratio in tumour and normal tissue and as a function of interval between treatments demonstrates that a simultaneous treatment yields the highest enhancement ratio. This is of the same magnitude in normal skin and tumour. Only by giving radiation followed by hyperthermia more than 4 hours later, is it possible to obtain a therapeutically acceptable difference in favour of the tumour response even though the enhancement is lower. This can be explained by the rapid recovery from radiation damage in normal tissue so that no normal tissue damage can be observed if the recovery time is more than 4 hours. The opposite sequence (heat before radiation) is less advantageous as repair of hyperthermic damage is more varying and longer than the recovery from radiation damage. If the heat cannot be selectively directed to the tumour, the use of hyperthermia as a cytotoxic agent directed towards radioresistant cells appears more logical and potentially useful than the use of the "hyperthermic radiosensitizing effect". However, it should be emphasized that the effect measured as thermal enhancement ratio is less than that observed when hyperthermia is given simultaneously with radiation.

In conclusion, it is possible to enhance the sensitivity of tissues to radiation when heat is applied simultaneous. This treatment is probably non-selective and should only be utilized if the tumour can be selectively heated. If both tumour and normal tissue has to be given the same heat dose, the hyperthermic cytotoxicity should be utilized.

APPLICATION OF HYPERTHERMIA

Before discussing the early experience with the use of hyperthermia in clinical practice the problems related to local heating of human tumours should be mentioned briefly. Whereas it is easy to uniformly heat experimental tumours (mostly by means of a water bath) heating of larger human tumours is a different and difficult problem.

The ideal heating source for use concurrent with radiation therapy should be able to localize the heat to a tumour without heating the surrounding normal tissue. Unfortunately, such a heating technique is not currently available, and the lack of sufficient technology in heating equipment is probably one of the greatest impediments in developing clinical experience with local hyperthermia.

Several techniques are of potential usefulness, including (1) water bath, (2) infra-red radiation, (3) hot air (hair-dryer technique), (4) low frequency current heating, (5) microwave radiation, (6) ultrasound, and (7) regional perfusion (Brenner and Yerushalmi, 1975; Kim and others, 1977; 1978, Johnson and others, 1977; U. and others, 1977; Luk and others, 1977; Sternhagen and others, 1978; Hornback and others, 1977; Har-Kedar and Bleehan, 1976; Cavaliere and others, 1967; Marmor and others, 1978; Gerner and others, 1975; Connor and others, 1977). Among these (1), (2) and (3) are suitable only for local tumours positioned closely to surfaces, and (7) is a technically difficult procedure with no selectivity and with a high risk of morbidity. Therefore we are left with ultrasound and electromagnetic heating at different frequences and with different application techniques (interstitial or external). Both of these modalities also have the theoretical possibility of focusing different beams and thereby to facilitate selective deep tissue heating. Only further research will show which of these techniques will be best suitable for a relatively selective local heating with the potential to penetrate sufficiently into the depth. However, even though the problems related to focusing hyperthermic treatment is not solved at the moment should one realize that not only by focusing the hyperthermia, but instead the radiation, may a similar effect by obtained. Such an effect can be obtained by giving a closely focused radiation treatment e.g. in the form of an interstitial implant (Manning and Miller, 1977).

Not only the heating technique, but also the temperature measurement is in an early and unsatisfactory state of development. Most temperature equipment includes different metals which interact with electromagnetic heating and partly also with the ultrasonic heat distribution. However, more optimism should be directed to the solution of these problems than to those of delivering the heat as several systems using fast or liquid crystals are under development and appear to have the properties needed for avoiding deformation of the electromagnetic field (Cetas, Connor and Boone, 1978). This new generation of thermometers are needed to test the heating equipment. But although such thermometers will not interact with the heating, they do not satisfy our ultimate desire for a temperature measurement source which should be a non-inversive technique.

CLINICAL EXPERIENCE

Most clinical experiences are at an early and preliminary phase I and phase II stage. For all studies account that they point to a certain effect of heat treatment added to other therapeutic modalities, but the effects cannot be accurately assessed due to the above-mentioned technical insufficiencies.

Nevertheless, several recent studies have provided us with valuable information, and some of the data will be summarized in the following. However, most observations are concerned with the technical possibilities of heating tumours and therefore only few studies include comparable tumours treated with one modality alone. Without such comparative information the evaluation of the modality and especially the thermal enhancement ratios is impossible.

Surface Heating

This group includes heating techniques where the heat is applied to the surface (e.g. the skin) and transferred by conduction to superficially situated tumours. The technique includes water bath, infrared and hot air heating. The principle of surface heating is far from optimal as it tends to cause higher temperatures in the skin than in the tumour, and the effect is therefore non-selective.

Brenner and Yerushalmi (1975) have used such a principle by applying hot air (partly combined with 2,450 Mhz microwave) to superficial metastases or recurrent tumours. With skin temperatures of about 47° C they observed complete regression in 3 of 6 treated tumours given radiation fractions of about 700 rad applied in the last part of the heating period. The treatment was given in 3 fractions of combined treatment with weekly intervals. Although the tumour response appears to be better than could be expected was also the skin reaction strong and all patients got significant moist desquamation in the heated and radiated areas. In contrast, in skin

TABLE 1 Recent Studies of Combined Local Hyperthermia and Radiation

References	Number of Treated Lesions/ Patients	Hyperthermia			Radiation Treatment	Response	
		Heating Method	Sequence and Interval	Temperature and Time		Tumour	Normal Tissue
Brenner and Yerushalmi	6	Hot air	Simultaneous	45-47° C/45-60 min	6-700 rad x 3	3/6 CR	Moist desquamation from combined treatment vs. moderate erythema from radiation only
Kim and co-workers	18	Water bath	Hyperthermia immediately after radiation	Tumour 40-41° C skin 43.5° C 30-60 min	300 rad/fx. (total 1200 - 2400 rad)	"Improved disease-free interval after combined treatment"	No unusual reactions
Kim and co-workers	37 (14 pts.)	27.12 Mhz radiofrequency	Heat immediately prior to radiation	Tumour 41-45° C skin 40-42.5° C 30-45 min	3-600 rad/fx. (total 3-4000 rad)	16/18 CR 1/8 CR after radiation alone	2 pts. with enhanced skin reaction
Johnson and coworkers	–	915 and 2450 Mhz microwave	Heat after radiation	41.5-42° C for 2 hours	–	Thermal enhancement 1.2-1.4	Thermal enhancement 1.2-1.4
U. and co-workers	14	915 and 2450 Mhz microwave (3 pts. water bath)	Radiation followed by heat	Tumour 42-43.5° C skin 35-42° C 45 min	2-600 rad/fx. (total 2000-4800 rad)	8/14 CR (3 pts. not evaluated)	no significant increase of normal tissue damage
Luk and co-workers	17 / 9	2450 Mhz microwave	Radiation prior to heat / Heat alone	Tumour 40-43° C skin 35-45° C 60 min	300 rad x 6-9 / –	6/17 CR 7/17 PR / 1/9 CR 4/9 PR	9 pts. with heat damage to skin
Sternhagen and coworkers	2 / 6	0.5-13 Mhz radiofrequency current (external or interstitial)	Radiation prior to heat / Heat alone	50° C/20 sec / 42-44° C/30-60 min (1-6 fx.)	150 rad/fx. / –	1/2 CR / 1/6 CR (all pts. palliative response)	Not evaluated

Fig. 3. Peak skin reaction in a patient treated with three fractions of combined radiation and hyperthermia. Note the increased reaction when no interval is allowed between the modalities

exposed only to a similar radiation treatment (outside the heat area) the peak reaction was moderate erythema. Thus, as both the tumour and skin reaction appeared to be markly enhanced, it appears unclarified whether any therapeutic advantage was obtained by using these simultaneous treatments.

Kim and coworkers (1977) used water bath heating at 43.5° C for 30 min immediately after radiation therapy which was applied with 300 rad per fraction. Total doses of 1200-2100 rad were given with 1 or 2 fractions per week to 18 patients with different malignant cutaneous lesions. All responses were compared to similar lesions treated with radiation only. In the total series of 26 patients which also includes 8 patients treated with 27.12 Mhz diathermia, 10 patients showed recurrence of the lesion and among these 7 appeared to have a longer disease-free survival after the combined treatment. In contrast to the observation by Brenner and Yerushalmi (1975), there was apparently no enhanced skin reaction in the patients treated with combined water bath heating and radiation. The difference in normal tissue reaction in the 2 studies should call attention to the observation that simultaneous treatment and higher temperatures certainly enhance the thermal enhancement ratio. The data by Kim and coworkers (1977) also showed how insufficient water bath heating is in penetrating to the depth.

Thus surface heating does not appear to be a technique with perspective in clinical practice.

TABLE 2 Effect of Combined Radiation and Hyperthermia on Metastatic Malignant Melanoma

Number of lesions	Radiation Dose	Time (days)	Hyperthermia 42-44° C/ 30 min time after radiation	Response Tumour	Normal tissue
4	900 x 3	8	4 hrs	3 CR 1 NE	Marked erythema with dry desquamation
4	900 x 3	8	-	1 CR 1 PR 2 NE	Marked erythema with dry desquamation
1	800 x 5	28	4 hrs	1 PR (60%) large tumour (10 x 10 cm)	Marked erythema with <25% dry desquamation
1	800 x 5	28	-	1 CR	Marked erythema (grafted area)
4	800 x 4	28	-	1 CR 1 PR 2 NE	Marked erythema (1 scrotum, 1 grafted area)
1	800 x 3	8	0 hr	NE	Moist desquamation ∼ 10% of area
1	800 x 3	8	4 hrs	NE	Marked erythema
2	700 x 3	8	0 hr	1 CR 1 NE	Marked erythema with dry desquamation
8	700 x 3	8	4 hrs	4 CR 3 PR 1 NE	Slight erythema
2	700 x 3	8	-	2 PR	Slight erythema (marked erythema in grafted area)
2	600 x 3	8	0 hr	NE	Slight and marked erythema
1	500 x 3	8	0 hr	PR	Slight erythema

CR: complete response. PR: partial response. NE: not evaluated.

Penetrating Heating

This includes first of all ultrasound and electromagnetic heating. Only pilot investigations are available at present (Table 1), but most of these point towards the preliminary conclusion that hyperthermia administered immediately before or after tumour irradiation appears to give better responses than radiation alone.

The normal tissue damage varies in these studies, probably due to the fact that the temperature varies markly on the skin when heated with microwave and ultrasound. In addition, some investigations have used surface cooling in order to facilitate a selective temperature increase in the tumours. However, when the modalities were given without any time interval enhanced skin reactions were seen (Fig. 3). Whether or not this improved skin reaction is less than the tumour enhancement cannot be evaluated at the present time, especially as there are unequivocal clear temperature variations which have not been overcomed as yet.

Certainly this heating technique needs much more evaluation before we can obtain reliable biological information, but the sparse present results are promising.

Sequence and Interval

In all mentioned studies on combined treatment the two modalities were applied, one immediately after the other. However, as experimental data indicate that it is doubtful whether such schedule has any advantages, it was decided to utilize only the direct cytotoxicity of hyperthermia and to avoid causing enhancement of the skin reactions in order to obtain a better therapeutic ratio. This was done by giving radiation 4 hours prior to hyperthermia. Eight patients with malignant melanoma and with a total of 31 lesions were treated with radiation alone or in combination with hyperthermia. The heat was applied with 27.12 Mhz radiofrequency current. Table 2 gives a summary of the result. Although the tumour data are few and difficult to interpret, they appear nevertheless to indicate an improved effect of the combined treatment especially when given with few, large fractions (e.g. 900 rad x 3). Such schedule may further be useful in the treatment of malignant melanomas (Habermalz and Fischer, 1976).

Splitting the treatment in no case caused stronger skin reactions than that seen after radiation alone. However, in a few fields which were treated with radiation followed immediately with heat, was a significant enhanced skin reaction observed (Fig. 3). Further studies of the split treatment are in progress and may yield more information on the potential use of combined hyperthermia and radiotherapy.

At a time where we cannot fully control the normal tissue temperature, it seems attractive to avoid causing stronger skin reactions than strictly necessary, in addition, the experimental data point towards a better therapeutic ratio when such split treatment is used. Therefore, in the future much more attention should be given to studies on questions like sequence and interval in the combined treatment.

CONCLUSION

The experimental biological observations on heat as a potentiator of radiation therapy appear to be confirmed when applied to human tumours. Here, enhancement ratios up to 1.4 can be obtained using temperatures of up to 44° C. Such temperatures can be tolerated locally without normal tissue damage. However, also normal tissue reactions may be enhanced by the combined radiation heat treatment, and the therapeutic gain - especially by simultaneous treatment - appears to be doubtful if the tumour cannot be selectively heated (or irradiated). Although sequential therapy produces less tumour enhancement, normal tissue damage can be avoided, and even better therapeutic ratios can be obtained.

The preliminary clinical results are encouraging and underline some of the problems which should be given high priority in the future research. This includes application of the heat where a focusing heating source allowing selective tumour heating is badly needed. Also the temperature measurement should be improved, ultimately resulting in a non-inversive technique. Biologically, parameters related to interval, sequence and fractionation of the two modalities should be explored and whenever possible both tumour and normal tissue reaction to combined hyperthermia and radiation treatment should be compared to a similarly treated structure given radiation alone in order to obtain the proper thermal enhancement ratios.

Much progress has been accomplished since Dewey and coworkers in 1974 gave a report similar the present one, and one may with optimism and expectations look forward to what will be presented at the next IUCC congress in 4 years.

ACKNOWLEDGMENTS

The present paper has been supported in part by Krista and Viggo Petersen's Foundation and the Danish Cancer Society. I wish to thank Bent Pedersen, M.D., Ph.D. for correcting the manuscript and Mrs. Lisa Wagner for skilful typing.

REFERENCES

Ben-Hur, E., M.M. Elkind, and E. Riklis (1978). The combined effects of hyperthermia and radiation in cultured mammalian cells. In C. Streffer (Ed.), Cancer Therapy by Hyperthermia and Radiation. Urban and Schwarzenberg, Baltimore and Munich. pp. 29-36.
Brenner, H.J., and A. Yerushalmi (1976). Combined local hyperthermia and X-irradiation in the treatment of metastatic tumours. Brit. J. Cancer, 33, 91-95.
Bronk, B.V. (1976). Thermal potentiation of mammalian cell killing: clues for understanding and potential for tumor therapy. In J.T. Lett and H. Adler (Eds.), Advances in Radiation Biology, Vol. 6. Academic Press, New York, San Francisco and London. pp. 267-324.
Cavaliere, R., E.C. Ciocatto, B.C. Giovanella, C. Heidelberger, R.O. Johnson, M. Margottini, B. Mondovi, G. Moricca, and A. Rossi-Fanelli (1967). Selective heat sensitivity of cancer cells. Biochemical and clinical studies. Cancer (Philad.), 20, 1351-1381.
Cetas, T.C., W.G. Connor, and M.L.M. Boone (1978). In C. Streffer (Ed.), Cancer Therapy by Hyperthermia and Radiation. Urban and Schwarzenberg, Baltimore and Munich. pp. 3-12.
Connor, W.G., E.W. Gerner, R.C. Miller, and M.L.M. Boone (1977). Prospects for hyperthermia in human cancer therapy. Part II: Implications of biological and physical data for applications of hyperthermia to man. Radiology, 123, 497-503.
Dewey, W.C., L.E. Hopwood, S.A. Sapareto, and L.E. Gerweck (1977). Cellular responses to combinations of hyperthermia and radiation. Radiology, 123, 463-474.
Gerner, E.W., W.G. Connor, M.L.M. Boone, J.D. Doss, E.G. Mayer, and R.C. Miller (1975). The potential of localized heating as an adjunct to radiation therapy. Radiology, 116, 433-439.
Gillette, E.L., and B.A. Ensley (1978). Response of C3H mouse mammary tumor and skin to simultaneous heat and irradiation. Radiat. Res. (in press).
Habermalz, H.J., and J.J. Fischer (1976). Radiation therapy of malignant melanoma. Experience with high individual treatment doses. Cancer (Philad.), 38, 2258-2262.
Har-Kedar, I., and N.M. Bleehen (1976). Experimental and clinical aspects of hyperthermia applied to the treatment of cancer with special reference to the role of ultrasonic and microwave heating. In J.T. Lett and H. Adler (Eds.), Advances in Radiation Biology, vol. 6. Academic Press, New York, San Francisco and London. pp. 229-266.

Hornback, N.B., R.E. Shupe, H. Shidnia, B.T. Joe, E. Sayoc, and C. Marshall (1977). Preliminary clinical results of combined 433 megahertz microwave therapy and radiation therapy on patients with advanced cancer. Cancer (Philad.), 40, 2854-2863.

Johnson, R.J.R., T.S. Sandhu, F.W. Hetzel, and H. Kowal (1977). A pilot study of hyperthermia as an adjuvant to radiotherapy. Radiat. Res., 70, 634-635.

Kim, J.H., E.W. Hahn, and N. Tokita (1978). Combination hyperthermia and radiation therapy for cutaneous malignant melanoma. Cancer (Philad.), 41, 2143-2148.

Kim, J.H., E.W. Hahn, N. Tokita, and L.Z. Nisce (1977). Local tumor hyperthermia in combination with radiation therapy. 1. Malignant cutaneous lesions. Cancer (Philad.), 40, 161-169.

Luk, K.H., D.G. Baker, P. Purser, J.R. Castro, and T.L. Phillips (1977). Preliminary report of local hyperthermia induced by 2450 megahertz microwave. Proceedings of 2nd R.T.O.G. Hyperthermia Group Meeting, November 28-29, 1977, Chicago.

Manning, M.R., and R.C. Miller (1977). Evaluation of the effects of localized heat and low dose rate radiation on human malignancies. Proceedings of 2nd R.T.O.G. Hyperthermia Group Meeting, November 28-29, 1977, Chicago.

Marmor, J.B., D. Pounds, T.B. Postic, and G.M. Hahn (1978). Treatment of superficial human neoplasms by local hyperthermia induced by ultrasound. Cancer (in press).

Müller, C. (1913). Die Krebskrankheit und ihre Behandlung mit Röntgenstrahlen und hochfrequenter Elektrizität bzw. Diathermie. Strahlentherapie, 2, 170-191.

Overgaard, J. (1977). Effect of hyperthermia on malignant cells in vivo. A review and a hypothesis. Cancer (Philad.), 39, 2637-2646.

Overgaard, J. (1978a). The effect of local hyperthermia alone, and in combination with radiation, on solid tumors. In C. Streffer (Ed.), Cancer Therapy by Hyperthermia and Radiation. Urban and Schwarzenberg, Baltimore and Munich. pp. 49-61.

Overgaard, J. (1978b). Combined hyperthermia and radiation in vivo. Effect of time, temperature, sequence and interval. Proceedings of the XI International Cancer Conference, Buenos Aires, 5-11 October 1978. Pergamon Press Ltd., Oxford (in press).

Overgaard, K., and J. Overgaard (1972). Investigations on the possibility of a thermic tumour therapy - I. Short-wave treatment of a transplanted isologous mouse mammary carcinoma. Europ. J. Cancer, 8, 65-78.

Robinson, J.E., M.J. Wizenberg, and W.A. McCready (1974). Radiation and hyperthermal response of normal tissue in situ. Radiology, 113, 195-198.

Schmidt, H.E. (1909). Zur Röntgenbehandlung tiefgelegener Tumoren. Fortschr. Röntgenstr., 14, 134-136.

Selawry, O.S., J.C. Carlson, and G.E. Moore (1958). Tumor response to ionizing rays at elevated temperatures. A review and discussion. Amer. J. Roentgenol., 80, 833-839.

Sternhagen, C.J., J.D. Doss, P.W. Day, W.S. Edwards, R.C. Doberneck, F.S. Herzon, T.D. Powell, G.F. O'Brien, and J.M. Larkin (1978). Clinical use of radiofrequency current in oral cavity carcinomas and metastatic malignancies with continuous temperature control and monitoring. In C. Streffer (Ed.), Cancer Therapy by Hyperthermia and Radiation. Urban and Schwarzenberg, Baltimore and Munich. pp. 331-334.

Stewart, F.A., and J. Denekamp (1978). The therapeutic advantage of combined heat and X rays on a mouse fibrosarcoma. Brit. J. Radiol., 51, 307-316.

Thrall, D.E., E.L. Gillette, and W.C. Dewey (1975). Effect of heat and ionizing radiation on normal and neoplastic tissue of the C3H mouse. Radiat. Res., 63, 363-377.

U., R., K.T. Noell, L.S. Miller, K.T. Woodward, B.T. Worde, R.I. Fishburn, and W.T. Joines (1977). Clinical pilot study of localized microwave hyperthermia and radiotherapy of malignant tumors. Proceedings of 2nd R.T.O.G. Hyperthermia Group Meeting, November 28-29, 1977, Chicago.

Combined Hyperthermia and Radiation In Vivo
Effect of Time, Temperature, Sequence and Interval

Jens Overgaard

*The Institute of Cancer Research and the Radium Centre,
Radiumstationen, DK-8000 Aarhus C, Denmark*

ABSTRACT

In order to optimize the therapeutic effect of the combined hyperthermia-radiation treatment of solid tumours the influence on the local tumour control and the damage to normal tissue has been studied during variations of heating time, temperature and the sequence and interval between the two modalities. As a model was used an isologous mouse mammary carcinoma transplanted into the feet of CDF_1 mice. Hyperthermia was performed in a controlled water bath. Radiation was given either before, during or after heating. An increasing thermal enhancement ratio was observed with increasing temperature or heating time between 41 and 43.5° C. This was similar in both tumour and normal tissue and no improvement in therapeutic gain could be found if treatment was given simultaneously. Different results were obtained by using an interval between the two modalities. Hyperthermic treatment given with varying intervals before radiation showed no improvement of the therapeutic ratio. However, if radiation was given before the heat, the normal tissue recovered within a few hours whereas a marked thermal enhancement was persistent in the tumour for more than 8 hours. Thus an increased therapeutic ratio could be obtained if radiation was given before heat and an interval of more than 4 hours was allowed. These data indicate that the optimal tumour effect may not be a hyperthermic radiosensitization, but rather a direct heat-killing of radio-resistant cells. This special heat sensitivity of radio-resistant tumour cells may be explained by the special environmental conditions (e.g. hypoxia and acidity) influencing such cells.

KEYWORDS

Hyperthermia, radiation, effect on tumour and normal tissue, therapeutic ratio.

INTRODUCTION

The interaction between hyperthermia and radiation on malignant and normal tissue is complex and involves several different biological mechanisms.

Firstly, heat has a direct radio-sensitizing effect (Bronk, 1976; Dewey and others, 1977). This interaction of heat and radiation occurs in several ways among which the most important are, (1) direct radio-sensitization (reduced D_0), (2) decreased repair of sublethal damage (reduced shoulder), and (3) sensitization of cells in relative radio-resistant phases of the cell cycle. Furthermore, heat may sensitize hypoxic cells more than well oxygenated cells - thus causing a decreased oxygen enhancement ratio (4). Finally, (5) heat is shown to reduce the repair of potentially lethal damage. This radio-sensitizing effect is often observed after relatively low heat doses (i.e. doses not causing significant damage when applied alone), and is strongly dependent on the time of application of the two modalities. In general optimal sensitization is obtained by simultaneous treatment, any time interval between the two components tending to reduce the sensitizing effect.

Besides its ability to act as a radio-sensitizer, heat also has a direct cytotoxic effect (Overgaard, 1977). This is mainly seen at higher heat doses (heat dose: combination of treatment time and treatment temperature), and may produce control of experimental solid tumours with an acceptable degree of normal tissue damage. This cytotoxicity is strongly enhanced by certain environmental conditions. Cells situated in areas of hypoxia, acidity and insufficient nutrition, (typical of large areas of solid tumours) are considerably more vulnerable to hyperthermic destruction. The fact that cells in such an environment are also the most radio-resistant may indirectly influence the response with combined radiation and hyperthermia as a smaller radiation dose may be adequate to control the better oxygenated peripheral tumour cells.

This cytotoxic effect, in contrast to the hyperthermic radio-sensitization, bears no time relation to the radiation treatment. It is therefore possible to assess the individual role of the radio-sensitizing effect versus cytotoxicity in experimental tumours and surrounding normal tissues. In order to find the maximum therapeutic ratio and thereby the greatest potential for the use of such treatment in the clinical situation, a series of experiments were performed to study the effect of heat and radiation with variations of heating time, temperature and the sequence and interval between the two modalities.

MATERIALS AND METHODS

A C3H mouse mammary carcinoma was used as a model. The tumour was transplanted to the feet of F1 hybrids between isologous C3H (♀) and D2BA (♂) mice (C3D2F1). Tumours were treated at a size of approx. 200 mm^2.

Hyperthermia was performed by placing unanaesthetized mice in a special jig which allowed immersion of the tumour-bearing foot into a controlled temperature bath. In certain tumour-bearing mice the temperature was measured at regular intervals with a fine-needle thermocouple and was found to rise rapidly and remain constant within 0.1^o C of the desired temperature. This procedure allows radiation to be given either before, during or after heating.

Radiation was given with a 250 KV clinical X-ray unit at 15 mA with 2 mm Al filtration and a HVL of 1.1 mm Cu. The dose rate was approximately 200 rads per minute.

The results of the different regimens were recorded as the radiation dose required for 50% tumour control at 120 days (TCD50) or the radiation dose required to achieve moist desquamation in 50% of the treated feet within 30 days (DD50). Based on the observed data the thermal enhancement ratio (TER) was calculated for a given treatment as the dose of radiation given alone compared to that dose of radiation needed to obtain the same response when given in combination with hyperthermia.

RESULTS

In order to study hyperthermic radiosensitivity, the effect on tumour and normal tissue response was first examined in a treatment schedule where heat and radiation were given simultaneously. Hyperthermia at temperatures between 41-44° C was applied for 1 hour and graded doses of radiation were given in the middle of the heating period. Fig. 1 showed the thermal enhancement ratios obtained with this treatment. The TER for local tumour control increased with increasing temperatures from about 1.2 at 41° C to about 5 at 43.5° C. However, a similar enhancement was observed for the radiation dose causing moist desquamation in 50% of animals, and therefore no gain in therapeutic effect was obtained with simultaneous radiation and hyperthermia when heat was given within a one hour period.

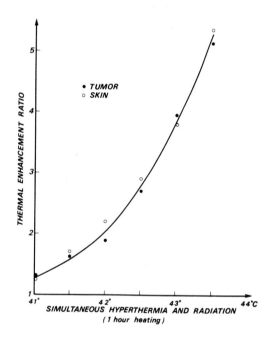

Fig. 1. Thermal enhancement ratio as a function of temperature. No difference was observed between the TER for skin and tumour response when hyperthermia and radiation were given simultaneously.

Since no therapeutic benefit was apparent with the previous schedule, a further experiment was performed with variations in heating time, still using a simultaneous application of the two modalities. Fig. 2 shows the variation in TER as a function of temperature and duration of heating. It can be seen that increasing the heating time at a given temperature results in an increasing thermal enhancement ratio. But again it was found that the effect was of the same magnitude for both tumour and surrounding normal skin. In this tumour system therefore, no significant improvement in the therapeutic ratio was seen by varying the period of heating in a simultaneously applied treatment schedule.

Since a simultaneous approach did not result in any significant improvement in therapeutic effect when compared to radiation alone the response with sequential therapy was investigated. Previous in vitro data point to a lack of direct hyperthermic sensitization with split times greater than 3-4 hours, suggesting that the direct cytotoxic effect of hyperthermia could be evaluated in combined treatment with radiation. For these studies heat treatment of 42.5° C was given for one hour. The heat was applied before or after radiation, and for both sequences the effect was investigated with varying time intervals from 30 minutes to 24 hours (Fig. 3). Both normal tissue damage and tumour response were increased most

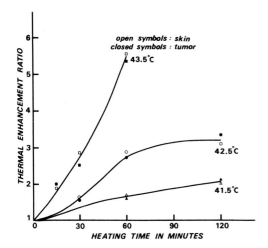

Fig. 2. Effect of heating time on the thermal enhancement ratio in tumour and surrounding skin by a simultaneous heat-radiation treatment.

by a simultaneous approach and decreased with increasing time intervals up to 4 hours. Hyperthermic treatment given at varying intervals before radiation showed (with exception of the 24-hour interval) no improvement in therapeutic ratio as the enhancement of normal tissue damage was similar to that of the tumour response. However, if radiation was followed by heating, the normal tissues showed a rapid recovery from excess thermal damage, and after 4 hours the normal tissue reaction did not exceed that observed after radiation alone. By contrast the tumour effect was still increased by a factor of about 1.4. This enhancement was lower than that seen after simultaneous application (about 2.6), but the difference between tumour and normal tissue response was now significant producing a real improvement in therapeutic ratio.

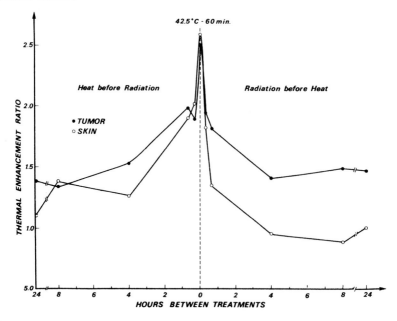

Fig. 3. Thermal enhancement ratio as a function of time interval and sequence between heat and radiation treatments.

DISCUSSION

The simultaneous application of heat and irradiation results in an increase of the thermal enhancement ratio with increasing temperature and/or duration of heating. However, as similar enhancement ratios were found in both tumour and surrounding normal tissue no improvement in therapeutic ratio was observed by using the combined treatment in this manner when compared to the results of radiation alone. With simultaneous treatment the hyperthermic effect is probably dominated by the direct radiosensitization, and it appears that the degree of sensitization was the same for both tumour and normal tissue. The gain which would be expected from increased hypoxic sensitization or from depressed repair of potential lethal damage in tumour cells was not found to cause any significant difference between tumour and skin responses. These results are in agreement with the data of several authors (Thrall; Gillette and Dewey, 1975; Hill and Fowler, 1978; Stewart and Denekamp, 1978) who also reported an increase of both normal tissue and tumour response. A few other studies (Robinson, Wizenberg and McCready, 1974; Gillette and Ensley, 1978) have found a slight enhancement of the tumour response compared to that seen in normal tissue, but this does not appear to be constant (Gillette and Ensley, 1978), and none of these studies used normal tissue adjacent to the tumour for assessing normal tissue damage. This is of some importance when considering the possible application of the treatment to the clinical situation.

Although allowing an interval between the two forms of treatment resulted in a decrease of the thermal enhancement ratio for both normal and tumour tissue, when radiation was given before heat this decrease in TER was most marked for the skin response which lost all heat potentiation with intervals of 4 hours or longer after radiation. In contrast the radiation dose required for local tumour control was still reduced if heated in this manner resulting in a real therapeutic gain (Fig. 4). The magnitude of the split in this sequence appears unimportant when greater than 4 hours, an interval corresponding to the time needed for repair of the sublethal radiation damage. The fact that hyperthermia resulted in an increased TER even with a split, is likely to be a result of the direct cytotoxicity which may be directed selectively towards the acidic and hypoxic cells in the tumour, whereas the skin in the more normal environment is unaffected (Overgaard, 1978).

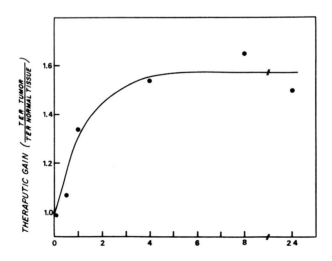

Fig. 4. Effect on therapeutic gain of the time interval between radiation and hyperthermia (42.5° C - 60 minutes).

When applying heat before radiation the tumour response appears to be similar to that with the opposite sequence. However, the skin reaction shows an increased TER of similar magnitude at intervals of at least 8 hours and therefore results in no enhanced therapeutic ratio. This may result from the slower repair of the hyperthermic damage that occurs after irradiation (Law, Ahier and Field, 1977). With an interval of 24 hours between heat and radiation some difference appears between the TER for tumour and skin and a slight improvement in the therapeutic ratio was found, but this was still less than that seen with radiation prior to hyperthermia.

Thus it seems that the most advantageous treatment schedule for combined heat and radiation therapy should involve irradiation given several hours before the hyperthermia. The data also indicate that the important tumour effect is the direct hyperthermic cytotoxicity rather than the hyperthermic radio-sensitization as the latter seems to occur to approximately the same degree in both normal and malignant tissue. As these results have been confirmed in other tumour systems (Hill and Fowler, 1978; Stewart and Denekamp, 1978) attention should be focused on such split treatment regimens when designing clinical protocols involving combined heat and irradiation.

ACKNOWLEDGMENT

The present paper has been supported in part by Krista and Viggo Petersen's Foundation and the Danish Cancer Society.

REFERENCES

Bronk, B.V. (1976). Thermal potentiation of mammalian cell killing: clues for understanding and potential for tumor therapy. In J.T. Lett and H. Adler (Eds.), Advances in Radiation Biology, Vol. 6. Academic Press, New York, San Francisco and London. pp. 267-324.
Dewey, W.C., L.E. Hopwood, S.A. Sapareto, and L.E. Gerweck (1977). Cellular Responses to combinations of hyperthermia and radiation. Radiology, 123, 463-474.
Gillette, E.L., and B.A. Ensley (1978). Response of C3H mouse mammary tumor and skin to simultaneous heat and irradiation. Radiat. Res. (in press).
Hill, S.A., and J.F. Fowler (1978). Combined effect of heat and X-rays on two types of mouse tumor. In C. Streffer (Ed.), Cancer Therapy by Hyperthermia and Radiation. Urban and Schwarzenberg, Baltimore and Munich. pp. 251-252.
Law, M.P., R.G. Ahier, and S.B. Field (1977). The response of mouse skin to combined hyperthermia and X-rays. Int. J. Radiat. Biol., 32, 153-163.
Overgaard, J. (1977). Effect of hyperthermia on malignant cells in vivo. A review and a hypothesis. Cancer, 39, 2637-2646.
Overgaard, J. (1978). The effect of local hyperthermia alone, and in combination with radiation, on solid tumors. In C. Streffer (Ed.), Cancer Therapy by Hyperthermia and Radiation. Urban and Schwarzenberg, Baltimore and Munich. pp. 49-61.
Robinson, J.E., M.J. Wizenberg, and W.A. McCready (1974). Radiation and hyperthermal response of normal tissue in situ. Radiology, 113, 195-198.
Stewart, F.A., and J. Denekamp (1978). The therapeutic advantage of combined heat and X rays on a mouse fibrosarcoma. Brit. J. Radiol., 51, 307-316.
Thrall, D.E., E.L. Gillette, and W.C. Dewey (1975). Effect of heat and ionizing radiation on normal and neoplastic tissue of the C3H mouse. Radiat. Res., 63, 363-377.

Unconventional Time-Dose-Fractionation in Radiotherapy of Cancer

Victor A. Marcial

*University of P. R. School of Medicine, Radiation Oncology Division,
P. R. Medical Center, San Juan, Puerto Rico*

ABSTRACT

American radiotherapists have traditionally favored therapeutic irradiation with five fractions per week, each ranging from 170 to 220 rads, but mainly 200 r, and total doses of 6000 r to 7000 rads, if no brachytherapy or surgical procedure forms part of the treatment plan. In the last twenty years some radiotherapists have challenged tradition by altering the standard time-dose-fractionation regimes. These variations may be: increase in total dose, reduced total dose, reduced fractionation (4 F/Wk, 3 F/Wk, 2 F/Wk, 1 F/Wk) increased fractionation (2 F/Day, 3 F/Day), continuous teletherapy irradiation, and split-course techniques.

Increased total dose has resulted in improvement of tumor control for certain sites, but at the expense of normal tissue damage. When chemotherapy and irradiation are combined, a reduction in the radiation dose is frequently needed. Reduced fractionation may have applicability for some clinical situations, such as treatment with four fractions per week, widely employed in several centers in the USA, with results that do not seem to vary from five fractions per week. Available controlled clinical data on three times per week fractionation for external irradiation of carcinoma of the uterine cervix and for advanced lesions of the laryngopharynx, show no difference to five fractions per week. Further study is needed in the treatment of early vocal cord lesions with three fractions per week.

Increased fractionation with two or three fractions per day shows promise in rapidly growing tumors of the Burkit's type, and in head and neck carcinomas, but has been of no value in glioblastoma multiformes.

Continuous teletherapy requires further investigation of its usefulness and practicality for the standard radiotherapy department.

Split-course technique can result in equal tumor control and normal tissue damage as conventional continuous therapy in carcinomas of the oropharynx and nasopharynx, using two treatment portions each consisting of ten 300 rad fractions, separated by a three weeks rest period. If no correction is introduced to compensate for the rest period, split-course irradiation may result in decreased curability.

Data from the author's department and/or the medical literature, and from the Radiation Therapy Oncology Group, are presented to support these statements.

Keywords: Radiation therapy, fractionation, dose-time relationships, neoplasms, carcinoma of cervix, carcinoma of oropharynx, carcinoma of laryngo-pharynx.

American radiotherapists have traditionally favored therapeutic irradiation with the following factors: five fractions per week ranging from 170 to 220 rad each (mainly 200), and total doses of the level of 6000 to 7000 rads, if no planned subsequent brachytherapy or surgical procedure is contemplated. The choice of the magnitude of the daily and total doses has been guided by the type, location and extent of the neoplasm, the size of the irradiated volume, and of no less importance, departmental traditions.

In the last twenty years some radiotherapists have challenged tradition by altering the standard time-dose-fractionation regimes. These variations are shown in Table 1. Although

Table 1 Time-Dose-Fractionation Variations

Increase in Total Dose
Reduced Total Dose
Reduced Fractionation:
    4 F/Wk., 3 F/Wk., 2 F/Wk., 1 F/Wk.
Increased Fractionation:
    2 F/Day, 3 F/Day
Continuous Teletherapy Irradiation
Split-Course Therapy

some mathematical formulae have been proposed to bring to a common denominator these variations in time-dose-relationships, such as Ellis' NSD (1971) and Orton's (1973) T.D.F. they have met with variable degrees of acceptance or skepticism by the radiotherapeutic community. As of this moment we have to conclude that sufficient evidence does not exist, based on clinical investigation, to completely adopt or reject these formulae. It is appropriate to review at this XII International Cancer Congress the status of the listed variations in standard therapy.

Following the concept that increased dose results in better local tumor control, some radiotherapists have felt inclined to increase the total dose to levels of 8000 rads or higher. Examples of this can be observed in the management of some brain tumors, in some head and neck carcinomas, particularly for the neck problem, and in soft tissue sarcomas. Increased dose has resulted in improved tumor control for some sites, but at the price of increased complications. One of these examples has been reported by Pierquin and co-workers (1966) for carcinoma of the tonsil. One should expect significant increase in complications with doses exceeding 7000 rads in 7 weeks with 200 rad fractions. Needless to say, with the usual daily fractions, a direct relationship exists between total dose, irradiated volume, and complications.

A reduction in total dose has been required with some frequency when radiotherapy has followed or has accompanied certain chemotherapy regimes. For example, some vigorous multidrug chemotherapy schedules utilized for head and neck cancer at the National Cancer Institute in Bethesda have necessitated a marked reduction in the total dose of irradiation as reported by Pomeroy (1975). It is to be expected that this will happen repeatedly as we increase the frequency of combined therapy. As of this moment we cannot accurately predict the magnitude of the toxicity resulting from these combined techniques.

Reduced weekly fractionation with four fractions per week has become standard therapy in some departments of the USA (Green, Vaeth and Lowy, 1968). This may be associated with reduction in the weekly or total dose, depending on how close to tolerance these departments may be operating. Basically I do not consider four times per week fractionation a significant deviation from standard therapy.

Three times per week fractionation has been the preferred regime in some departments (Botstein and Dalinka, 1968). Byhardt, Greenberg, and Cox (1977) in a retrospective study of three versus five fractions per week in carcinoma of the oral cavity and oropharynx concluded that five fractions per week gave better tumor control than three fractions. The authors cautioned against the use of three fraction per week schemata except on a controlled study basis. The authors quoted Fletcher, et al. as having had the same experience of fewer local controls with three fractions per week in laryngeal carcinoma, despite relatively equal NSD's as in the five fraction group.

In Puerto Rico, we conducted a randomized prospective trial of three versus five fractions per week in the external whole pelvis cobalt irradiation of carcinoma of the cervix (Marcial and Bosch, 1968) which began in July 1963. Consecutive cases were randomized to 3 F/Wk. versus 5 F/Wk. with the same weekly and total doses. External irradiation was followed by the same intracavitary irradiation in both groups. An initial publication referred to the first 200 cases registered in the study. An error of randomization resulted in an unequal distribution of cases with 108 in the 3 F/Wk. and 92 in the 5 F/Wk. The case distribution by stage in this group of 200 cases revealed a slight tendency to more advanced cases in the 3 F/Wk. group. The results of this analysis show no significant difference in the tumor control at 2 years, nor in the rectal and bladder complications (Table 2). No significant difference was noted in the 2 year survival achieved in the two

Table 2 Three F/Wk. vs. 5 FTWk. External Whole Pelvis Irradiation* Carcinoma of Cervix 200 Cases

	Cases Completing Therapy	Rectal and Bladder Complications
3 F/Week	98	10 (10%)
5 F/Week	81	9 (11%)

*Followed by Intracavitary Therapy.
From: Marcial, V.A., et al.

groups (Table 3). A second analysis presented by Marcial (1969) at the Carmel Conference referred to 260 patients with a minimum follow-up period of 3 years. No significant difference was observed in the reactions or three year survival in the two fractionation schedules (Table 4).

The British Institute of Radiology has been conducting a multi-institutional trial of 3 F/Wk. versus 5 F/Wk. radiotherapy in carcinomas of the laryngo-pharynx. The patients in the group receiving 3 F/Wk. had a reduction in total radiation dose of 11 to 13%. The last available report contains follow-up information up to 6 years (Wiernick and co-workers, 1978). A total of 732 cases were registered with 687 of them available for analysis. The

Table 3 Three F/Week vs. 5 F/Week External Whole Pelvis Irradiation* Carcinoma of Cervix - 200 Cases - 2 Year NED Survival

Stage	3 F/Week	5 F/Week
I	87%	95%
IIA	86%	67%
IIB	65%	59%
III	42%	21%
IV	38%	20%
Total	65%	59%

*Followed by Intracavitary Therapy.
From: Marcial, V.A., et al.

Table 4 Three-Year Survival

Stage	3 F/Week		5 F/Week	
	No.	%	No.	%
I	23/26*	88	23/24	96
II-A	20/28	71	20/30	67
II-B	18/30	60	17/31*	55
III	12/35*	34	7/24*	29
IV	4/17	24	3/11	27
Total	77/136	57	10/120	58

*Excluding one case lost to follow-up.

recurrence free rates and survival were found the same when all sites were analyzed as a group. No significant difference was noted in the reactions, though there was a tendency to have a higher percentage of mucosal changes in the 5 F/Wk. group (Table 5). Early

Table 5 Three F/Week vs. 5 F/Week Laryngo-Pharynx B.I.R. Study

	Acute Reactions	
	3 F/Wk.*	5 F/Wk.
	331 Cases	356 Cases
Marked mucous membrane reaction	15%	20%
Edema	14%	18%
Dysphagia	22%	27%

*Total Dose in 3 F/Wk. Reduced 11 to 13%.
From: Sixth Interim Progress Report - 1978.

lesions of the vocal cords showed better tumor control with 5 F/Wk.; however, the survival was the same in the two groups, but at the expense of more laryngectomies in the 3 F/Wk. patients. The conclusion of the British study was that 3 F/Wk. can be given safely to patients with advanced disease. The authors of the last report suggested the possibility that the doses used for the 3 F/Wk. schedules had a lower biological effectiveness than those used for 5 F/Wk., perhaps 3 to 10% lower.

The evidence for twice-a-week fractionation is less abundant than for the three or four times per week fractionation. We have used twice-a-week fractionation, with fractions of 400 rads for a total dose of 4000 rads, in selected cases with chest wall recurrence of carcinoma of the breast, with resulting satisfactory tumor control. The Radiation Therapy Oncology Group plans to begin soon a trial of 400 rads twice-a-week up to 4800 rads versus conventional radiotherapy for advanced carcinomas of the head and neck. This is in preparation for the administration and testing of misonidazole, an electron affinic compound, which is toxic and is best used twice-a-week.

Once-a-week fractionation has been used by several departments for palliation. We have used it for inoperable breast carcinomas with weekly 500 rad fractions up to 5000 rads with local results comparable to standard fractionation. However, advanced breast cancer may not be the most suitable test object. Ellis (1977) has advocated this type of fractionation and has used variable fraction and total doses. I am not familiar with any clinical trial where this regime has been employed.

Hyperfractionation with two of three fractions per day has had increased utilization in the recent years. This is expected to be useful for rapidly growing tumors where tumor regeneration during therapy is minimized by reducing the inter-fraction time (Norin and Onyango, 1977; Suit, 1977). Simpson in Canada has used three fractions per day for glioblastoma multiformes without demonstrable improvement in the results (Bush, 1974). Richard Marks, with our collaboration, recently completed a pilot project under the RTOG in which he tested various combinations of twice-a-day fractionation for carcinoma of the head and neck and esophagus. He has concluded that 125 rads can be administered twice daily to a dose level of 6000 rads in 5 weeks with adequate acceptance by patients, without severe acute or delayed reactions, and with acceptable ultimate local tumor control. Marks has finished the draft of an RTOG protocol for the comparison of a regime of two fractions of 120 rads each, separated 4 to 6 hours, for a total dose of 6000 rads in 5 weeks versus conventional fractionation with one fraction per day of 180 to 200 rads with total doses of 6 to 7000 rads, in advanced carcinomas of the head and neck. In the study protocol, Marks makes reference to a personal communication by Lester Peters who, at the M.D. Anderson Hospital, has investigated ultrafractionation. In a group of 55 patients with massive primary or nodal disease, or both, of the head and neck region treated with incremental doses of 6000 to 8000 rads, and a 3 hour inter-fraction interval, complete regression of the tumor mass(es) was obtained in 33 (60%) and of those followed for at least one year, 15 out of 37 (42%) remained disease-free.

The introduction of planned rest periods during the course of irradiation - called split-course technique - has become popular in recent years, following the publications of Sambrook in Britain, Scanlon in the USA, and Holsti in Finland (Marcial, 1972). We have used this technique in our Center for the last eighteen years. It has been claimed that during the rest period the patients' normal tissue reactions heal and that perhaps the tumor is more susceptible to radiation damage in the second half of therapy, for the radiation induced tumor regression evidenced after the rest period should lead to better oxygenation of the remaining tumor cells. Since 1971 we have directed a prospective collaborative trial of this technique for the Radiation Therapy Oncology Group in carcinomas of the nasopharynx, tonsillar fossa, base of tongue, urinary bladder stages C and D_1, and uterine cervix stages IIB, IIIA, IIIB, and IVA. Preliminary analyses of data for base of tongue, tonsillar fossa, and uterine cervix have been presented at recent meetings and will be published by Marcial and co-workers (1978a, 1978b).

The first site in which initial analysis was available is base of tongue. A total of 141 patients were randomized by institution, T and N stage, and treatment schedule. The treatment plan is shown in Table 6. The split cases received two courses each of ten 300 rad fractions separated by a three-week rest period. The continuous therapy patients received 30 fractions of 220 rads for total doses of 6600 rads. Case distribution by stage revealed that these were very advanced cases with 89% being T-3 and T-4, and with 74% positive nodes. The results in terms of acute and late normal tissue reactions are not significantly different in the two groups. The primary tumor control achieved initially (which means at the end and in the first three months after therapy) and at 1 year, reveals no difference. The survival was not different in the two groups.

Tonsillar fossa is the second site in which initial analysis was available. A group of 130 cases was randomized for the same treatment schedules mentioned for base of tongue (Table 6). As in the base of tongue part of this study, most cases were advanced with 90%

Table 6 Treatment Options

	Fraction Dose	No. of Fractions	Total Dose	NSD	TDF
Split	1st part 300 r	10	3000 r	2010	118
	3 wks. rest				
	2nd part 300 r	10	3000 r		
Continuous	200 r	30	6600 r	1940	115

T-3 and T-4 tumors and 69% with clinically positive nodes. The acute and late tissue reactions were not significantly different in the two groups, but a tendency to more completion of therapy and less interruptions was noted in the split group. The tumor control initially and at 1 year were the same in both groups as shown in Table 7. The survival appears the same in both groups.

Table 7 Primary Tumor Control by T-Stage

	Continuous		Split		Both	
	Initially**	1 Yr.*	Initially**	1 Yr.*	Initially**	1 Yr.*
T_1	4/5 80%	80%	1/1 100%	100%	5/6 83%	83%
T_2	3/3 100%	100%	5/5 100%	80%	8/8 100%	88%
T_3	19/26 73%	60%	23/29 79%	68%	42/55 76%	64%
T_4	11/33 33%	27%	9/21 43%	43%	20/54 37%	33%
All T's	37/67 55%	47%	38/56 68%	60%	75/123 61%	53%

*Actuarial estimation (Kaplan, E. and Meier, P.: Non-parametric estimation from incomplete observations. J. of the Am. Statistical Assoc., pp.457-481, June 1958).

**At the end of therapy or in the first three months of follow-up.

Primary tumor control means no evidence of tumor at the primary site.

The third site related to carcinoma of the cervix. External irradiation which preceded intracavitary brachytherapy, was administered by split technique or continuous course. As of June 1978 a total of 158 cases had been entered in the study with no significant difference in tumor control, survival or complications.

With the knowledge available, one can conclude that split-course irradiation can give equal results as continuous therapy, but some correction is necessary to compensate for the rest period. The tested split schedule reduced the number of fractions to two thirds of the continuous therapy regime with corresponding increase in fraction dose. Attempts to give split-course irradiation by simply adding a rest period, without compensating for it, have only resulted in diminished tumor control (Million and Zimmermann, 1975). In the case of lymphomas, Johnson (1974) has claimed that no correction is needed for the rest period.

Pierquin, Mueller and Baillet (1978) in Paris have attempted to treat cancer with continuous low dose rate teletherapy resembling the conditions of brachytherapy with apparent satisfactory initial results. As an experiment, this technique should be of interest to the radiotherapeutic community, but the prospects of application in the usual radiotherapy department are nil.

CONCLUSIONS:

Conventional time-dose-fractionation schedules in present cancer radiotherapy are the product of many years of clinical observation and evolution. The introduction of variations on these standard treatment regimes, simply by the application of the available mathematical formulae, involves considerable risk.

Certain variations from the traditional techniques such as three fractions per week and some split-course regimes are acceptable in view of the information available which is based on prospective clinical investigation.

Additional controlled clinical research is needed to test other variations of the conventional schedules and the application of the available mathematical formulae. This need is urgent as the introduction of combined modalities and the use of electron affinic compounds increases the uncertainties in this field.

REFERENCES

1. Botstein, C. and M.K. Dalinka (1968). Nine Years Experience with Reduced Fractionation. Front. Radiation Ther. Onc., 3, 212-219, (Karger, Basel/New York).

2. Byhardt, R.W., M. Greenberg, and J.D. Cox (1977). Local Control of Carcinoma of Oral Cavity and Oropharynx with 3 Vs. 5 Treatment Fractions Per Week. Int. J. Radiation Oncology Biol. Phys., 2, 415-420.

3. Bush, R. (1974). Information on page 204 of Proceedings of the Conference on the Time-Dose Relationships in Clinical Radiotherapy. Edited by Caldwell, W.L. and Tolbert, D.D., Wisconsin Center.

4. Ellis, F. (1971). Nominal Standard Dose and the Ret. Brit. J. Radiol., 44, 101-108.

5. Ellis, F. and A. L. Goldson (1977). Once a Week Treatments. Int. J. Radiation Oncology Biol. Phys., 2, 537-548.

6. Green, J.P., J. Vaeth, and R.O. Lowy (1968). Never on Sunday, Saturday, or Wednesday. Front. Radiation Ther. Onc., 3, 229-237.

7. Johnson, R. (1974). Personal Communication.

8. Marcial, V.A. and A. Bosch (1968). Fractionation in Radiation Therapy of Carcinoma of the Uterine Cervix: Results of Prospective Study of 3 Vs. 5 Fractions per Week. Front. Radiation Ther. Onc., 3, 238-249.

9. Marcial, V.A. (1969). Studies on the Relationships Between Dose-Time, Fractionation, and Rest Period in Radiation Therapy. Proceedings of NCI-AEC Conference on Time and Dose Relationships in Radiation Biology as Applied to Radiotherapy, Brookhaven National Laboratory - Publication 50203(C-57), 280-285.

10. Marcial, V.A. (1972). Split-Course Radiation Therapy Project. Cancer, 29, 1463-1467.

11. Marcial, V.A., J. Hanley, L. Davis, F. Hendrickson, and H. Ortiz (1978). Split-Course Radiation Therapy of Carcinoma of the Base of the Tongue: Preliminary Results of a Prospective National Collaborative Clinical Trial of the Radiation Therapy Oncology Group. Pending publication.

12. Marcial, V.A., J. Hanley, M. Rotman, L. Brady, and J. Ubiñas (1978). Split-Course Radiation Therapy of Carcinoma of the Tonsillar Fossa: Preliminary Results of a Prospective National Collaborative Clinical Trial of the Radiation Therapy Oncology Group. Pending publication.

13. Million, R.R. and R. C. Zimmermann (1975). Evaluation of University of Florida Split-Course Technique for Various Head and Neck Squamous Carcinomas. Cancer, 35, 1533-1536.

14. Norin, T. and J. Onyango (1977). Radiotherapy in Burkitt's Lymphoma - Conventional or Superfractionated Regime - Early Results. Int. J. Radiation Oncology

Biol. Phys. 2, 399-406.

15. Orton, G.G. and F. Ellis (1973). A Simplification in the Use of the NSD Concept in Practical Radiotherapy. Brit. J. Radiol., 46, 529-537.

16. Pierquin, B., M. Raynal, A. Ennuyer, and P. Bataini (1966). Etude Comparative des Résultats Concernant les Epithéliomas de la Région Amygdalienne Traités a L' Institut Gustave-Roussy et á la Fondation Curie. Annales de Radiologie, 9, 815-824.

17. Pierquin, B.M., W.K. Mueller, and F. Ballet (1978). Low Dose Rate Irradiation of Advanced Head and Neck Cancers: Present Status. Int. J. Radiation Oncology Biol. Phys., 4, 565-572.

18. Pomeroy, T. (1975). Communication at the time of the 57th Annual Meeting of the American Radium Society, San Juan, Puerto Rico.

19. Suit, H.D. (1977). Superfractionation (Editorial), Int. J. Radiation Oncology Biol. Phys., 2, 591-592.

20. Wiernik, G., N.M. Bleehen, J. Brindle, J. Bullimore, I.F.J. Churchill-Davidson, J. Davidson, J.F. Fowler, P. Francis, R.C.M. Hadden, J.L. Haybittle, N. Howard, I.F. Lansley, R. Lindup, D.L. Phillips, and D. Skeggs (1978). Sixth Interim Progress Report of the British Institute of Radiology Fractionation Study of 3 F/Week Versus 5 F/Week in Radiotherapy of the Laryngo-Pharynx. Brit. J. Radiol., 51, 241-250.

Radiotherapy of Subclinical Disease: Are Small Amounts of Radiation Effective for Small Amounts of Cancer Cells? Special Reference to the So-called Prophylactic Irradiation of Lung and Brain Occult Metastases

M. Tubiana

Institut Gustave-Roussy, 94800 Villejuif, France

ABSTRACT

Clinical and experimental data have demonstrated the high radiosensitivity of microfoci of cancer cells. These observations lead to prophylactic irradiation of occult metastases. In osteosarcoma a controlled clinical trial has demonstrated that 2 000 rads lung irradiation significantly reduces the incidence of lung metastasis and seems to have a favorable effect upon survival.

Prophylactic lung irradiation, eventually associated with chemotherapy, deserve to be tried in other cancers and for the treatment of patients with lung metastases.

Prophylactic brain irradiation gives excellent results in acute lymphocytic leukemia and encouraging data have been reported in oat cell lung cancer. Prophylactic irradiation of liver is possible and deserves to be investigated.

Keywords : Prophylactic irradiation, occult metastases, osteosarcoma.

INTRODUCTION

For more than half a century, the possibility of eradicating small cancer foci with relatively limited doses tolerated even when delivered on large target volumes has been used in the prophylactic post-operative irradiation.

This type of adjuvant radiotherapy is widely used for tumors of the upper respiratory and digestive tract and provides good results after doses of 5 000 rads in 5 weeks. Many papers have shown that for these cancers the incidence of new disease in areas of the neck initially clinically uninvolved is significantly reduced by a systematic irradiation (Fletcher).

Similarly, in breast cancer, a few randomized trials and numerous studies have demonstrated that systematic irradiation with moderate doses of the thoracic wall and of the axillary and supraclavicular lymph node areas significantly reduces the incidence of local and regional recurrences.

A few studies have been performed on the clinical dose-response curve of occult residual disease. For adenocarcinomas of the breast, 3 000 to 3 500 rads control about 60 % to 70 % of subclinical disease. For 4 000 rads this proportion rises to

80-90 % and is greater than 90 % for 5 000 rads. For squamous cell carcinomas of the upper respiratory and digestive tract, the data are similar ; 3 000 to 4 000 rads control about 60 to 70 % of occult residual disease and 5 000 rads more than 90 %. Using Cohen's mathematical model, Fletcher calculated that 4 000 rads (given in 4 weeks and 20 fractions) control 90 % of aggregates containing 10^6 cells and 99 % of 10^5 cell aggregates.

Similar calculations suggest that a dose of 2 000 rads, well tolerated by normal tissues, could control most tumor deposits up to about 10^5 cells.

EXPERIMENTAL DATA

The results of animal studies add relevant data. For example, in Lewis lung carcinoma, Shipley and others have measured the survival of the clonogenic tumor cells after in vivo irradiation. The cell radiosensitivity is much greater in a small tumor than in a large tumor. Anoxia is the main factor of radioresistance and this difference in radiosensitivity is probably mostly due to a rapid increase of the hypoxic fraction in experimental tumors when the diameters reach 1 to 2 mm as suggested by the available data on tumors which have been studied (Table 1).

TABLE 1 Proportion of Anoxic Cells in Small and Large Experimental Tumors

TYPE OF TUMOR	SIZE OF THE TUMOR	HYPOXIC FRACTION	REF.
C3H mammary carcinoma	0.6 mm^3 250 mm^3	0.001 0.15	Suit and Maeda
EMT6 tumor	diam. 0.5 to 2 mm 1 g	very low 0.3	Fu and others
Lewis lung carcinoma	0.5 mm^3 500 mm^3	0.005 0.36	Shipley and others

This observation is important not only for radiotherapy but also for chemotherapy because better tumor microvasculature also improves drug delivery and can explain the good results of systemic chemotherapy on micrometastases (Steel and others). In this connection it should also be noted that the amount of drug necessary to control experimental tumors increases very rapidly with their size.

However, besides good oxygenation, other factors may contribute to explaining the high radiosensitivity of small tumors. The mean lethal dose (D_o) of a well-oxygenated suspension of single cells originating from a tumor of 500 cu mm and irradiated in vitro is 110 rads. It is only 89 rads for cells originating from a tumor of 0.5 cu mm (Shipley). The difference, although small, is significant and might be explained by differences in cell kinetics. We have been able to demonstrate a decrease of growth fraction during growth of experimental tumors (Frindel and others) ; subsequently this has been confirmed (Simpson-Herren). The age-density distribution is therefore different in small and large tumors. Furthermore the repair of potentially lethal damage is larger for quiescent cells ; we have shown on NCTC fibrosarcoma that the amount of repair is greater for large than for small tumors (Little and others). This has been recently confirmed for another type of experimental tumor, the EMT6 tumor.

PROPHYLACTIC TREATMENT OF OCCULT METASTASES

Tumor cell kinetic studies have taught us during this past decade that metastases which become clinically detectable during the years following treatment of the primary tumor were in fact already present as occult metastases at the time of initial treatment. This finding stimulated much research on the so-called prophylactic treatment of metastases.

Rationale of Prophylactic Irradiation of Lung Metastases
Systematic whole lung irradiation is justified in cancers with a high incidence of pulmonary metastases. In osteosarcoma, choriocarcinoma and kidney carcinoma, about 75 % of the patients ultimately die of lung metastases. The proportion is smaller but still important for breast carcinoma (about 60 %) and prostate ca. (40 %).

Normal lungs tolerate bilateral irradiation of doses up to 2 000 rads in 2 weeks and 10 sessions or 1 500 rads in 4 sessions of 375 rads over 7 days. The analysis of the effects of such doses (Abbatucci and others, 1978) has shown that the functional pulmonary deficit is very small and does not increase during up to 3 years of follow-up. Probably slightly higher doses, for example 2 500 rads in 2.5 weeks, would still be tolerated, but there is no clinical data for such doses. Furthermore 2 000 rads permit association with chemotherapy. However in patients with respiratory deficiency pulmonary irradiation is contra-indicated.

A dose of 2 000 rads is probably able to sterilize microcolonies of up to 10^4-10^6 well-oxygenated tumor cells. As we have seen, experimental data suggest that larger metastases are much more difficult to control.

The size of the metastases depends on their age and their growth rate. The proportion of potentially curable patients therefore varies with the natural history of the primary tumor and its size.

Osteosarcoma
The prognosis for patients with osteosarcoma is very poor. The 5-yr survival rate does not exceed 20-25 %. Irrespective of the method of treatment of the primary tumor, prognosis depends almost entirely on the appearance of hematogenous metastases. Although metastases may appear in different organs such as the skeleton, metastases in the lungs are the main cause of death. 75 % of the lung metastases developping in patients with osteosarcoma are detected during the first 8 months after treatment of the primary tumor and 80 % during the first year.

The doubling times of osteosarcoma lung metastases have been reported in a few papers (Abbatucci and others, 1970; Breur; Combes and others; Rambert and others) and range from 10 to 60 days. This means that the size of the metastases at detection, e.g. when its diameter reaches 6 mm, corresponds to an age ranging from 10 months to 4 years. Taking into account 1) the distribution of the time intervals between the treatment of the primary tumor and the moment when the metastases become detectable, and 2) the distribution of the doubling times, one can calculate the distribution of the lung metastase sizes at the time of the treatment of the primary tumor. The results predict that at that time 20 % to 30 % of patients with subvisual lung metastases are potentially curable by whole lung irradiation because there are no more than 10^5 cell in any of their lung metastases (Abbatucci, 1972; Breur and others, 1978).

On the basis of those estimations, in 1970 a controlled clinical trial was started by the E.O.R.T.C. Radiotherapy Cooperative Group. The aim was to evaluate the effect on the development of lung metastases of an adjuvant irradiation of the lungs after "radical" treatment of primary osteosarcoma of the limbs. Five treatment centers in France and two in The Netherlands have been participating in this trial. Early in 1975 the trial was stopped. At that time 86 patients had been randomized into a group of 42 patients without any adjuvant therapy and a group of 44 patients with adjuvant lung irradiation to a dose of 2 000 rads in 2 weeks. The results show that this dose of radiation can decrease the percentage of cases in which metastases develop (Breur and others, 1978).

In both groups no metastase appears after a follow-up of 2 years ; the ultimate metastase-free percentage is 43 % for the lung irradiated group as compared with 28 % in the control group. The difference is significant at the 6 % level ($p = 0.059$). The results are relatively better in younger age groups. For patients under 17 years of age the percentage of patients remaining free of metastases are respectively 48 % and 28 %. The difference is statistically significant ($p = 0.02$).

The incidence of metastases is practically unchanged during the first 10 months after irradiation, whereas it is considerably reduced thereafter. The metastases which become detectable during the months following irradiation were probably just below the threshold of visibility and irradiation was relatively inefficient on these large subvisual metastases whereas it was much more efficient on smaller volume cell foci. This correlates well with experimental data.

The important contribution of patients by the Institut Gustave-Roussy (Villejuif, Paris) made it possible to compare this group of 38 patients with the group of 48 patients treated by the other participating centers. This comparison revealed remarkable differences. The metastase-free period was markedly shorter for the Paris Group than for the others. The ultimate metastase-free percentage for patients treated in Paris is 33 % in the adjuvant treatment group, which is hardly different from the 27 % found in the control group. For the patients in the other centers, the percentages are 50 % in the treated group and 30 % in the controls ; the difference is statistically significant ($P_1 = 0.021$). In order to find an explanation for the deviating results observed in Paris, all available criteria have been examined. No difference in site, size, histology or radiological characteristics of the tumor was found. Furthermore the average time interval between first symptoms and treatment is the same. One marked difference, however, was that in 12 out of 49 patients initially registered in Paris, lung metastases became demonstrable during the first weeks of treatment of the primary tumor, while in the other centers this was the case in only 3 out of 52 patients. The percentage of patients with metastases in just subvisual size at the time of the first treatment is therefore larger in Paris than at the other centers. This suggests that a selection of patients with an unfavourable prognosis is being referred to Villejuif, which is understandable since this center acts as a national referring center.

Although a longer observation time is needed for a final evaluation of the survival rates, present data indicate a better survival-rate in the group treated by adjuvant radiotherapy. For patients under 17 yrs of age the difference after 4 yrs between the treated group (59 %) and the controls (35 %) is almost statistically significant ($P_1 = 0.052$).

The survival curves after the appearance of metastases are identical for the treated group and the controls which suggests the absence of a differential effect of the treatment for metastases between the groups.

During the past 5 years, several reports have been published on the initial results of various chemotherapy programs as adjuvant treatment in osteosarcomas (Table 2). Most of these showed a marked postponement of the appearance of metastases, suggesting the possibility of an increase in the final cure rate. Unfortunately these trials were not performed as randomized studies and data from historical series have been used for comparison. There are many possible biases in such historical comparisons. 1) The earlier treatment of primary tumors may improve the survival-rate without any adjuvant therapy. This is illustrated in particular by the data of Taylor and others. 2) The selection of patients may differ during two subsequent periods. 3) Prognostic factors (age, tumor size, histology) may introduce bias when the number of patients is small. 4) When the follow-up is insufficient, actuarial methods may introduce uncertainties. This is illustrated by the wide differences in the successive sets of results reported for a same series.

Furthermore, the duration of adjuvant chemotherapy is generally of one to 2 years. When the observation period is short it is thus difficult to distinguish between a delay in the appearance of metastases or a true control. Furthermore, in some of the reports a few patients were still under chemotherapy at the time of the evaluation of the results.

TABLE 2 Relapse Free Survival observed after Various Types of Adjuvant Therapy (N.C.I. Meeting, March 78 - G.E.R. Meeting, September 78)

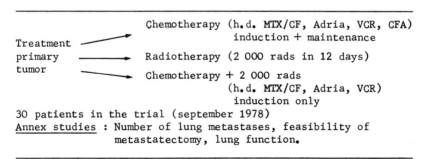

Despite these difficulties which tend to artefactually improve the apparent results of adjuvant chemotherapy when the follow-up is short, Table 2 shows that the results of 2 weeks of radiotherapy compare favorably with the results of 2 years of chemotherapy. Today it is still difficult to describe an optimal treatment for patients with osteosarcoma. The possibilities of a combination of lung irradiation and chemotherapy should be seriously investigated. This is the aim of a new controlled trial undertaken by E.O.R.T.C. and described in Table 3. By september 1978, 30 patients had already been introduced in the trial.

TABLE 3 E.O.R.T.C. Second Osteosarcoma Trial (Adjuvant Therapy)

Treatment primary tumor
- Chemotherapy (h.d. MTX/CF, Adria, VCR, CFA) induction + maintenance
- Radiotherapy (2 000 rads in 12 days)
- Chemotherapy + 2 000 rads (h.d. MTX/CF, Adria, VCR) induction only

30 patients in the trial (september 1978)
Annex studies : Number of lung metastases, feasibility of metastatectomy, lung function.

Curative Treatment of Lung Metastases

Bilateral pulmonary irradiation has been also proposed for patients with lung metastases after surgical resection of a solitary metastasis or in complement of a localized irradiation of a few grouped metastases.

This method has been investigated in particular by Abbatucci and others (1973) and Wharam and others (1974). The results are encouraging, especially when the primary tumor is radiosensitive. When the nodules are scattered in both pulmonary fields, combination of chemotherapy and bilateral lung irradiation also deserves systematic investigation. Association of 2 000 rads on the 2 lungs plus a boost of 1 000 - 1 500 rads on large nodules and chemotherapy has given encouraging results for nephroblastoma, Ewing's sarcoma and germinal cell testis carcinoma.

Prophylactic Irradiation of Brain Metastases
The brain appears to be a sanctuary for cancer cells during polychemotherapy due to brain-blood barrier. The remarkable results obtained by systematic cranial irradiation in the prevention of meningeal central nervous system involvement have considerably influenced the period of complete remission and survival in acute lymphocytic leukemia and have made it possible to cure this disease (Pinkel and others, 1977). A dose of 2 400 rads in 10 days is generally used. This achievement has led to other systematic brain irradiations, particularly in immunoblastic malignant lymphoma.

During recent years a few studies have been undertaken to assess the ability of prophylactic cranial irradiation to prevent CNS metastasis in small cell carcinoma of the lung. In patients treated by chemotherapy, relapses within the brain occur in about one-third of the patients ; they are generally associated with severe neurological complications for which there is no effective treatment.

A dose of 3 000 rads to the entire brain is generally used and is delivered in 10 fractions over a 2 week period. This dose is well tolerated and no complications have been reported (Jackson and others). The results suggest the efficacy of prophylactic cranial irradiation to prevent the development of brain metastasis (Table 4). However the survival rate was not substantially improved. Cranial irradiation deserves to be an integral part of the initial therapy of lung small cell carcinoma but is insufficient by itself to modify the poor prognosis of this cancer.

TABLE 4 Results of Various Trials on Prophylactic Cranial Irradiation in Small Cell Cancer of the Lung. Incidence of Brain Metastasis

	NO IRRADIATION	IRRADIATION
Maurer	16/81 (20 %)	2/66 (3 %)
Choi	17/58 (29 %)	1/15 (6 %)
Glatstein	4/22 (18 %)	0 % (?)
Jackson	4/15 (25 %)	0/14

Liver Metastasis
Prophylactic irradiation of the liver is possible, provided the total dose and the dose per session are not too high. A dose of 2 550 rads given by fractions of 150 rads over 3 to 4 weeks is well tolerated. We have used it in the routine treatment of oat cell carcinoma, unfortunately without noticeable improvement of the remission or of the survival periods. Nevertheless the usefulness of liver irradiation deserve to be assessed in other cancers.

REFERENCES

Abbatucci, J. S. (1972). L'irradiation pulmonaire de principe. Bases théoriques et expérience acquise. In A. Trifaud and R. Meary (Ed.), Pronostic et Traitement des Sarcomes Ostéogéniques, Masson and Cie. pp. 55-68.
Abbatucci, J. S., D. Fourré, R. Quint, A. Roussel, M. Urbajtel, and D. Brune (1973). Possibilités de la radiothérapie dans les métastases pulmonaires. Ann. Radiologie, 16, 385-392.

Abbatucci, J. S., A. Boulier, J. Fabre, and J. C. Lozier (1978). Functionnal evaluation of pulmonary bilateral irradiation effects. Europ. J. Cancer, 14, 781-785.

Breur, K. (1966). Growth rate and radiosensitivity of human tumours. Europ. J. Cancer, 2, 157-171 and 173-188.

Breur, K., P. Cohen, O. Schweisguth, and A. M. M. Hart (1978). Irradiation of the lungs as an adjuvant therapy of osteosarcoma of the limbs. An EORTC randomized study. Europ. J. Cancer, 14, 461-471.

Combes, P. F., J. Douchez, M. Carton, and A. Naja (1968). Etude de la croissance des métastases pulmonaires humaines comme argument objectif d'évaluation du pronostic et des effets thérapeutiques (basée sur 90 observations). J. Radiol. Electrol., 49, 893-902.

Fletcher, G. H. (1972). Clinical dose-response curve of subclinical aggregates of epithelial cells. J. Radiol. Electrol., 53, 201-206.

Fu, K. K., T. L. Phillips, and M. D. Wharam (1976). Radiation response of artificial pulmonary metastases of the EMT-6 tumor. Intern. J. Radiation Oncol. Biol. Physics, 1, 257-260.

Frindel, E., E. P. Malaise, E. Alpen, and M. Tubiana (1967). Kinetics of cell proliferation of an experimental tumor. Cancer Res., 27, 1122-1131.

Jackson, D. V., F. Richards, M. R. Cooper, C. Ferree, H. B. Muss, D. R. White, and C. I. Spurr (1977). Prophylactic cranial irradiation of the lung. J. Am. Med. Assoc., 237, 2730-2733.

Little, J. B., G. M. Hahn, E. Frindel, and M. Tubiana (1973). Repair of potentially lethal radiation damage in vitro and in vivo. Radiology, 106, 689-694.

Pinkel, D., H. O. Hustu, R. J. A. Aur, K. Smith, L. D. Borella, and J. Simone (1977). Radiotherapy in leukemia and lymphoma in children. Cancer, 39, 817-824.

Rambert, P., E. Malaise, A. Laugier, M. Schlienger, and M. Tubiana (1968). Données sur la vitesse de croissance de tumeurs humaines. Bull. Cancer, 55, 323-342.

Shipley, W. U., J. A. Stanley, and G. G. Steel (1975). Tumor size dependency in the radiation response of the Lewis lung carcinoma. Cancer Res., 35, 2488-2493.

Simpson-Herren, L. (1977). Growth kinetics as a function of tumor size. In B. Drewinko and R. M. Humphrey (Ed.), Growth Kinetics and Biochemical Regulation of Normal and Malignant Cells, Williams and Wilkins, Baltimore. pp. 547-559.

Skipper, H. E. (1971). Kinetic behaviour versus response to chemotherapy. Nat. Cancer Inst., Monograph, 34, 2.

Steel, G. G., K. Adams, and J. Stanley (1976). Size dependence of the response of Lewis lung tumors to BCNU. Cancer Treat. Rep., 60, 1743-1748.

Taylor, W. F., J. C. Ivins, D. C. Dahlin, and D. J. Pritchard. Osteogenic sarcoma experience at Mayo Clinic 1963-1974. To be published.

Wharam, M. D., T. L. Phillips, and E. M. Jacob (1974). Combination chemotherapy and whole lung irradiation for pulmonary metastases from sarcoma and germinal cell tumor of the tests. Cancer, 34, 136-142.

Use of Radiosensitizers in Radiotherapy

S. Dische

Marie Curie Research Wing for Oncology, Regional Radiotherapy Centre,
Mount Vernon Hospital, Northwood, Middlesex, U.K.

ABSTRACT

Clinical work with the hypoxic cell sensitizers began in 1973 with metronidazole and in 1974 with the more promising compound, misonidazole. It has been shown that there is a satisfactory penetration of the drug into tumours. Sensitization of hypoxic cells in man has been demonstrated as well as increased effect in tumours when single doses are employed. Neurotoxicity restricts the total dose which may be given. Randomized controlled clinical trials using this promising method for improving radiotherapy are now underway.

Key words: hypoxic cell sensitizers, clinical radiotherapy, clinical trials.

As a method to improve cancer treatment, the chemical agents which specifically sensitize hypoxic tumour cells in radiotherapy have arrived after a long period of laboratory study. It was Professor Adams in 1963 who first suggested that chemical agents might replace oxygen in its vital function in determining radiation injury (Adams and Dewey, 1963). There were many years of tests with many compounds using hypoxic bacteria as the test system. Some success lead to work with mammalian cells in culture and then to the in vivo situation in animals; first with artificially hypoxic normal and tumour tissue and then with tumours known to contain hypoxic cells (Adams and colleagues, 1976).

When in 1973 metronidazole was first given to patients and then in 1974, Ro-07-0582, now called misonidazole, improved tumour control had been demonstrated in many animal tumours (Adams and colleagues, 1976). No other method introduced so far into clinical radiotherapy has been better based upon laboratory study.

In our early clinical work in 1974 and 1975 using misonidazole we were able to show that the drug is well absorbed and that plasma concentrations reach a peak in 1-2 hr (Gray and colleagues, 1976). There is a good plateau period extending to 4 or 5 hr and then the drug concentration falls off with a half-life of about 12 hr. Plasma concentrations rise proportionately with the dose given. Now, in our studies of over 160 patients with some 4000 estimates of plasma concentration, we can show that a higher concentration is usually achieved in women than in men, but that this is compensated for as regards tissue exposure and, therefore, possible toxic effects by a shorter half-life (Dische and colleagues, 1978a).

Levels in normal tissues and in tumour rise with the plasma concentration and are usually between 60 and 100% of the plasma concentration at the time of sampling (Dische and colleagues, 1977). A core was taken by drill biopsy through a carcinoma of the breast 4 hr after 2.5g of misonidazole were given. At the very edge of the tumour there was a high concentration of fully viable tumour cells; these became fewer within the depths of the tumour and cells only seemed to survive around blood vessels. A tumour such as this can be expected to have a high concentration of hypoxic cells. A sample taken from the very centre of the core was submitted for estimation of misonidazole concentration; the level was 79 $\mu g/g$ - 99% of that in the plasma at the time of biopsy.

We were able to show in our first series of patients given the drug that misonidazole could sensitize hypoxic cells in man as in animals (Dische, Gray and Zanelli, 1976; Dische and Zanelli, 1976). We did this under oxic and under hypoxic conditions in order to see whether the drug could restore radiosensitivity. We used a complex system to make the skin hypoxic. An Esmarch's bandage was wound around the limb, a sphygmomanometer cuff applied at high pressure and then after uncoiling the Esmarch's bandage the limb was encircled in a bag of nitrogen. Under hypoxia with this system usually double the dose of radiation is required in order to achieve the same degree of pigmentation after treatment as that achieved when radiation is given under normal oxic conditions. Using hypoxia and the radiosensitizing drug, however, much enhanced reaction was achieved, in fact in this experiment when 800 rad were given under oxic conditions, we reduced our dose in hypoxia to 1200 rad and even then the response was greater than when 1600 rad were given under these hypoxic conditions, without sensitizer. In this situation, therefore, the radiosensitivity of artificially hypoxic skin in man has been taken more than halfway back to that of oxic skin by the addition of the radiosensitizing drug.

We are, of course, concerned not about normal skin made hypoxic but hypoxic tumour cells. In our original series of eight patients with multiple deposits of tumour we gave a single large dose of misonidazole combined with radiotherapy and compared the result with radiotherapy alone. In three of four assessable results we were able to demonstrate some increase in radiation response with the use of the sensitizer (Thomlinson and colleagues, 1976).

Unfortunately, misonidazole may give certain toxic effects. The important one which really limits the amount of drug which can be given is neurotoxicity, principally peripheral neuropathy (Dische and colleagues, 1977). However, if a dose of 12 g/m^2 of surface area is set at maximum when the drug is given over a minimum period of 17 days or greater there is a low and acceptable incidence of peripheral neuropathy. We have found that neurotoxicity can be reduced further by monitoring the blood concentrations achieved (Saunders and colleagues, 1978). This is because neurotoxicity is directly related to the plasma concentration reached. We have found that about 3% of patients do show unusually high serum levels for their calculated dose and so are particularly at risk of neurotoxicity. In others by monitoring the plasma concentration we can pick out those tending to show high values in the normal range and so with a slight reduction in dose we can keep the risk of neurotoxicity to a low level. In our most recent group of 60 patients the incidence of neurotoxicity was reduced to 11% and in nearly all this has been of mild severity, of short lasting and really of little clinical significance (Dische and colleagues, 1978a).

It might be that if the course of radiotherapy is a long one, and particularly if the dose is given in single amounts weekly, that some elevation of the total dose of misonidazole may be allowed. We await further evidence on this subject. A continuing collaboration between centres and a pooling of data should enable us to advance knowledge as to safe tolerance at a rapid rate (Dische and colleagues, 1978b).

How should we administer this dose? With a single treatment as for palliation the sensitizer would certainly improve the response. When we move to the multi-fraction techniques used in radiotherapy for cure there are many possible patterns of administration. First it can be given in a few large doses, each combined with a high radiation dose. Secondly, it can be given once or twice a week when daily radiotherapy is employed. In this situation when the misonidazole is given the radiation dose is commonly higher than on the other occasions in the week when the radiation is given alone. The third pattern shows a normal daily fractionated course of radiotherapy with misonidazole being given in small doses with each.

The first two techniques recognise the suggested advantage of high radiation dose and high sensitizer concentration. What is the evidence that small daily doses of misonidazole may also be helpful? We know that the relationship between sensitization and drug concentration is not linear. There is a steep rise at low concentration of drug, but as these become higher the curve flattens out. When, however, we look at the survival curves for oxic and hypoxic cells we can see that the curves tend to come together at the 150-200 rad dose level. Applying this biological data, therefore, it may be that no sensitization is achieved and, in fact, there might be reduced cell kill. It is, however, extremely difficult for our biological colleagues to mimic the clinical situation and assess the oxygen effect at this low level of dose. We now have the results of the Medical Research Council's hyperbaric oxygen trials and in the cases treated for carcinoma of cervix, where 75% were treated at around the 200 rad per day dose level, considerable benefit has been achieved thus showing the oxygen effect is relevant at this dose level (Watson and colleagues, 1978).

Finally there is the independent cytotoxic effect of the nitroimidazoles (Hall and Biaglow, 1977). It remains to be seen whether at the level of dose possible in man this effect is to be an important one with misonidazole but if it is then because of the importance of the duration of exposure, daily administration of misonidazole may lead to the greatest cell kill of hypoxic cells. We need careful biological experiments in man to increase our knowledge.

It is important that we explore all the possibilities in the combination of misonidazole with radiotherapy. However, we must be careful not to have too many trials and too many regimes when none may be adequately tested. Collaboration between radiotherapy centres at national and international level is essential if we are to learn the value of the drug at an early time.

It is too early to give an estimate as to the likely impact of misonidazole upon clinical radiotherapy. We have our favourable impressions based upon the 160 patients to whom we have given the drug, but only randomized controlled clinical trials can give the answer. The experience of Dr. Urtasun in a trial using metronidazole, a less potent radiosensitizer, in the treatment of glioblastomas gives us encouragement that we will achieve better with misonidazole (Urtasun and colleagues, 1977).

Professor Sealy in his trial in advanced mouth cancer in Capetown has reported that currently there is a margin of benefit in favour of those given misonidazole and we in our studies of advanced carcinoma of cervix have seen improved immediate response in those given the drug.

In many laboratories work is in progress to develop new radiosensitizing drugs for hypoxic cells which may give greater effect and be better tolerated by man. This is one of the most encouraging and most exciting fields of advance in radiotherapy. We have good expectation that misonidazole or one of the new compounds will be shown to significantly improve the results of radiotherapy and to have a place in the routine treatment of patients in every cancer treatment centre.

REFERENCES

Adams, G. E., and D. L. Dewey (1963). Hydrated electrons and radiobiological sensitization. Biochem. Byophys. Res. Comm., 12, 473-477.

Adams, G. E., J. F. Fowler, S. Dische and R. H. Thomlinson (1976). Increased radiation response by chemical sensitization. Lancet, i, 186-188.

Dische, S., A. J. Gray and G. D. Zanelli (1976). Clinical testing of the radiosensitizer Ro-07-0582. II. Radiosensitization of normal and hypoxic skin. Clin. Radiol., 27, 159-166.

Dische, S., and G. D. Zanelli (1976). Skin reaction - a quantitative system for measurement of radiosensitization in man. Clin. Radiol., 27, 145-159.

Dische, S., M. I. Saunders, M. E. Lee, G. E. Adams and I. R. Flockhart (1977). Clinical testing of the radiosensitizer Ro-07-0582: Experience with multiple doses. Br. J. Cancer, 35, 567-579.

Dische, S., M. I. Saunders, I. R. Flockhart, M. E. Lee and P. Anderson (1978a). Misonidazole. A drug for use in radiotherapy and oncology. (Awaiting publication).

Dische, S., M. I. Saunders, P. Anderson, R. C. Urtasun, K. H. Kärcher, H. D. Kogelnik, N. Bleehen, T. L. Phillips, and T. H. Wasserman (1978b). The neurotoxicity of misonidazole. The pooling of data from five centres. Letter to Br. J. Radiol. (Awaiting publication).

Gray, A. J., S. Dische, G. E. Adams, I. R. Flockhart, and J. L. Foster (1976). Clinical testing of the radiosensitizer Ro-07-0582. I. Dose tolerance, serum and tumour concentration. Clin. Radiol., 27, 151-157.

Hall, E. J., and J. Biaglow (1977). Ro-07-0582 as a radiosensitizer and cytotoxic agent. Int. J. Radiation Oncology, Biol. Phys., 2, 521-530.

Saunders, M. I., S. Dische, P. Anderson, and I. R. Flockhart (1978). The neurotoxicity of Misonidazole and its relationship to dose, half life and concentration in the serum. Br. J. Cancer, 37, Suppl. III, 268-270.

Thomlinson, R. H., S. Dische, A. J. Gray, and L. M. Errington (1976). Clinical testing of the radiosensitizer Ro-07-0582. III. Regression and regrowth of tumour. Clin. Radiol., 27, 167-174.

Urtasun, R. C., P. Band, J. D. Chapman, M. L. Feldstein, B. Mielke, and C. Fryer (1976). Radiation and high dose Metronidazole in supratentorial glioblastomas. New Engl. J. Med., 294, 1364-1367.

Watson, E. R., K. E. Halnan, S. Dische, M. I. Saunders, I. S. Cade, J. B. McEwen, G. Wiernik, D. J. D. Perrins, and I. Sutherland (1978). Hyperbaric oxygen and radiotherapy. A Medical Research Council trial in carcinoma of the cervix. Br. J. Radiol. (Awaiting publication).

Symposium No. 9 — Chairman's Closing Remarks

L. R. Holsti

Department of Radiotherapy, University Central Hospital, Helsinki 29, Finland

If we look at the factors affecting the treatment result, we can specify the tumour size, the proportion of anoxic cells in the tumour, and the treatment.
Big tumours probably contain more anoxic cells than small tumours, and especially the subclinical tumour cell aggregates.

It has been shown that 5000 rads in 5 weeks sterilizes regional, subclinical lesions in head and neck tumours and in pelvic tumours. Prof. M. Tubiana (France) indicated that 2000 rads in 2 weeks to both lungs decreases significantly the percentage of lung metastases in cases of osteosarcoma provided that the treatment is given immediately in connection with the treatment of the primary tumour. The result is the same or better than that achieved with chemotherapy, and radiation causes less side effects. Prof. M. Tubiana discussed tumour-cell kinetics as a base for elective treatment of subclinical lesions. The subvisual size of lung metastases at the time of irradiation is critical from the point of view of end results. The effect of whole-brain irradiation in acute leukaemia is well established, but data concerning solid tumour are still insufficient.

The time-dose relationship and the normal tissue tolerance are evergreens of radiotherapeutic problems. Prof. V. Marcial (Puerto Rico) reviewed the status of variations in standard therapy. One important fact can be recorded: it is now clear that a similar local control of tumours can be achieved with many different fractionation methods. No regimen seems to be definitely superior to others. Normal tissue damage seems however, to be enhanced with reduced number of fractions. With the knowledge available, prof. V. Marcial concluded that split-course irradiation and three fractions per week are acceptable from the point of view of normal tissue reactions. It has to be noted that the development in split-course therapy has been towards shorter and more concentrated courses separated by the rest period.
Fractionation two and three times per day is a new and interesting approach from which data is now accumulating.

It is generally accepted that human tumours contain anoxic and hypoxic cells. Since the discovery of the oxygen effect in tumours, radiobiologists and radiotherapists have searched for ways to overcome the anoxic problem. Hyperbaric oxygen, high-LET radiation, radiosensitizers and hyperthermia are the methods available. Hyperbaric oxygen has been used for more than 20 years and we can begin to evaluate the data accumulated. Prof. R. Sealy (South Africa) reviewed data showing improved results in the head and neck and the cervix, especially when large fractions are used.

There has been uncertainly among radiotherapists as to whether the favourable results are due to oxygen or to the fractionation. Prof. R. Sealy had data from comparative studies indicating that there is no difference in survival between cases treated with 27 fractions in air or oxygen. But there is a possible favouring of oxygen over air at 1 and 2 years when 10 fractions are used. Prof. R. Sealy also drew attention to the combination of neutron and oxygen as well as metranidazole and hyperbaric oxygen.

Dr. J.M. Sala (USA) and dr. M. Kligerman (USA) reviewed data on treatment with high-LET radiations. About 3000 patients throughout the world have been treated with neutrons. There are encouraging results with head and neck tumours, cervix and soft-tissue tumours; discouraging results with brain, breast and gastro-intestinal tumours. It is quite clear now that the injury rate is higher than with conventional radiation. Adjustment of dosage seems essential. Experiences with pions and heavy ions are accumulating from pilot studies. Normal tissue reactions are carefully studied. At the dose level used, no serious reactions have been found with pions at 15 months follow-up. There are some promising results concerning tumour response but it is premature to evaluate them now.

The development of radiosensitizing drugs for hypoxic cells is an exciting advance in radiotherapy. According to dr. S. Dische (United Kingdom), it is also one of the most encouraging advances. As dr. S. Dische pointed out "no other method introduced so far into clinical radiotherapy has been better based upon laboratory studies". Dr. S. Dische has shown that misonidazole can sensitize hypoxic cells in man. There seem to be several successful techniques to combine radiation and misonidazole; few large doses, small daily doses, and combination of these two. I would like to underline the important statement made by dr. S. Dische: "We must be careful not to have too many trials and too many regimens when none may be adequately tested". Only randomized, controlled clinical trials can give us the answer we need. The impression of the use of radiosensitizers is favourable but it is too early to evaluate their role in radiotherapy.

Dr. J. Overgaard (Denmark) discussed the biological background to experimental studies and clinical experience with hyperthermia. It was clearly stated that both the heating technique and the temperature measurement are in an early and unsatisfactory state of development. Heat has a radiosensitizing ability and appears to be selectively cytotoxic to radioresistant hypoxic cells. Enhancement ratios of 1.2 to 1.4 can be obtained using temperatures below $44^{\circ}C$. Heating is most effective when given after irradiation, and a 4 hour interval seems to be preferable. However, also normal tissue reaction may be enhanced.

Fractionation has been studied scantly in combination with heat. Dr. J. Overgaard has some interesting data on metastatic melanoma indicating improved treatment results when radiation and heat were given with few large fractions. The preliminary clinical results are encouraging but many more studies have to be done. Dr. J. Overgaard listed the problems which should be given high priority in future research. In addition to those mentioned, studies on tumour blood flow seem to be relevant.

Dr. J. Barragué (Argentina) discussed combined radiotherapy and chemotherapy. The objectives of chemotherapy in combination with radiotherapy were listed as follows: to diminish tumour volume, to treat sub-clinical disease, and to enhance the effect of radiotherapy. The use of different types of drugs in specific clinical situations was discussed.

In summing up, there are exciting developments taking place in radiotherapy. Further work is necessary and we can look forward to having interesting sessions when we meet again in four years time.

Radiation Therapy Role in Testicular Germinomas

M. A. Batata*, F. C. H. Chu*, B. S. Hilaris*, A. Unal*, W. F. Whitmore, Jr.**, H. Grabstald** and R. Golbey***

*The Department of Radiation Therapy, **Urologic Service and ***Solid Tumor Service, Memorial Sloan-Kettering Cancer Center, New York, New York, U.S.A.

ABSTRACT

Of 1000 testicular cancer patients treated mainly at the Memorial Sloan-Kettering Cancer Center between 1949 and 1974, 963 patients had germinal testicular tumors. Histologic types on orchiectomy were pure seminoma in 304 patients, embryonal carcinoma in 329, teratocarcinoma in 310, and pure choriocarcinoma in 20 patients. Overall survival without evidence of tumor at 5 years was 80% in pure seminoma and 45% in germinal carcinomas. Incidence of recurrence of testicular cancer in 5 or more years was 28% in pure seminoma and 59% in germinal carcinomas.

Survival and recurrence results in patients treated with radiation therapy mainly to the regional lymphatics, correlated with the radiation procedure. Five year survival rates with relatively adequate irradiation were 92% in pure seminoma and 63% in germinal carcinomas, and with inadequate dose-volume irradiation were 47% in pure seminoma and 14% in germinal carcinomas. The recurrence rates with relatively adequate versus inadequate irradiation were 11% versus 77% in pure seminoma, and 40% versus 91% in germinal carcinomas.

INTRODUCTION

The literature indicates that the prognosis of testicular germinomas is determined primarily by the histologic type and anatomic stage, and that treatment has a lesser effect on prognosis.[1-14] However, the rapidly developing multimodal therapy, requires a reappraisal of the prognostic significance of the various therapeutic modalities, especially radiation therapy. The purpose of this report is to emphasize the value of radiation therapy not only in pure seminoma but in germinal carcinomas, and to underscore the usefulness of relatively adequate irradiation in all stages of the disease.

MATERIALS AND METHODS

During the 25 year interval extending from 1949 to 1974, 1000

patients with a testicular cancer on orchiectomy were treated
mainly at the Memorial Sloan-Kettering Cancer Center. Nongerminal
testicular tumors either reticulum cell sarcomas (22) or soft part
sarcomas (15) were excluded in this study of 963 patients with
germinal tumors. Histologic types were pure seminoma in 304
patients, embryonal carcinoma in 329, teratocarcinoma (teratoma
with embryonal carcinoma) in 310, and pure choriocarcinoma in 20
patients. The tumors were staged clinico-pathologically according
to the initial gross and microscopic extent into three stages:
Stage I, tumor confined to testis and adnexa; Stage II, para-aortic
and/or pelvic (regional) lymph node involvement; and Stage III,
distant spread.[17]

Orchiectomy was done in virtually all patients, and was usually
inguinal and occasionally abdominal for undescended testes.
Bilateral para-aortic with ipsilateral pelvic (regional) node
dissection was performed in one-half (331/659) of germinal carcinoma
patients and in a few (15) pure seminoma patients. Adjuvant chemotherapy was administered more frequently in germinal carcinoma
patients (318) than in pure seminoma patients (13). External
irradiation was delivered to the regional lymphatics and/or other
regions in 98% (298) of pure seminoma patients and 46% (300) of
germinal carcinoma patients, with or without prior regional lymphadenectomy or laparotomy and/or adjuvant chemotherapy.

Treatment was understandably not uniform in this large series of
patients, due to the variations not only of cancer histology or
stage and patients' age or condition, but also of oncologists over
the 25 year review period. The patients were followed up for
periods of 5 or more years after treatment or until death, with the
exception of nine patients who are alive after treatment less than
5 years ago. The median interval from treatment until last follow-
up or death was 8 years in pure seminoma and 3 years in germinal
carcinomas. Autopsy data were obtained in 129 patients, 110 with
germinal carcinomas and 19 with pure seminoma.

Clinico-pathologic Findings:

All patients were male caucasians or orientals except a few blacks
(7), with an age range of 1 to 80 years and a peak incidence in the
third decade for germinal carcinomas and in the fourth decade for
pure seminoma. Nine-hundred patients presented with an intrascrotal
swelling usually painful, 134 patients with palpable nodes in lower
neck or groin or abdominal mass, and 45 patients with unilateral or
bilateral cryptorchidism. Initial investigations revealed para-
aortic or pelvic lymphadenopathy on pedal lymphangiography and/or
ureteral displacement on excretion urography in 336 patients,
pulmonary or mediastinal lesions on chest roentgenograms or tomograms
in 152 patients, and elevation of urinary or serum chorionic gonado-
tropin titers in 127 patients.

Pathological stage in 422 patients with germinal carcinomas (376)
or pure seminomas (46), was determined by regional lymph node
dissection or biopsy either negative (Stage I) in 175 patients or
positive (Stage II) in 177 patients, and by histologically proven
distant metastases (Stage III) in 70 patients. The initial clinical
stage was the same as the pathological stage in 326 stage I-III
patients (77%), was lower than the pathological stage in 87 stage II

patients (21%) with false negative lymphangiograms or excretion urograms, and was higher in 9 stage I patients (2%) with false positive radiological findings (Table 1). Pure seminoma (269) and germinal carcinoma (272) patients who did not undergo regional lymphadenectomy or metastatic biopsy, were staged only clinically: stage I in 300 patients, stage II in 105 patients, and stage III in 136 patients.

SURVIVAL AND RECURRENCE RESULTS

Overall tumor-free survival determined at 5 years was 56% (538/963), including 41 five year survivors with recurrences in less than 5 years resolved by subsequent treatment and remained recurrence-free at 5 or more years from initial treatment. Five year survival decreased with increasing malignancy: 80% in pure seminoma, 54% in teratocarcinoma, 39% in embryonal carcinoma, and nil in pure choriocarcinoma. Survival rates decreased also with increasing stage and were higher in pure seminoma than in germinal carcinomas across all stages (Table 2).

In pure seminoma and germinal carcinomas treated mainly by regional lymphatic irradiation, higher survival rates were observed across all stages (Table 3) with relatively adequate irradiation encompassing the entire para-aortic with at least the ipsilateral pelvic lymph nodes, including the orchiectomy scar and adjoining inguinal lymphatics, to tumor doses of 2500 to 3500 rads in 3 to 4 weeks in pure seminoma and 4000 to 4600 rads in 5 to 6 weeks in germinal carcinomas, with mediastinal-supraclavicular irradiation in stage II pure seminomas and/or other regions in stage III germinal tumors. Inadequate irradiation with lower radiation doses and/or smaller irradiated volumes, was associated with lower 5-year survival rates in all stages of pure seminoma and germinal carcinomas. In germinal carcinomas treated with or without irradiation, survival across all stages was higher with and lower without regional lymph node dissection or adjuvant chemotherapy (Table 4).

Regional and/or distant cancer recurrence or persistence occurred in about half of testicular cancer patients; 451 in less than 5 years and 21 after 5 years from the initial treatment. The incidence of recurrence increased with increasing malignancy: 28% in pure seminoma, 47% in teratocarcinoma, 67% in embryonal carcinoma, and 100% in pure choriocarcinoma. The frequency of recurrence increased with increasing stage and was lower stage for stage in pure seminoma than in germinal carcinomas (Table 5). Incidence of recurrence or persistence was also lower with relatively adequate irradiation and higher with inadequate irradiation in all stages of pure seminoma and germinal carcinomas (Table 6). In germinal carcinomas, the recurrence incidence across all stages was lower with and higher without regional lymph node dissection or adjuvant chemotherapy (Table 7).

DISCUSSION

The literature indicates that testicular neoplasms are usually malignant germinal tumors which may occur in any age from infancy to old age, but develop more frequently in caucasians than in blacks during the third and fourth decades of life.[1-14] This is in accord with the fact that 963 of 1000 testicular cancer patients, mostly caucasians, had malignant germ cell tumors which developed after puberty and peaked during early manhood age from 21 to 40 years in germinal carcinomas (442/659) and pure seminoma (188/304).

About half of the testicular cancer patients had stage I tumors predominantly (48%) pure seminomas, and the remainder had stage II or III tumors usually (84%) germinal carcinomas. Tumor stage (Table 1) was determined pathologically by regional lymphadenectomy or distal biopsy in 422 patients whose majority (376) had germinal carcinomas, and only clinically without histological determinations of regional or distant spread in 541 patients usually (390) with pure seminoma or metastatic germinal carcinomas. Since the clinical tendency is to underestimate rather than overestimate the extent of the cancer, the pathological stage was higher than the clinical stage in 21% and lower in 2% of the 422 pathologically staged cases.

The survival and recurrence results in this series of testicular germinoma patients, correlated favorably with the initial treatment modalities mainly radiation therapy in pure seminoma and retroperitoneal lymph node dissection and/or irradiation and adjuvant chemotherapy in germinal carcinomas (Tables 2-7). The favorable treatment results, though apparent in all stages of pure seminoma and germinal carcinomas, were also influenced by the tumor histology and stage. Although the results were correlated with each mode of therapy in order to underscore its prognostic significance, post-orchiectomy multimodal therapy was more frequently used in germinal carcinomas with extensive regional or distal spread and/or mixed elements of choriocarcinoma or seminoma.

The results here reported also indicate that regional irradiation can be effective as regional lymphadenectomy, but the radiation dose-volume techniques in pure seminoma and germinal carcinomas should be adequately used. The procedure of retroperitoneal node dissection was feasible only in technically and medically operable patients, whereas regional irradiation was used in inoperable and operable patients without the loss of ejaculation consequent to radical surgery. The favorable effects of adjuvant chemotherapy clearly demonstrated in all stages of germinal carcinomas, warrant routine administration in such cases. A previous report describes the various chemotherapeutic agents used, individually or in combinations, for this group of patients.[6]

External megavoltage irradiation of the regional lymphatics should encompass the entire bilateral para-aortic area and at least the ipsilateral pelvic region, including the orchiectomy scar (Figure 1). With tumor involvement of the spermatic cord, tunica albuginea or scrotal skin, the anterior field is extended to include the entire ipsilateral inguinal lymphatics which are boosted to the level of the regional tumor dose. The mediastinal and supraclavicular lymphatics should be irradiated in pure seminoma with involved regional lymph nodes (Figure 2). The recommended radiation doses are 3000 rads in 3 to 4 weeks in pure seminoma and 4000 to 4500 rads in 5 to 6 weeks, with adjuvant chemotherapy, in germinal carcinomas.

REFERENCES

1. Batata, M.A., Whitmore, W.F., Hilaris, B.S. and others, (1976). Cancer of the undescended or maldescended testis. Am. J. Roentgenol. Rad. Ther. Nucl. Med., 126, 302-312.
2. Boctor, Z.N., Kurohara, S.S., Badib, A.O. and others, (1969). Current results from therapy of testicular tumors. Cancer, 24, 870-875.
3. Dixon, F.J., and Moore, R.A., (1953). Testicular tumors. Cancer, 6, 427-454.
4. Earle, J.D., Bagshaw, M.A., and Kaplan, H.S. (1973). Supervoltage radiation therapy of the testicular tumors. Am. J. Roentgenol. Rad. Ther. Nucl. Med., 117, 653-661.
5. Fayos, J.V., and Kim, Y.H. (1978). Treatment of testicular tumors. Radiology, 128, 471-475.
6. Golbey, R.B. (1970). Place of chemotherapy in treatment of testicular tumors. J.A.M.A., 213, 101-103.
7. Horn, Y., and Hochman, A. (1967). End results in treatment of malignant testicular tumors. Oncology, 21, 72-79.
8. Kurohara, S.S., Badib, A.O., Webster and others, (1968). Prognostic factors in the common testis tumors. Am. J. Roentgenol. Rad. Ther. Nucl. Med., 103, 827-836.
9. Maier, J.C., and Sulak, M.H. (1973). Radiation therapy in seminoma. Cancer, 32, 1212-1216.
10. Maier, J.G., and Mittemeyer, B. (1977). Carcinoma of the testis. Cancer, 39, 981-986.
11. Quivey, J.M., Fu, K.K., Herzog, K.A. and others, (1977). Malignant tumors of the testis. Cancer, 39, 1247-1253.
12. Varkarakis, M., Merrin, C., Gaeta, J. and others, (1974). Nonseminomatous testicular tumor. Urology, 111, 684-688.
13. Werf-Messing, V.D. (1976). Radiotherapeutic treatment of testicular tumors. Int. J. Radiation Oncology Biol. Phys., 1, 235-248.
14. Whitmore, W.F., Jr. (1973). Germinal testis tumors in adults. In, Proc. Seventh Nat. Cancer Conf. Philadelphia, J.B. Lippincott Co., 793-801.

TABLE 1 CLINICAL VERSUS PATHOLOGICAL STAGE

Initial Clinical Stage		Pathological Stage			
		None	I	II	III
I	553	300	166*	87+	–
II	204	105	9++	90*	–
III	206	136	–	–	70*
Total	963	541	175	177	70

*Same clinical pathologic stage.

+Clinically understaged; 66% (57/87) with microscopic and/or one lymph node involvement.

++Clinically overstaged.

TABLE 2 FIVE YEAR SURVIVAL ACCORDING TO HISTOLOGY AND STAGE

	Five Year Survival			
	Pure Seminoma	Embryonal Carcinoma	Terato-Carcinoma	Pure Choriocarcinoma
Stage I	200/227 (88%)	64/90 (71%)	124/156 (79%)	0/2
Stage II	33/53 (62%)	47/127 (37%)	38/100 (38%)	0/2
Stage III	10/24 (42%)	16/112 (14%)	6/54 (11%)	0/16 (0%)
Total	243/304* (80%)	127/329* (39%)	168/310* (54%)	0/20 (0%)

*Each includes three patients alive with follow-up shorter than 5 years.

TABLE 3 FIVE YEAR SURVIVAL VERSUS RADIATION PROCEDURE

	Five Year Survival			
	Adequate Irradiation		Inadequate Irradiation	
	Seminoma	Carcinomas	Seminoma	Carcinomas
Stage I	176/187 (94%)	42/45 (93%)	21/35 (60%)	15/43 (35%)
Stage II	18/21 (86%)	20/49 (41%)	14/31 (45%)	10/79 (13%)
Stage III	9/13 (69%)	4/10 (40%)	1/11 (9%)	2/74 (3%)
Total	203/221 (92%)	66/104 (63%)	36/77 (47%)	27/196 (14%)

TABLE 4 FIVE YEAR SURVIVAL WITH AND WITHOUT REGIONAL
LYMPHADENECTOMY AND ADJUVANT CHEMOTHERAPY IN
GERMINAL CARCINOMAS

Five Year Survival

	Regional Lymphadenectomy	No Regional Lymphadenectomy	Adjuvant Chemotherapy	No Adjuvant Chemotherapy
Stage I	139/160 (87%)	49/88 (55%)	65/70 (93%)	123/180 (68%)
Stage II	82/147 (56%)	3/82 (4%)	65/116 (56%)	20/112 (18%)
Stage III	9/24 (37%)	13/158 (8%)	18/132 (14%)	4/49 (8%)
Total	230/331 (69%)	65/328 (20%)	148/318 (47%)	147/341 (43%)

TABLE 5 RECURRENCE ACCORDING TO HISTOLOGY AND STAGE

	Incidence of Recurrence or Persistence			
	Pure Seminoma	Embryonal Carcinoma	Terato-Carcinoma	Pure Choriocarcinoma
Stage I	38/227 (17%)	35/90 (39%)	35/156 (22%)	2/2
Stage II	30/53 (57%)	88/127 (69%)	63/100 (63%)	2/2
Stage III	16/24 (67%)	99/112 (88%)	48/54 (89%)	16/16 (100%)
Total	84*/304 (28%)	222+/329 (67%)	146++/310 (47%)	20/20 (100%)

*12, +8 and ++1 died with recurrences after 5 years.

TABLE 6 RECURRENCE VERSUS RADIATION THERAPY PROCEDURE

	Incidence of Recurrence or Persistence			
	Adequate Irradiation		Inadequate Irradiation	
	Seminoma	Carcinomas	Seminoma	Carcinomas
Stage I	15/187 (8%)	4/45 (9%)	23/35 (66%)	34/43 (79%)
Stage II	4/21 (19%)	31/49 (63%)	26/31 (84%)	71/79 (90%)
Stage III	6/13 (46%)	7/10 (70%)	10/11 (91%)	73/74 (99%)
Total	25/221 (11%)	42/104 (40%)	59/77 (77%)	178/196 (91%)

TABLE 7 RECURRENCE WITH AND WITHOUT REGIONAL LYMPHADENECTOMY AND ADJUVANT CHEMOTHERAPY IN GERMINAL CARCINOMAS

	Incidence of Recurrence or Persistence			
	Regional Lymphadenectomy	No Regional Lymphadenectomy	Adjuvant Chemotherapy	No Adjuvant Chemotherapy
Stage I	24/160 (15%)	48/88 (55%)	8/70 (11%)	66/180 (37%)
Stage II	72/147 (49%)	81/82 (99%)	55/116 (47%)	97/112 (87%)
Stage III	15/24 (63%)	148/158 (94%)	116/132 (88%)	46/49 (94%)
Total	111/331 (34%)	277/328 (84%)	179/318 (56%)	209/341 (61%)

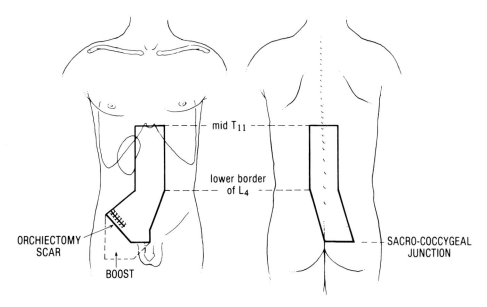

Fig. 1 FIELD PLACEMENT IN TESTICULAR CANCER

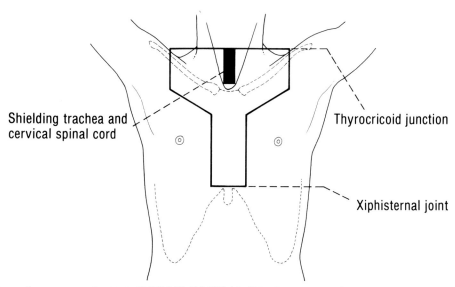

Fig. 2 MEDIASTINAL-SUPRACLAVICULAR IRRADIATION IN SEMINOMA WITH INVOLVED REGIONAL LYMPH NODES

Results of 8,056 Cases of Carcinoma of Cervix Uteri Treated by Irradiation: Clinical Use of Peking-type Applicators

People's Republic of China

*Department of Gynecology, Tumor Hospital,
Chinese Academy of Medical Sciences, Peking*

ABSTRACT

8,056 cases of carcinoma of the uterine cervix were treated in the department of gynecological oncology, during a relatively short period of time (1958-1972). Brief descriptions of the three modalities of irradiation used were given with particular reference to the use of Peking-type applicators. Analysis of the treatment results revealed that cases treated by intracavitary irradiation combined with transpelvic external irradiation fared better than those treated by the other two modalities. Complications and recurrences were also noted. All of the patients were followed up for more than five years, 3,570 of whom for over ten years. The overall 5- and 10-year survival rates are 68.7% and 62.2% respectively, those for Stage I being 93.4% and 89.7%; Stage II, 92.7% and 74.0%; Stage III, 63.6% and 52.7%; Stage IV, 36.6% and 9.5%.

INTRODUCTION

Progress in radiation physics and dosimetry, in radiobiology and in irradiation technics has brought about a steady improvement of cure rates and remission rates for all malignant tumors in general, including carcinoma of the cervix. We present here the detailed results of treatment in 8,056 cases of carcinoma of the cervix.

CLINICAL MATERIAL

Of the 8,056 cases of carcinoma of the cervix 70.8% belonged to clinical stages III and IV. Treatment was given by the following irradiation modalities:

Modality (1): intracavitary irradiation with use of Peking-type applicators alternating with transpelvic external irradiation chiefly by telecobalt therapy, 4 portals.

4,639 cases of all stages of carcinoma of the cervix were treated by this modality. Included were 113 cases of Stage I (2.5%) treated by intracavitary irradiation alone.

Modality (2): intravaginal cone irradiation combined with telecobalt transpelvic irradiation, 4 portals.

564 cases were treated by this modality. Of these, 294 cases were given supplementary intracavitary irradiation.

Modality (3): external telecobalt rotation 300° or bilateral 160°. 2,853 cases of all stages were irradiated by the rotation technic. Of these, 2,248 cases (78.8%) were given supplementary intracavitary irradiation.

It is shown in Table 1, more than half of the patients (57.6%) were treated by modality (1).

TABLE 1 Clinical stages and their distribution with reference to irradiation modalities

Stage	Cases	%	Modality (1) Cases	%	Modality (2) Cases	%	Modality (3) Cases	%
I	320	4.0	300*	93.8	10	3.1	10	3.1
II	2,028	25.2	1,556	76.7	204	10.1	268	13.2
III	5,509	68.3	2,721	49.4	336	6.1	2,452	44.5
IV	199	2.5	62	31.2	14	7.0	123	61.8
I-IV	8,056	100.0	4,639	57.6	564	7.0	2,853	35.4

*The 300 cases include those treated by intracavitary irradiation with and without transpelvic external irradiation.

TREATMENT METHODS

The three modalities of irradiation used in the treatment of carcinoma of the cervix are as follows:

Modality (1): Combination of intracavitary and external irradiation.
a. Intracavitary irradiation. Before 1961, "palisade-pattern" vaginal applicators were used. Since 1962, we have been using the "Peking-type applicators", which have retained the advantages of the "palisade-pattern" applicators and have been found to be more adaptable for individualized treatment, with better versatility and an additional shielding effect. The Peking-type vaginal applicator is made up of 1) unit-receptor, 2) radoactive unit, and 3) dud shielding unit (Fig. 1). To adapt the applicator to the needs of various morphological aspects of tumor growths, suitable sizes of unit-receptors are used, carrying 2 to 6 radioactive units. During application, the radioactive as well as the dud shielding units may be arranged as desired for the zone to be irradiated. The dud shielding units protect the rectum from overdosage. Wherever situations call for avoidance of irradiation, they may serve as shields, minimizing unnecessary excess irradiation. Intracavitary irradiation is given by fractionation method. In general, 4-6 applications are given in a complete course. Some cases may occasionally require 7-8 applications. The first, second and third applications are given at 1-week intervals, thereafter the intervals may extend over to 2 weeks. The duration for each application is 20-22 hours. In most of our cases, the total dose ranged between 6,000 and 10,000 mg hr, with 3,000-4,000 mg hr delivered intrauterine.

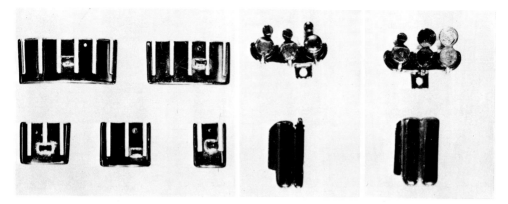

Unit-receptor (5 sizes)   Radoactive unit   Dud shielding unit

Fig. 1. "Peking-type applicator".

b. Transpelvic external irradiation. This is given by telegamma therapy, using the 4-portal method, each 8 x 15 cm in size. Tumor dose given to zone "B" ranges from 3,000 to 5,000 rad in 6-8 weeks.

Modality (2): Intravaginal cone X-ray irradiation combined with telecobalt transpelvic irradiation, X-rays of HVL 1 mm Cu - 37.5 cm FSD.

Modality (3): Rotation therapy. Rotation radius of 75 cm and a field area of 8 x 15 cm. In one set-up, the center rotation is at the cervix, and the Co^{60} machine rotates round 300°, avoiding the rectum for 60°, 7,000-8,000 rad being delivered to the cervix in 6-8 weeks. Another set-up is irradiation by way of two lateral arcs; the center of each arc of rotation is at the so-called point "B" in the pelvis, total dose delivered to the cervix ranges between 5,000 and 6,700 rad in 6-8 weeks.

RESULTS

<u>Survival</u>. All of the 8,056 cases were followed up for more than 5 years after completion of radiation treatment and 3,570 of them over 10 years. The 5- and 10-year survival rates obtained by the three modalities of radiation are shown in Table 2. Patients lost to follow-up are classified as dead.

DISCUSSION

Apart from the meticulous planning and the principle of individualization in radiotherapy, the modality of delivering the radiation is also an important factor.

1) Figure 2 gives a diagrammatic illustration of the "intensity" of doses in the various zones of the pelvis obtained through intracavitary and through transpelvic irradiation. The heavy-lined zone includes the vagina, fornices, cervix, uterine body, a large section of zone A and a smaller section of zone B. The aim of intracavitary irradiation is to deliver an adequate dose to the heavy-lined zone.

TABLE 2 Five- and ten-year survival rates for 8,056 cases of carcinoma of the cervix by radiation treatment

Stage	No. of patients	Modality (1)		(2)	(3)	5-year survival rate	10-year survival rate
		Intra-cavitary alone	Intra-cavitary & trans-pelvic			(1)(2)(3)	(1)(2)(3)
I	Survived /treated %	110/114 96.5	170/186 91.4	9/10	10/10	299/320 93.4	243/271 89.7
II	Survived /treated %		1,298/ 1,556 83.4	173/204	206/268	1,677/ 2,028 82.7	960/ 1,298 74.0
III	Survived /treated %		1,817/ 2,721 66.8	185/336 55.1	1,500/ 2,452 61.2	3,502/ 5,509 63.6	1,022/ 1,907 53.7
IV	Survived /treated %		13/62 18.6	3/14 21.4	37/123 30.1	53/199 26.6	9/94 9.5
I-IV	Survived /treated %	110/114 96.5	3,298/ 4,525 72.9	370/564 65.6	1,753/ 2,853 61.4	5,531/ 8,056 68.7	2,234/ 3,570 62.6

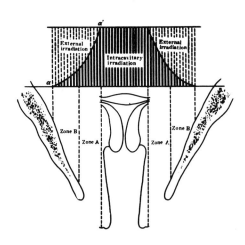

Fig. 2. Diagrammatic illustration of "intensity" of doses in the various zones as delivered by intracavitary and transpelvic irradiation.

The slope of the curve a'a indicates that the ideal intracavitary irradiation would be to raise the curve a'a so that the level of point a approaches maximally that of a', increasing thus the relative volume of heavy-lined zone.

Irradiation (radium, Co^{60} or Cs^{137}) directed to the zone of primary growth, on the principle of multiple doses by intracavitary method, yields the best radiation effect with minimal integral dose. The shape of dose distribution closely correlates to the zone of the primary growth and its extensions to the parametria. The incidence of local recurrence and postirradiation complications in our patients treated by intracavitary irradiation, was lowest as compared with that in patients treated by the other two modalities of irradiation. With its versatility and protective device, the applicator we used "fits" the local lesions as if "tailored" and at the same time gives appropriate protection to the surrounding normal tissues.

2) Intravaginal X-ray cone irradiation maybe effective but, due to its isodose distribution, satisfactory and adequate dosage can only be delivered to the relatively superficial portion of the primary growth. For deeper tissues and zone A, the dosage is inadequate. In our patients treated by this method, even though 52.7% of these were given supplementary intracavitary irradiation, local recurrence still amounts to 22.1%. Thus, it may be said, aside from early cases with supperficial lesions, intravaginal X-ray cone therapy is not an ideal procedure for treatment of carcinoma of the cervix.

3) Rotation therapy of carcinoma of the cervix is a satisfactory modality to deliver sufficient dosage of irradiation, but it entails a large integral dose. It is mostly indicated for late cases. In 78.8% of our patients treated by this method, intracavitary irradiation was being given as supplementary treatment, and for these cases the results were fairly good. Nevertheless, the incidence of local recurrence was 9.8%, being twice that in patients treated by intracavitary irradiation. The incidence of postirradiation injury was also higher. However, the very late cases and those unfit for the intracavitary method, such as patients with senile atrophic vagina and patients with small vagina, rotation therapy with additional intracavitary irradiation may sometimes yield good results.

4) A combination of radiation and chemotherapy by intraarterial or by intra-lymphatic perfusion may also be an avenue for improving the present results.

# Preoperative Irradiation of T3-Carcinoma in Bilharzial Bladder

H. K. Awwad, H. Abd El Baki, M. N. El Bolkainy, M. V. Burgers,
S. El Badawy, M. A. Mansour, O. Soliman, S. Omar and
M. Khafagy

National Cancer Institute, Kasr El Aini Street, Fom El Khalig, Cairo, Egypt

ABSTRACT

The present report deals with a prospective randomized study investigating the value of preoperative telecobalt irradiation in the management of T3-carcinoma in Bilharzial bladder.

A total preoperative dose of 4000 rad was split into two equal courses with a gap of one week. Two dose-time regimens were compared: conventional fractionation, 200 rad/day (SC) and hyperfractionation (HF). In the latter the daily dose amounted to 1000 rad and was divided into 17 hourly acute fractions, 60 rad each.

The SC and HF regimens produced equivalent local tissue reactions. HF, however, was associated with a somewhat higher incidence of radiation sickness. Both regimens resulted in an increase of the two-year disease-free survival rate from 19 ± 10 (3/16) to 53 ± 9 (17/32) without added surgical hazard and both HF and SC appeared to be equally effective.

INTRODUCTION

Radical cystectomy is the standard treatment adopted in the management of carcinoma in Bilharzial bladder in Egypt, since most cases are seen in advanced stages. Radiation therapy in such stages is handicapped by a compromised local tolerance as well as by a poor radiation response principally due to bulky disease. (Awwad et al 1970).

Radical cystectomy effers a 25-38% 5-year disease-free survival rate and local pelvic recurrence accounts for about 75% of failures (El Sebai 1977). A need therefore exists for improving the end results.

The present study aims at investigating the therapeutic value of moderate doses of preoperative telecobalt irradiation.

Two dose-time factors were employed: conventional fractionation and hyperfractionation. The latter is a simulation of continuous low dose rate irradiation of radium treatment, hoping to obtain the radiobiological advantages thought to be associated with this from of therapy particularly as regards improving the therapeutic margin by increasing normal tissue tolerance.

MATERIAL AND METHODS

Forty eight previously untreated cases of histologically verified carcinoma in Bilharzial bladder were included. According to the UICC system of clinical staging (1974) they belonged to the T_3-category and they were eligibible for radical cystectomy.

They included 40 males and 8 females with an age range of 30 to 56 years and a median age of 45 years.

The association of bladder cancer with Bilharzial cystitis was confirmed by clinical, radiological and histological criteria (El Bolkainy et al 1972). The group included squamous cell carcinoma: 31 transitional cell carcinoma: 12 and adenocarcinoma: 5 cases.

Therapeutic Groups: Patients were allocated at random to one of three groups:
a) Radical cystectomy alone.

b) Split-course (SC) pre-operative radiotherapy group: two courses were given with a gap of one week. Each course comprised 10 fractions, 200 rad each delivered in 14 days. Cystectomy was done 15 to 20 days after the end of the second course.

c) Hyperfractionation (HF) pre-operative radiotherapy group: two courses were also given with a gap of one week. Each course consisted of a two days treatment and on each day 17 fractions, 60 rad each were delivered at a rate of a fraction every hour i.e. a total daily dose of 100 rad. Cystectomy was performed 15 to 20 days after the end of the last course.

Radiotherapy Technique: Telecobalt therapy was used at an SAD of 80 cm. The target volume included the entire pelvis and extended upwards to the level of the middle of the body of the fifth lumbar vertebra and laterally for 1.5 cm beyond the inlet of the true pelvis to cover the external iliac nodes. The proximal 1.5 cm of the prostatic urethra was also included.

One direct anterior portal and two lateral 45° wedge fields were used. Since a rectal bladder was the most frequently used urine diversion procedure, the rectum was spared as much as possible. In most cases the anterior rectal wall only was included in the target volume and the dose to the posterior wall did not exceed 50% of the prescribed dose.

Scoring of the Acute and Chronic Radiation Effects and Reactions:

For scoring of the acute skin reactions the system proposed by Berry, Wiernick & Patterson (1974) was used. For chronic skin and subcutaneous tissue reactions the scoring system proposed by Arcangel, Friedman & Pauluzi (1974) was employed.

Acute bladder and rectal reactions were also scored using an arbitrary scale 0, 1, 2, 3 for no, mild, moderate and severe rectal reactions, respectively.

Blood loss during surgery, the period of hospitalization, time of wound healing, and any post-operative complications were also recorded.

RESULTS

Tolerance and Normal Tissue Reactions: All patients could tolerate the entire prescribed pre-operative radiotherapy regimes. No significant differences were noted amongst the two pre-operative radiotherapy groups as regards the incidence and severity of radiation sickness, maximum skin, rectal and bladder reactions.

Clinical End Results: Clinical and radiological examination performed 1-2 days before cystectomy showed that some degree of tumor shrinkage was achieved in 26 out of the 32 patients of the two pre-operative radiotherapy groups. In six patients no tumor could be detected and in two of them complete tumor disappearance was confirmed in the cystectomy specimens (Table 1).

TABLE 1. PATHOLOGICAL STAGING

	CONTROL	HF	SC
P_0	0/16	1/16	1/16
P_2	2/16	2/16	4/16
P_{3a}	5/16	5/16	5/16
P_{3b}	4/16	5/16	3/16
P_4	5/16	3/16	3/16
Downstaging	2/16	8/32	
Upstaging	5/16	6/32	

Radical cystectomy including pelvic lymphadenectomy could be performed in all except seven patients (control group: 2, SC: 1 and HF: 4 patients). In four of them the common iliac nodes could not be completely removed due to fixity to iliac vesses and in one of them the lower aortic nodes were also involved. In two patients peritoneal malignant nodules were found and only palliative cytectomy was performed. In the seventh patient the rectum and part of the anterior abdominal wall had to be excised since they were infiltrated by tumor.

Pre-operative radiotherapy did not seem to increase the operative mortality or the incidence of the common post-operative complications. The overall operative mortality of the entire series amounted to 10%, peritonitis due to leakage through anastomotic lines being the most common cause (Table 1).

Pathological examination of the cystectomy specimens showed the same degree of upstaging and downstaging errors (Table 2).

The one-year and 2-year disease free survival rates in patients receiving preoperative radiotherapy amounted to $59\pm9\%$ (19/32) and $53\pm9\%$ (17/32) respectively (Table 1). In the surgery alone group the corresponding rates were significantly lower, being $25\pm11\%$ and $19\pm10\%$ respectively (P 0.05). The survival rates did not differ significantly in the SC and HF groups, however.

Local pelvic recurrence accounted for all treatment failures in patients that died of the malignant disease. In one patient inguinal nodal metastases also developed.

Two patients of the SC group died during the second post-cystectomy year of causes other than the malignant tumor (Table 2).

TABLE 2. CLINICAL END RESULTS

	Control	HF	SC
No. of patients:	16	16	16
Operative mortality:	1	3	1
cause	peritonitis	-peritonitis -c. haemorrhage -liver failure	Peritonitis
Alive and NED:			
1 Year	4	8	11
2 Year	3	8	9
Died of disease Within 2-year:	12	5	4
Died of other causes:	0	0	2
			- Haematemsis (portal hypertension) - Coronary heart disease.

DISCUSSION

The present data confirm the previously reported finding that local pelvic recurrence accounts for nearly all of the failures of surgery of carcinoma in Bilharzial bladder (El Sebai- Ghoneim 1976). Moreover moderate pre-operative radiation doses seemed to reduce significantly the risk of local recurrence and to improve the disease-free survival rate without increasing the hazards of surgery. In this respect the beneficial effect of preoperative irradiation seems to be similar to that reported in case of transitional cell bladder cancer occurring in Europe or North America. In the latter form of the disease pre-operative irradiation increased significantly the local control rate in those cases with deep muscle infiltration (T_3) whereas no benefit could be demonstrated in cases with only a superficial degree of bladder wall infiltration (Whitmore et al 1977 - Van der Werf-Messing 1975). Pre-operative irradiation can therefore , cause reduction of the fraction of viable cells prior to surgery particularly in the smaller cell burdens in the infiltrating deep margins of the tumor which constitute the highest risk of residual disease after surgery . Such foci expected to show a significant radiation response after moderate doses of preoperative irradiation since they are composed of relatively small cell numbers in a better state of oxygenation.

REFERENCES

Arcangeli, G., Friedman, M., Paoluzi, R. (1974). A quantitative study of late radiation effect on normal skin and subcutaneous tissues in human beings. Brit. J. Radiol., 47: 44-50.

Awwad, H. K., Massoud, G. E., Ghorab, M., Tahan, W. T. (1970). Behaviour of carcinoma of the urinary bladder associated with urinary Bilharziazis. GANN. 9: 195-203.

Berry, R. J., Wiernik, G., Patterson, T. J. S. (1974). Skin tolerance to fractionated X-irradiation in the pig-How good a predictor is the NSD formula. Brit J. Radiol. 47: 185-190.

El Boulkainy, M. N., Ghoneim, M. A., Mansour, M. A. (1972). Carcinoma of the Bilharzial bladder in Egypt. Brit. J. Urology 44: 561-570.

El Sebai, I. (1977). Parasites in the Etiology of Cancer. Cancer of the Bilharzial bladder CA-A, Cancer J. for Clinicians 27: 100-112.

Ghoneim, M. A., El-Hamady, S., El Bolkainy M. N., Ashamallah, A. G., Mansour, M. A., Soliman, E. H. (1976). Radical cystectomy for carcinoma of Bilharzial bladder. Urology 8: 547-552.

Whitmore, W. F., Batata, M. A., Hilaris, B. S., Reddy, G. N., Unal, A. Ghoneim, M. A., Grabstald, H., Chu, F. (1977). A comparative study of two preoperative radiation regimens with cystectomy for bladder cancer. Cancer 40: 1077-1086.

UICC: TNM classification of malignant tumours. (1974). 2nd ed. Geneva, 79-83.

ACKNOWLEDGEMENT

This work is part of the Project "promotion of the Radioresponsiveness of carcinoma of the bilharzial bladder". The Project is supported by the Dutch Government and is run in collaboration with the Radiotherapy Department of the National Netherlands Cancer Institute, Chief: Professor K. Breur.

The Value of Post-operative Irradiation in Stage II Breast Cancer: the Pattern of Appearance of Metastases and the Value of C.E.A. as a Predictor of Metastases. Preliminary Report of a Randomised SASIB Breast Study

S.A.S.I.B. Breast Study Group

Paper read by R. Sealy

Radiotherapy Department, Groote Schuur Hospital & University of Cape Town, South Africa

ABSTRACT

A multicentre randomised trial on the value of post-operative radiotherapy in Stage II breast cancer is reported. 377 patients were entered and carefully stratified for age, degree of malignancy, periglandular growth and number of involved nodes. Prior to entry and throughout follow-up, a rigorous scheme of investigation was undertaken to exclude and detect metastases. Preliminary results would indicate that post-operative radiotherapy does not affect survival but may be associated with the earlier appearance of metastases. Prospective C.E.A. estimations suggest that when positive, it may be a useful predictor of further disease where bone or mixed metastases are present. Follow-up for 4 years is required to allow final conclusions to be drawn.

Key words: Breast cancer,   post-operative radiotherapy,
             lymphopaenia,   carcinoembryonic antigen.

BACKGROUND AND PATIENT MATERIAL

The objectives of the trial were to firstly, investigate the value of post-operative radiotherapy in operable breast cancer, to investigate the effects of therapeutic irradiation on the pattern of recurrence and metastases in breast cancer, and discover the prognostic importance of any observed effect. It was also hoped to correlate lymphopaenia with metastatic pattern and to study the value of carcinoembryonic antigen as a predictive marker for metastases or recurrence. The participating centres were the Ludwig Institute for Cancer Research, Lausanne, Groote Schuur Hospital, Cape Town, The Radium Hospitalet, Oslo, The Jubileumskliniken, Götenborg, The Institute of Oncology, Ljublana, with the International Agency for Cancer Research, Lyon, as a statistical office.

The trial was commenced on 1st January 1974, and terminated on 31st May 1977. 377 Patients were entered in the trial, which was designed to stand a 90% chance of detecting distant metastases or local recurrences varying by a factor of 3 or more at the 5% level of significance.

It was felt that an examination of the pattern of metastases rather than the pattern of survival (about 1000), would allow a smaller number of patients (about 400) to be entered into the study. The patients accepted into the trial were all those in each centre, staged T_1 N_1 or T_2 by the U.I.C.C. Classification of 1968. Exclusion factors included patients over 75 years of age, associated pregnancy, mental unsuitability, co-existence of gross chronic disease, previous treatment by irradiation, hormones or cytotoxic drugs, and Paget's Disease.

Patients who were clinically acceptable for the trial had an extensive work-up before surgery and during follow-up, which included careful assessment of the clinical status, X-ray of lungs and bone, bone and liver scans, liver enzymes, sedimentation rate, and a full blood count, including differential count. It is hoped thereby that occult Stage IV disease patients were excluded and also that the recurrence or metastatic pattern was elucidated in individual patients as efficiently and at an early stage as possible.

Surgical treatment was either a Halsted type of radical mastectomy or a modified radical mastectomy. Patients were not considered eligible if local mastectomy, partial mastectomy or excision biopsy only were to be performed. Pathological evaluation of the surgical specimen was then undertaken, which included assessment of the degree of malignancy as described by Bloom and Richardson (1957). The number of involved axillary nodes was recorded, any reactive changes in the lymph nodes noted, and the presence or absence of any periglandular growth. Patients were then formally reviewed to ensure that they had Stage II (T1 or T2 N1) carcinoma of the breast and were strictly eligible for the trial. They were then classified into one of four groups on the basis of malignancy grade (1, 2 or 3), and periglandular growth (present or absent) and then randomised to receive radiation or no radiation.

The two groups were comparable with regard to age (below or about 50 years), degree of malignancy, extra-capsular tumour growth, and with regard to the number of axillary lymph nodes which were found to be positive. (Fig.1).

	Per cent	
	Surgery	Radiotherapy
Age: Less than 50 years	37	40
50 years or more	63	60
Degree of malignancy 1	11	13
2	53	51
3	36	36
Periglandular growth Absent	58	57
Present	42	43
Positive axillary nodes 1-3	67	63
4	33	37

Fig. 1. Balance of prognostic variables between the two therapeutic groups.

Each clinic followed its own irradiation schedule, treating the ipsilateral internal mammary, supraclavicular and axillary lymph nodes. Not all clinics irradiated the skin flaps. All data was registered with the Central Statistical Office at the International Agency for Cancer Research at Lyon.

FOLLOW-UP

The follow-up of patients was undertaken at least every 3 months. During the first year, complete pre-operative work-up was performed every 3 months and thereafter every 6 months. The value of the rigorous follow-up schedule in the detection of metastatic pattern was clear in many patients, whose first evidence of disease was found only by special investigations rather than by clinical presentation, which tended to appear weeks or months later. It is felt that the rigorous follow-up undertaken enabled evaluation of the evolution of metastatic pattern in a critical fashion which would not otherwise have been possible.

The results presented here are preliminary ones and are made without a complete and final intercentre review of the radiotherapy dosage and techniques, the absolute lymphocyte counts and peer review of all X-rays and scans of all the clinics.

RESULTS

There is no difference in the percentage of patients who received radiotherapy or not, who remain well, but a high percentage of those who had surgical treatment only, developed local disease as the first event, whereas of those patients who received radiotherapy, a higher percentage developed metastases as the first event. These findings are at present significant in the case of local disease at the 7% level and in the case of metastases, at the 3% level of significance. (Fig. 2).

Event	Absolute	%	S	%	RT	%
Healthy	288	76,4	152	77,6	136	75,1
Contralateral breast	2	0,5	1	0,5	1	0,6
Local Met.	26	6,9	18	9,2	8	4,4
Local & Distant (same time)	11	2,9	6	3,1	5	2,8
Distant Met.	49	13,0	18	9,2	31	17,1
Dead	1	0,3	1	0,5	-	-
Total	377	100,0	196	100,0	181	100,0

Fig. 2. First event suffered.

There is no significant difference at the present time in the overall mortality of the two groups of patients treated (Surgery 8/167, and Radiotherapy 12/163), nor would there appear to be any difference in the eventual (as opposed to the first event suffered) local metastasis pattern in both groups. Whereas it is possible, although at present not statistically significant, that the eventual distant metastatic pattern will be higher in the irradiated group. (Fig. 3).

Therapy	Year 1	Year 2	Year 3	Total
S	13/187 (7%)	14/105 (13%)	0/26 (0%)	27
S + RT	22/175 (13%)	10/102 (10%)	4/24 (17%)	36
	35/362 (10%)	24/207 (12%)	4/50 (8%)	63

Fig. 3. Distant metastases.

It would therefore appear, at this preliminary stage of our investigation, that radiotherapy given prophylactically to Stage II cancer of the breast after radical mastectomy, does not improve the mortality but that it possibly decreases the incidence of local disease and increases the appearance of distant metastases, as the first evidence of disease progression. This would suggest that it is of no material value in the prophylactical management of this disease at this stage, but were it to be used for some particular indication, there is a strong suggestion that systemic prophylactic chemotherapy be considered.

CARCINOEMBRYONIC ANTIGEN

During the course of the investigation, blood was taken for carcinoembryonic antigen, stored and estimated by Dr. J.P. Mach. The data so far available concerns 333 patients, 48 of whom developed metastases. In 16 of these 48 patients, C.E.A. was predictive for recurrence before other tests, but in 32 patients, not. There was no correlation between the degree of malignancy and the percentage of positive values, but it would appear that C.E.A. estimations are of little value in predicting lung or liver metastases, but may be useful for bone, regional or metastases at mixed sites. (Fig. 4).

Positive		and localization of metastases		
Lung	Liver	Bone	Loco-regional	Mixed
0/8	0/3	6/22	2/3	8/12

Fig. 4. C.E.A. - Breast cancer T.1/2 N(+) M"0"

The complete data in this trial however, will not be finally evaluable until 4 years from now. We hope to report more fully on this trial at that time.

REFERENCES

T.N.M. Classification of Malignant Tumours (1968), U.I.C.C., Geneva.
Bloom, H.J.G., Richardson, W.W. (1957). Histological grading and prognosis in breast cancer. Brit. J. Cancer, 11, 359.

Comparison of Three Modalities of Treatment for Carcinoma of Breast Stage III. Results from a Prospective Randomized Clinical Trial

E. Caceres, F. Tejada*, M. Zaharia, M. Lingan, M. Cotrina, L. Leon, M. Moran, O. Castro and A. Solidoro

Instituto Nacional de Enfermedades Neoplasicas, Lima, Peru
*Comprehensive Cancer Center for the State of Florida.
Supported in part by PAHO and National Cancer Institute
(USA) Latin American Program Protocol INEN-74-01

Radiotherapy alone is of definite but limited value in the treatment of Stage III Breast Carcinoma. In a prospective study recently reported from our Institution (1) the effect of a fixed dose of radiation therapy (5000 rads) for different tumor sizes, and patient's age, delivered in 5 weeks, achieved a 61 per cent response rate in patients with locally advanced cancer of the breast.

The duration of response was from three to more than 25 months, with an average duration of 13.1 months. The five year survival in the patients who responded to radiotherapy was 30 per cent in contrast to 5 per cent survival for non responders.

These results indicate that radiation therapy, when used alone, has been found to be of limited usefulness for the treatment of Stage III (locally advanced) breast cancer. Although around 50 per cent of our patients had recurrence in the irra - diated field within 24 months after onset of treatment, most of the patients had relapse of disease at distant sites from the irradiated fields. This fact emphasizes the idea that early dissemination of the disease may occur before the initiation of treatment for locally advanced cancer of the breast.

MATERIALS AND METHODS

On this basis, in January 1975 we started a prospective randomized and controlled clinical trial, on patients with locally advanced breast cancer or stage III (UICC). The patients were randomly allocated in three groups: 1) radiotherapy alone, 2) radiotherapy followed by chemotherapy (Cyclophosphamide, Methotrexate and 5-Fluorura cil) and 3) radiotherapy followed by total mastectomy and dissection of the lower axilla.

The patients were stratified by age, under 49 years vs 50 years or more, tumor fixed to fascia or extended to skin vs none.

The dose of radiation therapy administered was 5000 rads to the whole breast in five weeks plus a boost to the residual mass of 1500 given in 5 days and 4500 to the axilla in 4 weeks. When the separation of the tangential fields was more than 18 cm. we added a 12 x 6 field to the internal mammary chain giving a dose of 4500 rads calculated to 4 cm. deep. If the node in the axilla was larger than 3 cm. through a direct reduced field we gave a boost of 1000 rads in one week.

Chemotherapy was administered in the following way:

Cyclophosphamide 100 mg/m^2 PO days 1 to 14
Methotrexate   40 mg/m^2 IV days 1 and 8
5-Fluoruracil   600 mg/m^2 IV days 1 and 8

Each cycle was repeated every 29 days.

Patients randomized to receive surgery had total mastectomy plus lower axillary lymph node dissection (level I and II). Surgery was done three to four weeks after finishing the course of radiation therapy.

A total of 103 patients were admitted to the study, 35 patients for the radiotherapy plus surgery, 33 patients for the radiotherapy and chemotherapy group and 35 patients in the radiation alone group. As of July 1978, there were 28, 24 and 29 patients evaluable respectively for each group.

Seven patients in the radiotherapy and surgery group, 9 patients in the radiotherapy and chemotherapy group and 6 in the radiation therapy alone group were not considered evaluable for treatment response because of intercurrent death (cardiovascular disease), failure to complete treatment or protocol violation.

No significant difference was found between pre and post menopausal patients, size of primary tumor or lymph nodes in the axilla, similarly there was no difference in reference to age, extension of the disease or radiation dosage to the primary or lymph nodes.

Patients were informed of the investigational nature of the study and a written consent was obtained from each patient.

RESULTS

The objectives of this clinical trial were to confirm which of the three modalities of treatment was superior in improving, 1) the length of the disease free interval and 2) survival.

There were 18 recurrence in the radiation plus surgery group or 64 per cent with an average recurrence time of 329 days, versus 12 patients in the radiation plus chemotherapy group or 50 per cent, average recurrence time 582 days and in the last group radiation alone relapsed 23 patients or 79 per cent, average recurrence time 393 days.

The statistical analysis indicated significant difference favoring radiation and chemotherapy versus radiation with or without surgery.

This indicate that more local failures were seen in radiation only while more metastatic disease was present in radiation plus surgery. Radiation plus chemotherapy gives the most local regional and metastatic control which was significantly different. There were 2 local only recurrence in radiation plus chemotherapy while there were 6 local only recurrence in radiation plus surgery, all the other were combined recurrence.

This seems to indicate that radiation, chemotherapy and surgery play a role in the treatment of stage III breast cancer although their appropriate sequence has not been determined.

When we analyzed the survival in the three groups, at the 15% percentile radiation plus chemotherapy has better survival which it is changed when we analyze the survival at 30% percentile, although this study has only 3 1/2 years of follow-up, therefore, we do not draw conclusions at this time as regards to survival.

BIBLIOGRAPHY

1. Zaharia, M., E. Caceres, M. Moran, and R. Celis (1978) Management of Locally advanced breast cancer. <u>Breast</u>, Diseases of the Breast, Vol. 4, Number 1: 35-39.

Index

The page numbers refer to the first page of the article in which the index term appears.

Acute lymphocytic leukemia 263
Adjuvant treatment 131
Alloantigens, histocompatibility 3
Allogeneic AKR lymphoma 29
Allogeneic bone marrow grafts 143
Alpha-fetoprotein (AFP) 3, 37, 143
Antibodies
    labelled 95
    localizing 95
    specific antitumour 3, 95
Antigens
    carcinoembryonic 3, 305
    histocompatibility complex (MHC) 13
    of chemically-induced tumours 3
    soluble 19
    specific lung tumour 7
    tumour associated 3, 77, 85
    tumour rejection 19
Antitumour antibodies, specific 95
Antitumour reactivity, natural cell-mediated 3

BCG 67, 77, 121, 137
    side effects and hazards 109
Bilharzial bladder, T3-carcinoma 299
Bladder, T3-carcinoma 299
Bone marrow transplantation 149
Brain, prophylactic irradiation 263
Breast cancer
    stage II 305
    stage III 311
Bronchial carcinoma, non-small cell 131

C-type oncornaviruses 3
Carcinoembryonic antigen 3, 305
Cell-mediated immune reactivity 3, 85
Cell proliferation 165
Cervix uteri, cancer 223, 293
Chemically-induced tumours, antigens of 3
Chemotherapy 183

Chemotherapy (cont'd)
    combination 121
Choriocarcinoma, testicular 279
Clinical trials 215, 271, 311
Combination treatment 121, 183
Corynebacterium parvum 131, 137
Cryopreservation 149
Cystectomy, radical 299
Cytolytic T lymphocytes 3
Cytophilic activity 29
Cytotoxic agents 95
Cytotoxic T cells 13
Cytotoxicity 3

Electron microscopy 201
Electrophoretic mobility, of macrophages 3
Embryonal carcinoma, testicular 279
Epithelial cell response 201
Experimental tumour systems, host immune responses 13

Fibrilar proteins 201
Fractionation 165, 173, 183, 191, 253, 299
    hyper- 299

Gene expression, viral 3
Germinomas, testicular 279
Guinea pig macrophages 29

Head and neck cancer 223
Hepatoma BW7756 37, 143
High linear energy transfer (High LET) 215
Histocompatibility alloantigens, major 3
Histocompatibility complex
    antigens 13
    I-region 3
HLA-D locus 3
Host immune responses, in experimental

tumour systems 13
Hyperbaric oxygen (HBO) 191, 223
Hyperfractionation (HF) 299
Hypersensitivity, delayed 3
Hyperthermia-radiation therapy 235, 247
Hypoxic cells 191, 223
  sensitizers 271

I-region, of major histocompatibility
  complex 3
Immunocomplexes 3
Immunodiagnosis 3
Immunogenicity 3, 13
Immunopotentiation 77
Immunosuppression 77, 143
Immunosuppressive drugs 109
Immunotherapy 3, 37, 131, 137
  mechanisms 37
  non-specific 109
  of cancer 47
  of lung cancer 77
  specific active 67, 85
Intercellular contact 173
Interferon 3
Intracavitary irradiation 293
In vitro cultivation 165

LET, high- and low- 215
Leukemia 149
  acute lymphocytic 263
Leukocyte migration inhibition 3
Levamisole 77, 101, 137
Liver, prophylactic irradiation 263
Lung cancer
  immunotherapy 77
  oat-cell 263
  prophylactic irradiation 263
Lymphocytic stimulation test 3
Lymphocytes
  cytolytic T 3
  subpopulations in tumour immunity
    139
  tumour-derived 67
Lymphoma, allogeneic AKR 29
Lymphopaenia 305

Macrophages 3, 13
  electrophoretic mobility 3
  guinea pig 29
Mammalian cells 173
Melanoma, malignant 121, 223
Metastases 95
  occult 263
Metastatic sarcomas 121
Meth A sarcoma 19
Metronidazole 271
MHC regions, human 3
Microcytotoxicity 67
Misonidazole 223, 271

Mouse hepatoma BW7756 37, 143
Multifraction irradiation 173

Natural killer (NK) cells 13, 149
Neurotoxicity 271
Nucleoproteins 85

Oncolysates, viral 121
Oncornaviruses, C-type 3
Orchiectomy 279
Osteosarcoma 263

Particle radiotherapy 215
Peking-type applicators 293
Pelvis 215
Plasma membranes 19
Prophylactic irradiation 263
Public health programs 85

Radiation 159
  fractionated 165, 173, 183, 191
  injury 201
Radioiodine 95
Radiosensitivity 263
Radiosensitizers 191, 223, 271
Radiotherapy 191, 271, 275
  combined with hyperthermia 235, 247
  dose fractionation 165, 183, 253
  of bladder carcinoma 299
  of breast cancer 305, 311
  of cervix uteri carcinoma 293
  of subclinical disease 263
  of testicular germinomas 279
  particle 215
  post-operative 305
Repair processes 173
RNA, immune 3

Sarcoma
  metastatic 121
  meth A 19
  osteo- 263
Seminoma 279
Sensitivity, differential 165
Sensitizers 191, 223, 271
Spheroids, multicell 173
Structural proteins, of viruses 3
Subclinical disease 263, 275
Suppressor cells 3, 143, 149

T cells
  activated helper- 3
  cytotoxic 13
  interaction system 3
  maturation and differentiation 3
  suppressor- 3
Telecobalt irradiation, preoperative 299
Teratocarcinoma, testicular 279
Testicular cancer 279

Therapeutic gain 215
Therapeutic ratio 235, 247
Thymectomy 29
Time-dose effect 173, 253, 275
Tolerance 37
Tolerances, normal tissue 215
Tonofilaments 201
Transfer factor 3
Transpelvic external irradiation 293
TSTA 19
Tumour-associated antigen (TAA) 3, 85
Tumour-derived lymphocytes 67

Tumour growth enhancement 109
Tumour rejection
  failure 143
  immune mechanisms 47
Tumour rejection antigens (TSTAs) 19
Tumour response 183

Viral gene expression 3
Viral oncolysates 121

Winn test 67